NEURAL NETWORKS IN TELECOMMUNICATIONS

NEURAL NETWORKS IN TELECOMMUNICATIONS

EDITED BY

Ben YUHAS
Bellcore
Morristown, NJ 07960

Nirwan ANSARI
Electrical and Computer Engineering Department
New Jersey Institute of Technology
Newark, NJ 07102

KLUWER ACADEMIC PUBLISHERS
Boston/London/Dordrecht

Distributors for North America:
Kluwer Academic Publishers
101 Philip Drive
Assinippi Park
Norwell, Massachusetts 02061 USA

Distributors for all other countries:
Kluwer Academic Publishers Group
Distribution Centre
Post Office Box 322
3300 AH Dordrecht, THE NETHERLANDS

Library of Congress Cataloging-in-Publication Data

Neural networks in telecommunications / edited by Ben Yuhas, Nirwan Ansari.
    p. cm.
Includes bibliographical references and index.
ISBN 0-7923-9417-8 (alk. paper)
  1. Telecommunications systems--Data processing. 2. Neural networks (Computer Science) I. Yuhas, Ben, 1959- . II. Ansari, Nirwan, 1958- .
TK5102.5.N47 1994   93-33665
621.382'0285'63--dc20   CIP

Copyright © 1994 by Kluwer Academic Publishers

All rights reserved. No part of this publication may be reproduced, stored in a retrieval system or transmitted in any form or by any means, mechanical, photo-copying, recording, or otherwise, without the prior written permission of the publisher, Kluwer Academic Publishers, 101 Philip Drive, Assinippi Park, Norwell, Massachusetts 02061.

Printed on acid-free paper.

Printed in the United States of America

CONTENTS

1 INTRODUCTION
Ben Yuhas and Nirwan Ansari — 1
1. What is a Neural Network? — 2
2. Computing with Neural Networks — 2
3. Building the Correct Neural Network — 6
4. When are Neural Networks Appropriate? — 8

2 NEURAL NETWORKS FOR SWITCHING
Timothy X Brown — 11
1. Introduction — 11
2. Switching — 12
3. Neural Circuit Design — 15
4. Neural Approaches to Switching — 24
5. Conclusion — 32

3 ROUTING IN RANDOM MULTISTAGE INTERCONNECTION NETWORKS
Mark W. Goudreau and C. Lee Giles — 37
1. Introduction — 37
2. The Communication System — 38
3. Exhaustive Search and Greedy Routing — 41
4. The Neural Network Router — 42
5. Performance Measures for Routers — 49
6. Simulation Results — 51
7. Analysis — 55
8. Conclusions and Future Work — 57

4 ATM TRAFFIC CONTROL USING NEURAL NETWORKS
Atsushi Hiramatsu — 63
1. Introduction — 63
2. ATM Traffic Control — 64
3. Neural Network Applications in the ATM Network — 70
4. Adaptive Call Admission Control — 72
5. Adaptive Link Capacity Control — 79
6. Further Research Directions — 87

5 LEARNING FROM RARE EVENTS : DYNAMIC CELL SCHEDULING FOR ATM NETWORKS
Daniel B. Schwartz — 91
1. Introduction — 91
2. ATM Networking — 92
3. On-Line Dynamic Programming — 93
4. Experimental Evaluation — 101
5. Simulations — 102
6. Conclusions — 105

6 A NEURAL MODEL FOR ADAPTIVE CONGESTION CONTROL IN ATM NETWORKS
Xiaoqiang Chen — 109
1. Introduction — 109
2. Problem Formulation — 110
3. Adaptive Learning Rate Algorithm — 111
4. Neural Network Control Scheme — 114
5. Simulation Results — 118
6. Summary — 124

7 STRUCTURE AND PERFORMANCE OF NEURAL NETS IN BROADBAND SYSTEM ADMISSION CONTROL
Phuoc Tran-Gia and Oliver Gropp — 127
1. Introduction — 127
2. Neural Networks for Connection Admission Control — 127
3. Example of a Neural Network for Admission Control — 132
4. Conclusions — 141

Contents

8 NEURAL NETWORK CHANNEL EQUALIZATION
William R. Kirkland and D. P. Taylor **143**
1 Introduction — 143
2 A Brief Introduction to Digital Transmission Theory — 145
3 Signal Detection and Estimation — 147
4 Two Dimensional Signalling and Complex Neural Networks — 158
5 Summary — 166

9 NEURAL NETWORKS AS EXCISERS FOR SPREAD SPECTRUM COMMUNICATION SYSTEMS
Richard Bijjani and Pankaj K. Das **173**
1 Introduction — 173
2 Time Domain Processing — 178
3 Conclusions — 188

10 STATIC AND DYNAMIC CHANNEL ASSIGNMENT USING SIMULATED ANNEALING
Manuel Duque-Antón, Dietmar Kunz and Bernd Rüber **191**
1 Problem Description — 191
2 Existing Approaches — 193
3 Applying Simulated Annealing to the Static CAP — 194
4 Simulated Annealing in a Running System — 205
5 Conclusions — 208

11 CELLULAR MOBILE COMMUNICATION DESIGN USING SELF-ORGANIZING FEATURE MAPS
Thomas Fritsch **211**
1 Introduction — 211
2 Mobile Network Planning — 213
3 Traffic Density Determination using a Self-organizing Feature Map — 216
4 The Main Algorithm — 219
5 Experimental Results — 228
6 Conclusions — 231

12 AUTOMATIC LANGUAGE IDENTIFICATION USING TELEPHONE SPEECH
Yeshwant K. Muthusamy and Ronald A. Cole — **233**

1. Introduction — 233
2. Background — 235
3. A Telephone Speech Corpus For Automatic Language Identification — 237
4. Human Listening Experiments — 241
5. A Segmental Approach to Automatic Language Identification — 243
6. Summary and Future Work — 249

13 TEXT-INDEPENDENT TALKER VERIFICATION USING COHORT NORMALIZED SCORES
David J. Burr — **255**

1. Introduction — 255
2. Talker Identification Background — 256
3. Related Work — 257
4. Cohort Normalization — 258
5. STIMIT — 259
6. Talker Identification System — 260
7. Vector Quantization — 260
8. Experiments on 90 Talkers — 262
9. Experiments on All 630 Talkers — 263
10. Multilayer Perceptron — 264
11. Fixed Versus Variable Training Sentences — 265
12. Preliminary Experiment on Real Telephone Speech — 266
13. Discussion — 267
14. Conclusion — 268

14 NEURAL NETWORK APPLICATIONS IN CHARACTER RECOGNITION AND DOCUMENT ANALYSIS
L.D. Jackel, et al. — **271**

1. Introduction — 271
2. The Character Recognition Process — 272
3. The Basic Recognizer: LeNet — 273
4. Segmentation — 275

	5	Normalization	278
	6	Finding the Region of Interest	280
	7	Additional Applications	280
	8	Conclusions	283

15   IMAGE VECTOR QUANTIZATION BY NEURAL NETWORKS
Rosa Lancini — **287**

1	Introduction	287
2	Neural Networks for Vector Quantization	288
3	Neural Networks for Adaptive Vector Quantization	295
4	Conclusions	301

16   MANAGING THE INFOGLUT: INFORMATION FILTERING USING NEURAL NETWORKS
Thomas John — **305**

1	Introduction	305
2	Previous work on filtering	306
3	Neural Networks for Filtering	311
4	Reinforcement learning - AIR	313
5	An Unsupervised Classifier - ART	315
6	Experimental determination of filter validity	316
7	Future Directions	320

17   EMPIRICAL COMPARISONS OF NEURAL NETWORKS AND STATISTICAL METHODS FOR CLASSIFICATION AND REGRESSION
Diane Duffy, Ben Yuhas, Arvind Jain and Andreas Buja — **325**

1	Introduction	325
2	Two Problems	326
3	Regression and Classification Methodologies	329
4	Results on Regression for Switch Memory	338
5	Results on Classification of DS0-Rate Traffic	342
6	Discussion	347

18	A NEUROCOMPUTING APPROACH TO OPTIMIZING THE PERFORMANCE OF A SATELLITE COMMUNICATION NETWORK	
	Nirwan Ansari	**349**
	1 Satellite Communications	349
	2 The Self-organization Method	350
	3 The Cost Minimization Method	355
	4 Discussion	363

Index   **367**

CONTRIBUTORS

Nirwan Ansari
Center for
Communications & Signal Processing
Electrical and Computer Engineering
New Jersey Institute of Technology
University Heights
Newark, NJ 07102
email: ang@hertz.njit.edu

Manuel Duque-Antón
Philips GmbH Forschungslaboratorien
Postfach 1980
D-5100 Aachen
GERMANY
email: duque@philfa.pfa.philips.de

H.S. Baird
AT&T Bell Laboratories
101 Crawfords Corner Road
Holmdel, NJ 07733

M.Y. Battista
AT&T Bell Laboratories
101 Crawfords Corner Road
Holmdel, NJ 07733

J. Ben
AT&T Bell Laboratories
101 Crawfords Corner Road
Holmdel, NJ 07733

Richard Bijjani
The Bear Group, Inc.
951 Mariners Island Blvd, Suite 230
San Mateo, CA 94404

J. Bromley
AT&T Bell Laboratories
101 Crawfords Corner Road
Holmdel, NJ 07733

Timothy X. Brown
Bellcore, MRE 2E-378
445 South Street
Morristown, NJ 07960
email: timxb@faline.bellcore.com

Andreas Buja
Bellcore, MRE 2E-330
445 South Street
Morristown, NJ 07960

C.J.C Burges
AT&T Bell Laboratories
101 Crawfords Corner Road
Holmdel, NJ 07733

David J. Burr
Bellcore, MRE 2B-225
445 South Street
Morristown, NJ 07960
email: djb@bellcore.com

Xiaoqiang Chen
AT&T Bell Laboratories, Rm 4F-614
101 Crawfords Corner Road
Holmdel, NJ 07733
email: xchen@attmail.com

Ronald A. Cole
Center for
Spoken Language Understanding
Oregon Graduate Institute
of Science and Technology
19600 NW von Neumann Drive
Beaverton OR 97006-1999

E. Cosatto
AT&T Bell Laboratories
101 Crawfords Corner Road
Holmdel, NJ 07733

Pankaj K. Das
Electrical, Computer & Systems Eng.
Rensselaer Polytechnic Institute
Troy, NY 12180-3590
email: das@ecse.rpi.edu

J.S. Denker
AT&T Bell Laboratories
101 Crawfords Corner Road
Holmdel, NJ 07733

Diane Duffy
Bellcore
445 South Street
Morristown, NJ 07960

Thomas Fritsch
Institute of Computer Science
University of Wuerzburg
AM Bubland, D-8700 Wuerzburg
GERMANY

C. Lee Giles
NEC Research Institute
4 Independence Way
Princeton,NJ 08540
email: giles@fuzzy.nec.com

Mark Goudreau
Department of Computer Science
University of Central Florida
Orlando, FL 32816-2362
email: goudreau@cs.ucf.edu

H.P. Graf
AT&T Bell Laboratories
101 Crawfords Corner Road
Holmdel, NJ 07733

Oliver Gropp
Institute of Computer Science
University of Wuerzburg
Am Hubland, D-8700 Wuerzburg
GERMANY

Atsushi Hiramatsu
NTT Communication
Switching Laboratories
3-9-11, Midori-cho
Musashino-shi
Tokyo 180
JAPAN
hiramatu@nttgoso.ntt.jp

Lawrence Jackel
AT&T Bell Laboratories, HO 4D-433
101 Crawfords Corner Road
Holmdel, NJ 07733
email: jdl@neural.att.com

Arvind Jain
University of California at Berkley

Thomas John
Southwestern Bell Technology Resources
550 Maryville Centre Dr.
St Louis, MO 63141
email: john@sbctri.sbc.com

H.P. Katseff
AT&T Bell Laboratories
101 Crawfords Corner Road
Holmdel, NJ 07733

William R. Kirkland
CRL Rm 205
McMaster University
1280 Main Street West
Hamilton, Ontario
CANADA L8S 4K1
email: kirkland@sscvax.mcmaster.ca

Dietmar Kunz
Philips GmbH Forschungslaboratorien
Postfach 1980
D-5100 Aachen
GERMANY

Rosa Lancini
CEFRIEL, Via Emanueli 15
20126 Milano
ITALY
email: rosa@mailer.cefriel.it

Y. LeCun
AT&T Bell Laboratories
101 Crawfords Corner Road
Holmdel, NJ 07733

Yeshwant K. Muthusamy
Texas Instruments
P.O. Box 655474
MS 238
Dallas, TX 75265
email: yeshwant@csc.ti.com

C.R. Nohl
AT&T Bell Laboratories
101 Crawfords Corner Road
Holmdel, NJ 07733

Bernd Rüber
Philips GmbH Forschungslaboratorien
Postfach 1980
D-5100 Aachen
GERMANY

E. Sackinger
AT&T Bell Laboratories
101 Crawfords Corner Road
Holmdel, NJ 07733

Daniel B. Schwartz
Motorola, Inc.
Phoenix Central Research Laboratory - EL 508
2100 E. Elliot Rd.
Tempe AZ 85284
email: a186aa@email.sps.mot.com

J.H. Shamilian
AT&T Bell Laboratories
101 Crawfords Corner Road
Holmdel, NJ 07733

T. Shoemaker
AT&T Bell Laboratories
101 Crawfords Corner Road
Holmdel, NJ 07733

C.E. Stenard
AT&T Bell Laboratories
101 Crawfords Corner Road
Holmdel, NJ 07733

B.I. Strom
AT&T Bell Laboratories
101 Crawfords Corner Road
Holmdel, NJ 07733

D.P. Taylor
CRL Rm 205
McMaster University
1280 Main Street West
Hamilton, Ontario
CANADA L8S 4K1

R. Ting
AT&T Bell Laboratories
101 Crawfords Corner Road
Holmdel, NJ 07733

P. Tran-Gia
Institute of Computer Science
University of Wuerzburg
Am Hubland, D-8700 Wuerzburg
GERMANY

T. Wood
AT&T Bell Laboratories
101 Crawfords Corner Road
Holmdel, NJ 07733

Ben Yuhas
Bellcore, MRE 2E-330
445 South Street
Morristown, NJ 07960
email: yuhas@bellcore.com

C.R. Zuraw
AT&T Bell Laboratories
101 Crawfords Corner Road
Holmdel, NJ 07733

1

INTRODUCTION

Ben Yuhas and Nirwan Ansari*

Bellcore

New Jersey Institute of Technology

This book provides an overview of the wide range of telecommunications tasks that are being addressed with neural networks. We will focus on specific applications, describe specific solutions and demonstrate the benefits that neural networks can provide. By doing this, we show that neural networks should be included as another tool available to telecommunications engineers. Neural networks can offer the computational power of non-linear techniques and can provide a natural path to efficient hardware implementation. In addition, their ability to learn allows them to be used on problems where there are no straightforward heuristic or rule-based solutions. Together these capabilities allow neural networks to offer unique solutions to problems in telecommunications.

The growth of telecommunications and its expansion into new areas is always accompanied by new challenges. In the past, these challenges have been met through the rapid integration of the latest technology into the industry. The technological advances of the recent past are pervasive through the telecommunications systems of today, from digital switches to the optical interconnections.

A striking aspect of this book is the wide range of tasks to which neural networks have already been applied—from the design and control of the underlying transport network to the filtering, interpretation and manipulation of the transported media. This introductory chapter gives a brief overview of the different ways that neural networks can be used and then references those applications in the book that demonstrate these different uses. This is not intended to be a primer on neural networks. The reader is referred to Hertz *et al.* [1], Lippmann [2] or Rumelhart *et al.* [3] for a more thorough treatment of neural networks.

1 WHAT IS A NEURAL NETWORK?

Neural networks encompass a diverse set of computational models which share a set of simple underlying characteristics. Inspired by the computational style of biological systems, a neural network can be viewed as an assembly of *simple*, interconnected processing units (or *neurons*) acting in parallel. These individual units communicate to each other using unidirectional connections (or *synapses*). Each of these connections has a value (or *weight*) associated with it. Since the individual processing units are usually identical, the computation is defined by the specific way in which they are connected together, along with the weights associated with these connections. The diversity of the computational models that can be constructed from these simple elements is demonstrated in the applications presented in this book.

A processor within a neural network receives input from the other processors to which it is connected. Its job is simply to add these inputs up and generate an output signal, which is then sent out to other units. For example, a unit i computes

$$o_i = f(\sum_j w_{ij} o_j - \theta_i), \qquad (1)$$

where o_j is the output associated with a unit j, w_{ij} represents the weight associated with a unidirectional connection between unit j and i, and θ_i is a threshold associated with unit i. The *transfer function*, f, can take on many forms, but for the purposes of this book f will usually be defined as the so-called sigmoid function, $f(u) = (1 + e^{-u})^{-1} = 0.5\,(1 + tanh(u/2))$. This is a monotonically increasing real function, approaching one and zero at $\pm\infty$, respectively. A slight change in f can create a function that will approach ± 1 at $\pm\infty$.[1]

Given this simple paradigm, two questions immediately arise: (1) What can be done with a network of such units? (2) How does one construct a network that will perform the desired task?

2 COMPUTING WITH NEURAL NETWORKS

In the previous section, we claimed that networks of simple processors can perform many types of computations. The specific computation is largely

---
[1] In some networks, the individual units may have unique transfer functions, in which case the functions are identified as f_i.

Introduction

defined by the network's *architecture*, i.e, the number of units there are and the way in which they are connected together. In this section, we describe some of the different types of computations that are performed by neural networks.

2.1 Regression and Classification

Neural networks are most commonly used on regression and classification problems. Regression and classification are estimation problems, where an output value is estimated from a set of input variables. In regression, one attempts to estimate an unknown quantitative variable, while classification implies the estimation of a class membership. In both problems, estimation is based on the known values of input variables.

The result of a regression analysis is a regression model, i.e., a function that maps a set of input variables to a value of some unknown quantitative variable (the estimate). The result of a classification analysis is a classification rule, typically implemented as a vector function of the covariates with as many components as there are classes. The components of the vector give the estimated (relative) probabilities of falling in the different classes. In either case—regression or classification analysis—the problem is that of finding functions of the covariate variables suitable for the given task.

The regression and classification models discussed in this book are usually limited to the *feed-forward* class of neural network architectures. Feed-forward networks have no cycles in their directed graphs, so the flow of information is non-recurrent. This implies that the neurons can be numbered such that for two neurons i and j, if $i > j$ then $w_{ji} = 0$. Often feed-forward neural networks are composed of *layers* of units that are stacked. There are no connections between units within a given layer, but all of the units in one layer are connected to the units in the layer above. A simple three-layer neural network is graphically represented in Fig. 1. To complete the graphical intuition behind this architecture, input variables are presented to *input units* at the bottom of the figure. The input units often do not perform any computational function other than holding these input values. The top layer of units consists of one or more *output units* which furnish regression or classification estimates in one or more dimensions. Those units layered between the input and output units are called *hidden units*.

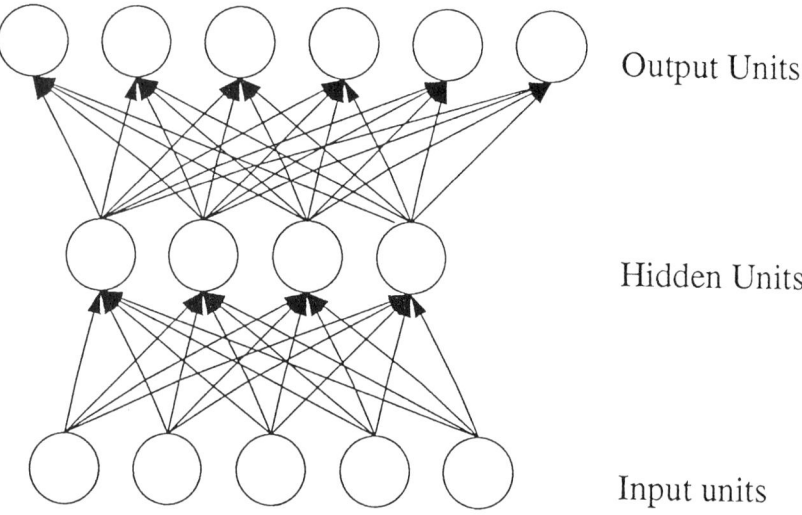

Figure 1 A three-layered feed-forward neural network.

A feed-forward neural network model produces a class of functions built up from building blocks of the form

$$f(w_0 + w_1 x_1 + w_2 x_2 + ...). \qquad (2)$$

The *input units* compute functions of the individual input variables and then pass these results on to other units. Additional layers of units cause these functional blocks to become nested resulting in more complicated functions. The weights w_{ij} in equation (1) are the free parameters that can be adjusted to fit training data. The process of adjusting these weights, called model fitting by statisticians, is referred to as *learning* or *training*. The algorithms that control these adjustments are called *learning algorithms* and are discussed in a later section. It is this learning capability that makes neural networks attractive for problems where the form of the solution is not known.

Many of the telecommunications problems described in this text are essentially estimation problems. In admission call control and adaptive congestion control in ATM (Asynchronous Transfer Mode) control systems, it is necessary to estimate the effect that available traffic can have on a given network. Various aspects of this problem are examined by Hiramatsu, Chen, and Tran-Gia and Gropp in Chapters 4, 6 and 7, respectively. In Chapter 8 Kirkland and Taylor discuss channel equalization, where the goal is to estimate the original signal that has been degraded by transmission. In Chapter 9, Bijjani and Das use neural networks to estimate the configuration of a notch filter for anti-jamming resistance in spread spectrum communications. In Chapter 12, Muthusamy and Cole describe a successful application in which neural networks are asked to classify a speech utterance into the appropriate language, while in Chapter 13, Burr designs a classifier to verify the identification of a speaker. In Chapter 14, Jackel *et al.* describe the success that neural networks are having at classifying handwritten digits. In all of these examples, the successes come from the power of neural networks to approximate the necessary regression or classification model. Finally, in Chapter 17, Yuhas *et al.* compare the performance of classical classification and regression techniques with feed-forward neural networks.

2.2 Constrainted Function Optimization

Neural networks are also used on functional optimization problems, where the problem is to find a solution that satisfies a set of constraints while minimizing some cost. The classic example of such a problem is the Traveling Salesman Problem, where it is not sufficient to simply visit all of the cities—it must be done in a way that the total distance traveled is minimized.

For these type of problems, neural networks provide a parallel method of quickly searching the space of possible solutions. In Chapter 2, Brown examines the problem of trying to satisfy the maximum call requests within the constraints of a given switch architecture. In Chapter 3, Goudreau and Giles use the Hopfield net to solve a similar switching problem, establishing point-to-point communication in an interconnection network.

Gradient-descent search techniques are commonly used to solve optimization problems, but they are susceptible to being trapped in local minima. Neural network implementations can augment gradient search techniques with *simulated annealing*. Simulated annealing reduces the probability of getting stuck in a local minima, by allowing occasional "uphill moves." That is, when searching for a lower cost solution, one occasionally accepts a higher cost solution in the hope that this will eventually lead to an even better solution. In Chapter 10, Duque-Antòn *et al.* describe a successful application using simulated annealing to assign channels to each cell of a mobile radio network. Also in Chapter 18, Ansari demonstrates how to increase the throughput of a circuit switched satellite communication network by simulated annealing.

For additional background in this area, the reader is referred to [4] and [5].

3 BUILDING THE CORRECT NEURAL NETWORK

Having decided to apply a neural network to a specific problem, the next step is to construct a network that will solve your problem.

A neural network model is parameterized not only by its weights but its architecture. A learning algorithm will find the weights for a given architecture, but it usually requires the *a priori* selection of an architecture. While the input and output variables may constrain the architecture somewhat, there are still many variations available. It is not uncommon to explore multiple architectures to determine an appropriate one.[2] The need to find a neural network architecture that can fit the training data well is balanced by the need to guard against too much flexibility, which can result in overfitting the training data. Overfitting the training data reduces the network's ability to perform on new examples. It is a problem arising from having too many free parameters in the model and not enough training samples. This problem is particularly acute when the number of training samples is small, and is treated more thoroughly in [6].

3.1 By Construction

When the relationship between the input variables and the task are well understood, it is sometimes possible to construct a network using heuristics. In Chapter 2, for example, Brown shows how the specific constraints of call requests and switch capacity can be directly translated into a network architecture. By directly embedding existing knowledge into a neural architecture, Brown does not need the learning capabilities of neural networks, but he does exploit the computation expediency of the architecture.

In most cases, either the solution or its translation into a neural architecture is not obvious, and it is necessary to find the correct architecture and/or weights through automated learning procedures.

---

[2]Fahlman and Lebiere (1990)[7] and others have explored learning algorithms that construct the architecture as they learn.

Introduction

3.2 By Supervised Learning

Learning algorithms use a training set of data to find a neural network that can be applied to future examples or previously unseen data.

If there is a specific output associated with a given set of input variables, one can use a *supervised learning* algorithm. Supervised learning requires that an output pattern be provided for every input training pattern. A learning algorithm modifies the weights in the neural network to minimize the difference between the *desired* output pattern and the *estimated* output pattern generated by the neural network. One popular algorithm is *back-propagation*, a version of gradient-descent adapted for neural networks.

A supervised learning method has an error measure to quantify the difference between the desired and estimated output values. A commonly used error measure is the sum of squared errors across the output units,

$$E = \sum_n \sum_j (y_j(n) - o_j(n))^2,$$

where $y_j(n)$ denotes the desired value of sample n and output unit j, and $o_j(n)$ is the corresponding output value generated by the network. The criterion E measures the overall deviation between desired and predicted values, and the network is trained to approximate the desired values $y(n)$ of the training sample as a function of the observed input values. One of the difficulties in applying supervised learning techniques is that they require *training* examples that provide the correct output for a given set of input variables. This is often referred to as *labeled data* and can often be a major obstacle.

There is also a class of learning algorithms that simply uses the binary decision as to whether the output is correct as the basis of modifying the weights. This type of learning is termed *reinforcement learning*, and is used in Chapter 5 by Schwartz for cell scheduling on ATM networks as well as by John in Chapter 16 for information filtering.

An issue common to these learning algorithms is the problem of sequential or *on-line* learning versus *batch* or off-line learning. By sequential learning one means the presentation of training samples one at a time and adjustment to the weights as the samples are presented; by batch learning one means that all data are available at once and weight adjustment can be based on the entire set of data.

3.3 By Unsupervised Learning

A second type of learning is *unsupervised learning*, where there is no teacher to provide the *correct answer* associated with a given set of input variables. Two tasks that lend themselves to this type of learning are clustering and coding.

In a clustering task, the goal is to identify natural groups in the data. In Chapter 11, Fritsch shows how the placement of base stations for cellular coverage can be approached as a clustering problem. Here a cluster is defined as a group of topographic points that can be serviced by the same radio base station. For background in clustering, the reader is referred to Kohonen [8].

A second application area for unsupervised learning is coding. The goal of coding is to define a more compressed representation for a set of input variables. In Chapter 15, Lancini shows how coding and clustering can be closely related. Other methods of using neural networks for coding are described by Hertz [1].

4 WHEN ARE NEURAL NETWORKS APPROPRIATE?

Having briefly described the various ways in which neural networks can be used, the question that remains to be answered is "When?"

- When one can exploit the parallel architecture by providing fast, compact solutions to existing problems
- When the underlying relationship among data may not be well understood, and an automated learning procedure may be necessary
- When there are inherent nonlinearities in the solution that are not well approximated by traditional linear techniques
- When the solution to a problem is constantly evolving and requires real-time adaptation

While each of these individual characteristics may be available from other tools, neural networks may be unique in that they can provide all of these in one computational paradigm.

REFERENCES

[1] J. Hertz, A. Krogh, and R. Palmer, *Introduction to the Theory of Neural Computation*. Reading, MA: Addison-Wesley, 1991.

[2] Lippmann, R.P., "An Introduction to Computing with Neural Nets," IEEE ASSP, April 1987, pp. 4-22.

[3] Rumelhart, D.E., J.L. McClelland and the PDP Research Group, Parallel Distributed Processing: Explorations in the Microstructure of Cognition. Cambridge, MA: MIT Press, 1986.

[4] Hopfield, J.J. and D.W. Tank, "Neural Computation of Decisions in Optimization Problems," *Biological Cybernetics 52*, 1985, pp. 141–152.

[5] van Laarhoven, P.J.M. and E.H.H. Aarts, *Simulated Annealing: Theory and Applications*. Holland: Reidel Publishing Co., 1987.

[6] MacKay, D.J., "Bayesian Interpolation," *Neural Computation*, *4*, May 1992, pp. 415-557.

[7] Fahlman, S.E. and C. Lebiere, "The Cascade-Correlation Learning Architecture," in D.S. Touretzky (ed.), *Advances in Neural Information Processing Systems 2*. San Mateo, CA: Morgan Kaufmann, 1990, pp. 524–532.

[8] Kohonen, T., *Self-Organization and Associative Memory*. Berlin: Springer-Verlag, 1984.

2

NEURAL NETWORKS FOR SWITCHING

Timothy X Brown

Bellcore

1 INTRODUCTION

Telecommunication networks hinge on several critical components; call admission into and routing through the network of nodes; data transport between nodes; and switching of data within nodes. We assume data can be transported into and out of a node and that data enters a node knowing which output it needs and focus on problems controlling the switching within nodes.

Switching nodes must successfully route the many arriving calls to the correct output without collisions of calls in the switching fabric. In broad terms, the switching control problem is characterized by:

- Many highly-interconnected simple elements comprise the switch.
- Many parameters describe the switch state and incoming data.
- Information is localized both within the switch and within the current time.
- Incremental but significant changes occur in the state as new data either enters or leaves the switch.
- A good solution from many potential solutions is required quickly.

Neural networks are noted for their ability to process large amounts of data quickly using a copious number of highly interconnected processors. The necessity to quickly find solutions; the multitude of readily available data; and the similar neural and switching network topologies all suggest the potential for exploiting neural networks in the switching control problem. In fact this has successfully been explored by a variety of researchers.

But, these results are based on a neural technique known as the Hopfield energy equation approach. In this chapter we present a framework for designing neural network solutions using extensions of the winner-take-all circuit. Unlike the Hopfield energy function approach that requires the researcher to first define the constraints of the problem and then go through an imprecise and obscuring energy function to define the weights, these networks have properties that can be directly defined and controlled. This direct approach allows efficient implementations that are scalable to large sizes and as long as the external inputs are within defined limits, the network will always satisfy the constraints embodied in the winner-take-all circuits.

We first describe more precisely a switch and look at two classes of problems, which we term loosely *circuit routing* and *packet arbitration*. We then give relevant background on neural networks developing the unifying framework. With this background, we cover the known results within this framework.

2 SWITCHING

2.1 Background

This section will introduce some basic concepts of switching theory. More detailed development of switching theory can be found in any of several references, [1, 2, 3]. We abstract a switch as a device that takes a set of N signal inputs and reproduces them in any permuted order at the output. We restrict our discussion to square switches, where the number of inputs and outputs is the same, although the concepts that we develop readily generalize to non-square switches. If the inputs and outputs of the switch are N distinct lines, then we refer to the switch as a *space* switch. Alternatively, the switch could have one line for the input and one line for the output. In this case, the inputs and outputs are N distinct blocks of data in which the order that the blocks are sent is permuted by the switch. Such a switch is referred to as a *time* switch. Since each time switch is equivalent to some space switch, we will restrict our discussion to space switches [2, pp. 114–117].

The basic switch is the $N \times N$ crossbar switch. Conceptually, it comprises a grid of N wires by N wires with a closable contact at each of the crosspoints as shown in Figure 1a (Figure 1b shows the schematic representation). A *legal*

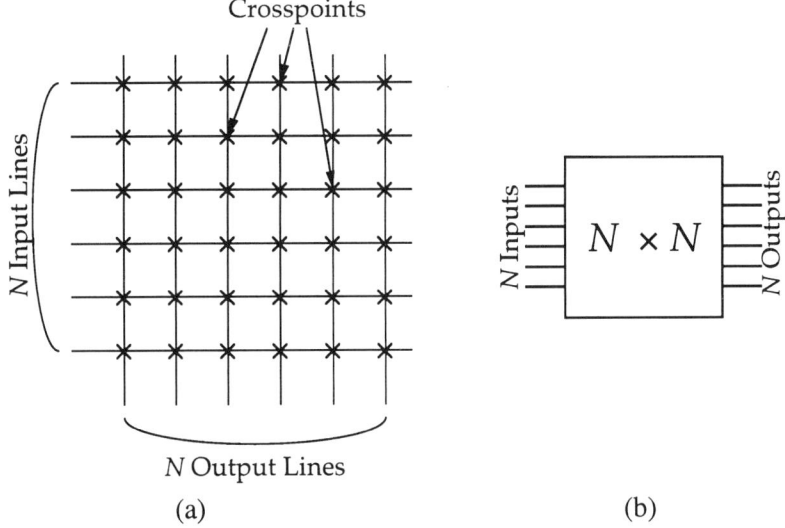

Figure 1 The $N \times N$ Crossbar Switch (a) and Its Schematic (b)

call request in a general switch is a request for a connection from one unused input to one unused output. A call request is *blocked* if the connection can not be put up through the switch. This occurs due to constraints from the architecture of the switch, or due to the current state of the switch. A switch is *non-blocking* if simultaneously given any legal set of calls (each input and output used at most once), the switch can put up all of the calls. A switch is *strictly non-blocking* if any sequence of legal call requests, with intervening call disconnects, can be put up as each request arrives, no matter what algorithm is used for routing calls. The $N \times N$ crossbar is the prototypical example of the non-blocking class. While such a switch is desirable, unfortunately it is at the expense of N^2 crosspoints.

The crosspoint count of a switch is often used as a measure of its cost or complexity. We desire to reduce the number of crosspoints. Usually this is achieved by building larger switches from stages of smaller crossbar switches. Figure 2a shows a general three-stage example built from smaller $r \times r$ and $n \times n$ crossbars. For an appropriate choice of n and r, such a multistage switch has a reduced number of crosspoints. For example, $r = n = 16$ uses more than 5 times fewer crosspoints than the equivalent 256×256 crossbar. Unfortunately, multistage switches are often not strictly non-blocking.

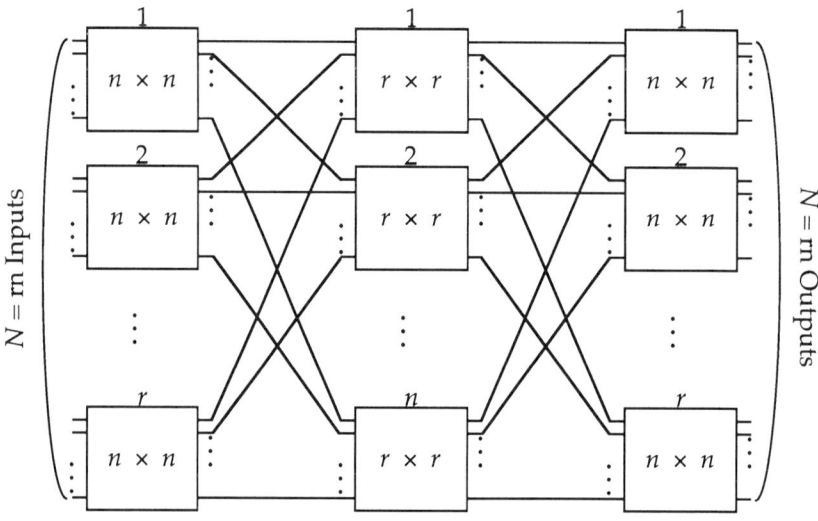

Figure 2 A General Three-Stage Switch

A switch can be blocking in various senses. Yet switches can be designed so that statistically the probability that a call is blocked in whatever sense is small, even when the call traffic being routed through the switch is high. Typically this requires several stages, and the provision of many alternative routes through which a call can be put up [3, pp. 526–552].

2.2 What questions will we address?

In circuit switching, a call, once it is put-up, occupies the circuitry on its path from input to output until the call is torn down at some (usually unspecified) time later. The state of the switch is then determined by what paths are being used. New calls that arrive must be routed from among the remaining paths.

In the case of packet switching, multiple packets can arrive simultaneously at different inputs destined for the same output and can not be routed to the same output at the same time without colliding. Due to the asynchronous nature of packet networks and the potential difficulty in coordinating distributed and disparate users, such so-called *output blocking* must occur. Since blocking can not be avoided, packets must be buffered. Depending on the buffer and switch architecture this can lead to multiple calls waiting at each input with conflicting output destinations.

We focus on two classes of questions one closely tied to circuit and the other closely tied to packet switching. First, given a switch with a set of calls already put up and a set of call requests how do we route the new calls (circuit routing)? Second, given a switch with a set of calls already put up and a set of call requests, how do we choose a non-blocking subset of the requests to put-up (packet arbitration)? The second question is a subset of the first, but is relevant when the routing is readily determined. Despite this simple classification, most switching control (as opposed to design) problems can be framed as either a circuit routing or packet arbitration question. Before we address these questions, we first develop the necessary neural tools.

3 NEURAL CIRCUIT DESIGN

3.1 The Hardware-Based Model

We start with the Hopfield continuous model [4], since it is defined in terms of electronic components, and therefore close to an actual implementation. In this model the neurons are high-gain amplifiers that are interconnected through resistors as shown in Figure 3 for a four-neuron network. The amplifiers have both positive and negative outputs. A connection from a positive output is known as *excitatory*, and a connection from a negative output is known as *inhibitory*. We define $W = \{w_{ij}\}$, the *connection matrix*. If R_{ij} is the value of the resistive connection from neuron j to neuron i, then we say that the strength of the connection, $|w_{ij}|$, is $1/R_{ij}$. The sign of w_{ij} is positive or negative depending on whether the connection is excitatory or inhibitory. Each neuron can have an external input, I_i. A threshold, t_i, is subtracted from all the inputs, and the result is the net input to the neuron.

The response or output of the neuron is the non-linear function $g(x) = f(\gamma x)$, where f is a "sigmoidal" function and γ is the gain. The results here are valid for virtually all definitions of monotonic continuous sigmoidal functions with outputs ranging from 0 to 1 (e.g. $g(x) = 1/(1 + e^{-x})$). Neurons with outputs close to +1 and 0 are said to be *on* and *off* respectively. In a system of N neurons, given a neuron i, its state variable, u_i, is governed by the differential equation:

16   CHAPTER 2

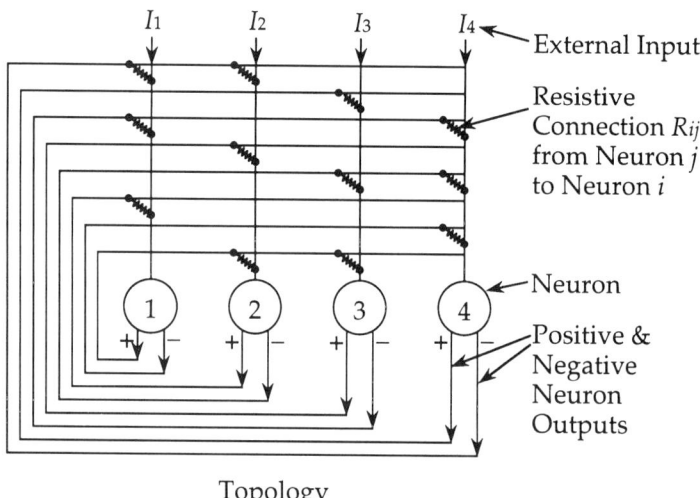

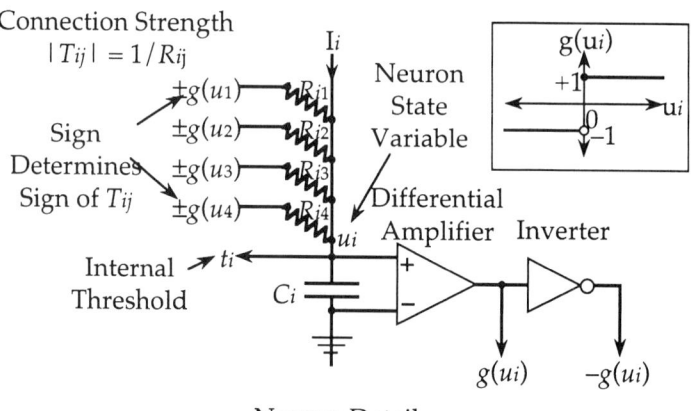

Figure 3 The Electrical Model of a Neural Network Showing Topology and Details of a Neuron

$$c_i \frac{du_i}{d\tau} = -\lambda_i u_i + \sum_{j=1}^{N} w_{ij} g(u_j) + I_i - t_i, \qquad (1)$$

where

$$\lambda_i = \sum_{j=1}^{N} |w_{ij}|.$$

This equation is derived from simple circuit analysis of each neuron. For simplicity, we assume that $c_i = C$ for all neurons. We also set the diagonal elements to 0 although the results here do not depend on this.

We define I_i^{tot} to be the sum of the last three terms in (1):

$$I_i^{tot} \triangleq \sum_{j=1}^{N} w_{ij} g(u_j) + I_i - t_i. \qquad (2)$$

We see that u_i evolves as a negative exponential with time constant c_i/λ_i, and that it decays toward the value I_i^{tot}/λ_i. In general neuron i is on or off depending on whether I_i^{tot} is positive or negative. We proceed now to define a series of useful results initially suggested by results in [5].

3.2 The Winner-Take-All Circuit

With the neural circuitry defined, we discuss a neural circuit that we will use repeatedly, the winner-take-all (WTA) circuit. As its name implies, it has the property that given N neurons all with the same initial internal state (i.e., at $t = 0$, $u_i = u_0$ for all i), only one neuron turns on, the one with the largest external input. This process of choosing the neuron with the largest input is known as a *competition*. By using external inputs, this circuit provides a method for selecting a neuron to turn on.

A generalization of this circuit is the K-winners-take-all (KWTA) circuit. This has the property that not just the neuron with the largest, but the K neurons with the K largest inputs, turn on. For an N-neuron KWTA, we define the network prototypically as:

$$w_{ij} \triangleq \begin{cases} -1 & \text{if } i \neq j \\ 0 & \text{if } i = j \end{cases}$$
$$t_i \triangleq 1 - K, \qquad (3)$$

We limit the inputs by:

$$0 < I_i < 1, \text{ or } I_i < 1 - K. \qquad (4)$$

Note that in general, if $I_i < 0$, neuron i will be off and out of the competition. The circuit is shown in Figure 4a. Using definition (3) in (1):

$$C\frac{du_i}{d\tau} = -\lambda u_i + g(u_i) - \sum_{j=1}^{N} g(u_j) + I_i + K - 1, \qquad (5)$$

where $\lambda = N - 1$. The competition is slightly complicated if fewer than K neurons have a positive input (i.e., $I_i > 0$), but it produces the obvious result:

Theorem 3.1 *Given a neural network defined by (3), satisfying (4), and at time $t = 0$ with all of the neurons starting at the same initial state u_0; let P be the number of neurons satisfying $I_i > 0$. Given these conditions, the neural computation results in the K' neurons with the K' largest inputs on, where $K' = \begin{cases} K & \text{if } P \geq K \\ P & \text{if } P < K \end{cases}$; the rest of the neurons are off. ([11, Theorem 2.2])*

3.3 The Multiple Overlapping Winner-Take-All Circuit

We describe an extension to the winner-Take-All circuit, the multiple overlapping winner-take-all (MOWTA) circuit. The idea is that we have a set of neurons, Θ, with subsets, S_i. These subsets represent constraints on the computation. Within each subset we restrict the neural computation to having at most one neuron on. The idea is shown in Figure 5. Note that the subsets are not necessarily disjoint. Intuitively, the neural network that will satisfy these

Neural Networks for Switching

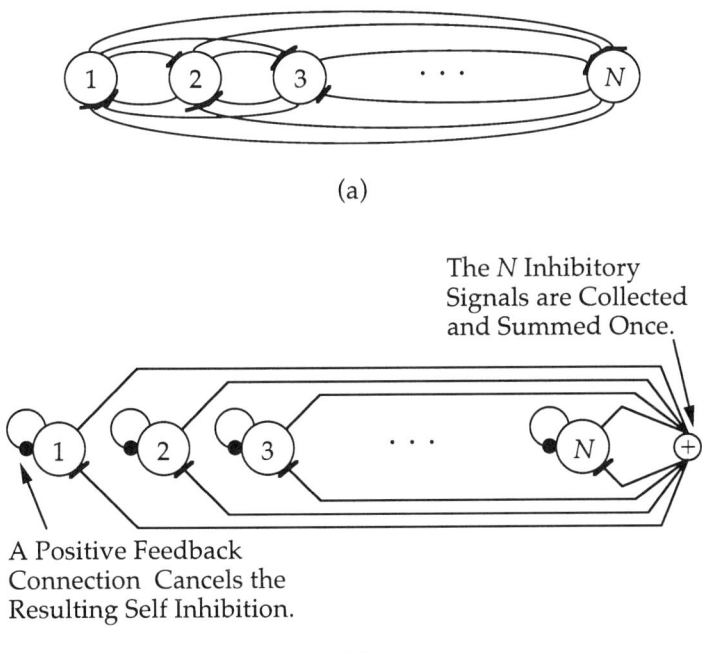

Figure 4 Winner-Take-All Mutual Inhibition Circuit (a), and Connection Saving Equivalent (b)

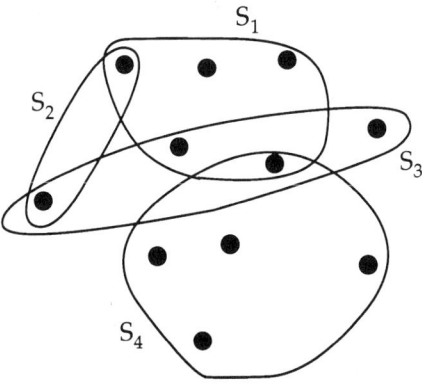

Figure 5 The Multiple Overlapping Winner-Take-All Concept

constraints is the one in which the neurons in each subset are connected in a separate WTA network. For a given Θ and $\{S_i\}$, we define the network as follows:

$$w_{ij} \triangleq \begin{cases} -(\text{number of } S_i \text{ that neurons } i \text{ and } j \text{ are jointly in}) & \text{if } i \neq j \\ 0 & \text{if } i = j \end{cases}$$
$$t_i \triangleq 0, \tag{6}$$

Note that if there is only one subset, $S_1 = \Theta$, then we reduce to the WTA. We limit the inputs as in the WTA case by:

$$0 < I_i < 1, \text{ or } I_i < 0. \tag{7}$$

We state the desired result as a theorem:

Theorem 3.2 *Given a neural network defined by (6), and satisfying (7), then the neural computation results in each subset, S_i, having at most one neuron on, the rest off.* ([11, Theorem 2.4][25, Appendix B])

It can be shown that the neural network never chooses the trivial solution of all neurons off, and in fact will always turn on a neuron if it does not violate any constraints. This is easy to see since if all of a neuron's neighbors are off, then $I_i^{tot} = I_i > 0$ for at least one neuron unless all of the neurons are constrained to be off ($I_i < 0$).

3.4 The Multiple Overlapping K-Winner-Take-All Circuit

We develop one final generalization to the winner-take-all circuit, the multiple overlapping K-winner-take-all circuit (MOKWTA). This is identical to the multiple overlapping winner-take-all circuit, except that in certain instances, we can define additional subsets, S_i', where not at most one neuron, but at most K_i neurons are allowed to turn on per S_i'. The only limitation on these subsets is that S_i' and S_j' are disjoint for all $i \neq j$, that is, each neuron can belong to no more than one set that allows more than one neuron to turn on.

For a given Θ and two sets of constraints $\{S_i\}$ and $\{(S'_i, K_i)\}$, we first define a MOWTA circuit using Θ and $\{S_i\}$ in (6). Using this as a basis, we incorporate the additional constraints.

$$w'_{ij} \triangleq \begin{cases} w_{ij} - \frac{1}{K_k} & \text{if } i \neq j \text{ and neuron } i \text{ and } j \text{ are in } S'_k, \\ w_{ij} & \text{otherwise}; \end{cases}$$

$$t'_i \triangleq \begin{cases} \frac{1}{K_k} - 1 & \text{if } i \text{ is in } S'_k, \\ 0 & \text{otherwise}. \end{cases} \quad (8)$$

Comparing this definition with (3) we see that we are simply adding a K-winner-take-all circuit to each neuron set, S'_i, that is scaled by a factor of $\frac{1}{K_i}$. Define $K_{max} \triangleq max_i\{K_i\}$. We limit the inputs by:

$$0 < I_i < \frac{1}{K_{max}}, \text{ or } I_i < 0. \quad (9)$$

This further reinforces the scaled KWTA concept. We now state the result:

Theorem 3.3 *Given a neural network defined by (8) and satisfying (9), then the neural computation results in each subset, S_i, having at most one neuron on, and each subset, S'_i, having at most K_i neurons on, the rest of the neurons being off*([11, Theorem 2.6]).

3.5 Designing with Winner-Take-All Circuits

Having defined these winner-take-all circuits, we show how to reduce their connectivity, and how we can incorporate neural networks into larger systems. To reduce the connectivity, we note that two neurons in Figure 4a, i and j, have the same input except that i does not connect to itself, but connects to j, and vice versa for j. Several researchers have noted that we can exploit this regularity [5, 6, 7]. They all fundamentally rely on the same principle. By making a weighted sum of all neurons only once for the whole circuit and providing a self connection in each neuron to negate the resulting feedback, we reduce the number of connections needed from $N(N-1)$ to $3N$ for a single

WTA. This modified circuit is shown in Figure 4b. We will assume that all mutually inhibitory connections are made in this manner.

For a multiple overlapping winner-take-all circuit, we could connect the network using the definition in (6). But every neuron is connected to every other neuron within each subset. This implies a total of $O(\sum_i |S_i|^2)$ connections. If instead we connect each subset in a separate WTA as described above, we produce a network which is mathematically equivalent to (6), but now there are only $3|S_i|$ connections per subset S_i. This results in a total of $3\sum_i |S_i|$ connections in the entire network, yielding a significant savings.

The definitions and analysis of these WTA circuits are all for a particular network scale. But as can be seen in (1) the network can be matched to the particular voltage and current levels appropriate for a particular implementation by scaling the connection resistors. From (1) we see that the threshold, t_i, and the external input, I_i, are fundamentally the same except for a change in sign. Since the only dependency on K in (5) occurs in the threshold, by adjusting the external input to all of the neurons we can change the threshold, and so use the same circuit as a K-winners-take-all circuit for any K. We also note that inputs from neurons outside of the WTA are equivalent to the input and threshold, therefore these can be used to modify the value of either I_i or t_i.

The definition of the KWTA allows for a range of values for the external input, I_i. These can be used to indicate various levels of "priority" of the neurons. The neurons with the highest priority will then be the neurons which win the competition.

The analysis of Section 3.2 relies on initially identical internal states. This requires outside circuitry that can reset the network every time new winners must be selected. If we relax the requirement that the internal states are all initially identical, we only lose the ordering on the internal states, otherwise the result is the same. This implies that we can use the KWTA in addition to the MOWTA in a completely asynchronous mode as a selector, or we can use the KWTA circuit as a discriminating selector. By using external inputs and neurons outside of the circuit as described previously, we can "program" the WTA to compute particular functions.

3.6 The Hopfield Model

Before we continue, it is instructive at this point to ask why use these techniques. They appear to be restricted to simply winner-take-all structures, and lack the apparent generality of the more popular Hopfield energy function approach [8]. We briefly explore the Hopfield energy function and show that whenever possible the direct WTA approach should be taken. Fortunately, this is often possible.

The Hopfield energy function is simply a quadratic function over a set of variables in the range $[0, 1]$. Hopfield devised a method for converting this equation to a neural network feedback system with one neuron corresponding to each of the variables and connections determined from the equation. As the neural network evolves over time, the neuron outputs change so as to continuously reduce the energy function until it reaches a minimum. Typically, a design begins by listing all of the constraints on the neurons and then devising quadratic energy function terms that are minimized when the constraints are satisfied. For example, given $V_i = g(u_i)$ and constants c_i $(1 \leq i \leq N)$:

$$E = A \sum_i V_i(1 - V_i) + B(\sum_i V_i - 1)^2 - C \sum_i c_i V_i$$

would represent an energy function that is minimum when all V_i are either 0 or 1 and only the V_i with largest c_i is on; i.e., a WTA. The parameters A, B, and C represent the relative weighting that the designer places on all the neurons being 0 or 1 (A), exactly one neuron turning on (B), and the influence of the c_i (C).

This example, while simple, embodies the method's deficiencies. Although reducible by choosing the correct A and B to (3) it is possible that even with A and B correct, no C exists that gives the WTA behavior (e.g. if all the c_i are negative). Further a group of parameters that works for this network may not work if the size of the problem is increased. These problems grow as the number of constraints (energy terms) grow. Often, the majority of simulation effort focuses on finding workable parameter combinations. From this we see that while it is easy to define networks that tend to compute what is desired (the lure of the Hopfield method), there are no guarantees that any of the constraints will be met. Even with a good set of parameters in the above example, the designer would still miss the circuit simplifications of the previous section resulting in N^2 instead of N connections.

In summary, the Hopfield approach requires careful balancing of parameters, does not always satisfy embodied constraints for all inputs, and results in complicated networks with many connections. In addition, it is an unnecessary step that obscures the structure of the neural solution. This is in stark contrast to the WTA design framework described above.

The Hopfield approach is more generally applicable, but still many optimization problems can be shown to fit within the WTA framework. For instance in the N-city traveling salesman problem, using a formulation similar to [9] and a direct application of the framework here, we can produce a neural network guaranteed to produce a valid tour with order N^2 connections as opposed to previous best of order N^3. As with virtually all neural solutions to optimization problems, this produces very good and but not necessarily optimal solutions. As we shall see, this is often more than sufficient. To understand the ease and breadth of the WTA frameworks applicability we return to the switching control problem.

4 NEURAL APPROACHES TO SWITCHING

In this section we use the winner-take-all framework from the previous section to solve a variety of problems from circuit switching. The problems are based on neural attempts to solve switching available in the literature. Typical approaches are based on using the problematic Hopfield energy function approach [8]. Interestingly we can show either directly or with a modified problem representation that the problems addressed can be solved with-in the WTA framework. Because of the rigorous nature of the previous WTA results, can gain guaranteed constraint satisfaction with a reduced number of connections defined using no free parameters.

4.1 Circuit Routing

As mentioned earlier, large multi-stage switches are typically blocking. But, by having a large number of routes available between any input and output of the switch, the probability that all these routes are blocked simultaneously is very low. The AT&T ESS #4 toll switch, for example, has 1,920 (16×120) possible routes between any input and output pair. The problem is that, in such a switch, finding an unblocked route from the many possible routes is a time consuming task, especially as the switch becomes loaded with calls. The problem can be generalized to the case when the outlet stage is not unique, as for toll switches where any one of several trunks between two offices would be acceptable.

The simplest parallel algorithm to find a route checks all possible routes at the same time, and then chooses one unblocked route if it exists. This algorithm was developed for arbitrary size multistage networks using a neural network by the author in [10][11]. The key element is a winner-take-all at each stage in order to choose a unique path that is influenced by signals propagating from both the desired input and desired output performing a full parallel search through available paths. The case of a three-stage Clos switch in Figure 2 is especially simple and was also addressed in references in [12] using the above mentioned Hopfield approach.

Whereas the above methods apply to the case where calls are set up and torn down one at a time, alternative approaches were developed in [13], [14] and [15] for the case where all calls are setup at once. The first is for arbitrary multistage switches and is described and analyzed in the next chapter by Goudreau and Giles. The second is designed specifically for three-stage networks. The third applies to the Beneš network [1], a non-blocking multistage switching network that has the property of using a minimal number of crosspoints. Reference [15] describes an interesting neural network that will always find a routing through the Beneš network for any set of legal call request in time complexity less than for any known algorithm.

Another circuit routing problem and neural solution is given in [16]. This applies to one sided switches that try to minimize noise by distributing calls evenly through the switch.

References [13]–[15] all use interesting variations on the approach of constraining each call to choose only one route, and each route to choose only one call—clearly an application for the WTA framework. But, instead they are all based on the Hopfield energy function with the result that several of these papers report occasional convergence to invalid states, while others have follow-on papers that experimentally define workable energy function parameters. Using the WTA framework directly in these designs would obviate this as well as bringing the benefit of reduced connectivity.

For [10] [12]–[15] the problem representations are given in terms of WTA-like constraints and lend themselves to direct applications of the WTA framework. Reference [15] is noteworthy in that it produces optimal solutions, and claims to have a proof that the neural network will always give the desired result.

Reference [16] does not map directly into the framework. But with an alternate representation, too lengthy to describe here, it can also be fitted.

4.2 Packet Arbitration

Model Definition

A typical packet routing situation is given in Figure 6. The packets arrive at an

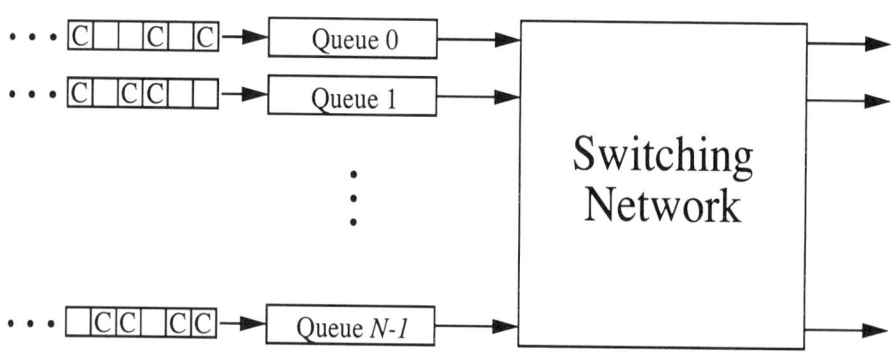

Figure 6 The Queueing Model

input according to some random process, and each packet is destined for one of the outputs according to a second process. For simplicity here we assume that the packets are fixed length and arrive synchronously at periodic time intervals (Such as in the asynchronoous transfer mode (ATM) protocol). Packets that are blocked and can not be sent are queued at each input. We also assume that an input can choose any packet in its queue to send. In this paper, we will not focus on the relative advantages and disadvantages of this model relative to alternative queueing and packet switching models (see [17]–[20]) but rather the neural approaches to this model.

This very model has been addressed directly for the case of a non-blocking switch in [21]–[24]. The first three take a pure Hopfield approach, although the first is notable in that its analysis are based on a VLSI circuit simulation. The fourth uses an alternative approach that we build upon in a later section.

In [25] and [11] a broader switch class is addressed denoted as *deterministic*. A switch is deterministic if:

A. The switch is composed of non-blocking switch nodes.

B. The nodes can be arbitrarily interconnected, but any external input or output has only a single link to one of these nodes.

C. Between each input and output there is exactly one route (defined in terms of links and nodes).

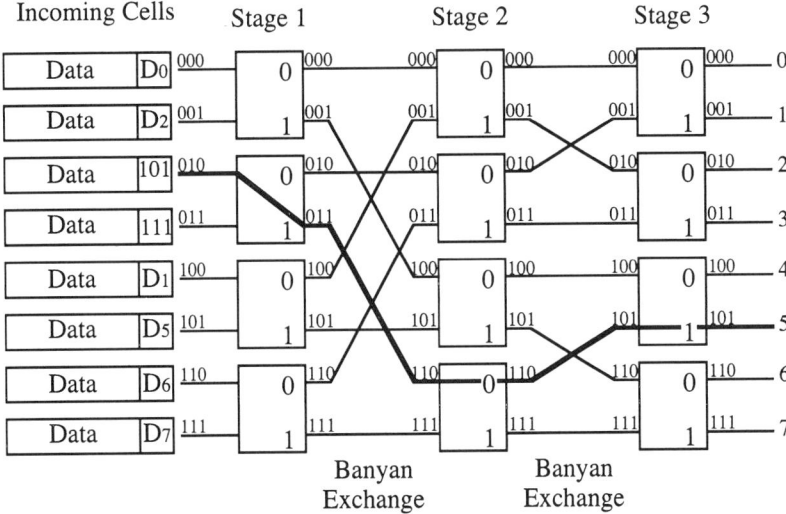

Figure 7 An 8 × 8 Example of a Banyan Switching Network

The well-known Banyan network in Figure 7 is an 8 × 8 deterministic switch composed of $log_2(N) = n$ stages of 2 × 2 nodes. A non-blocking switch is a deterministic switch with a single node—the switch—that has links to the inputs and outputs. Omega networks, baseline networks, and flip networks are all topologically equivalent to the Banyan network and therefore also deterministic switches [26].

A packet waiting to be sent from input I to output destination D we denote with the ordered pair (I, D). For a given switch architecture, let $S = \{(I, D)\}$ be a set such that every packet in this set is mutually blocking, that is, given any two packets in S, these packets collide somewhere in the switch. Such a set we call a *constraint set*. For deterministic switches, these constraints are simple to define, and can be used to completely define blocking.

To define the constraint sets, we note that whenever two packets both attempt to use the same link between two switches, there is a collision, and thus blocking. These two packets will always collide, since they have only one choice for routes, and both routes are through this link. Each of the switching elements are non-blocking, as long as only one packet arrives per input and leaves per output. Therefore, none of the packets are blocked if and only if there is no link used by more than one packet.

Let $\mathcal{L}^A = \{l_j\}$, be the set of links in a switch with architecture A. Each link, l_j, in the switch defines a constraint set,

$$S_j = \{(I, D) |\text{ the route from inlet } I \text{ to outlet } D \text{ uses link } l_j\}.$$

Using these constraint sets, our definition of non-blocking is clear:

Definition: A set of packets, $P = \{(I, D)\}$, is *non-blocking* if and only if each set, S_j, contains at most one $(I, D) \in P$.

For the Banyan switch, $\mathcal{L}^{\text{Banyan}} = \{l_j^k\}, 0 \leq k \leq n, 1 \leq j \leq N$, where l_j^k is the link connected to the jth outlet of the kth stage of switches. In the case of the non-blocking switch, the only links are the input and the output links. In terms of the Banyan definition of l_j^k, $\mathcal{L}^{\text{NB}} = \{l_j^k\}$, where $k \in \{0, n\}$, and $0 \leq j \leq N$.

A more graphical interpretation can be gleaned from looking at an $N \times N$ matrix of all input output combinations. Using this matrix, Figure 8 shows the $log_2(N) + 1 = 4$ different levels of constraint sets for an $N = 8$ input Banyan.

Given a set of packets $P = \{(I, D)\}$ queueing at the inputs of the switch, we create a binary $N \times N$ *packet matrix* by placing a 1 at entry (I, D), $\forall (I, D) \in P$ as shown in Figure 9.

The controller then uses the blocking constraints on the subsets S_j to choose a subset P' that is non-blocking. Optimally the network controller will choose a P' that has the maximal overlap with P. For traffic uniformly distributed on

Neural Networks for Switching

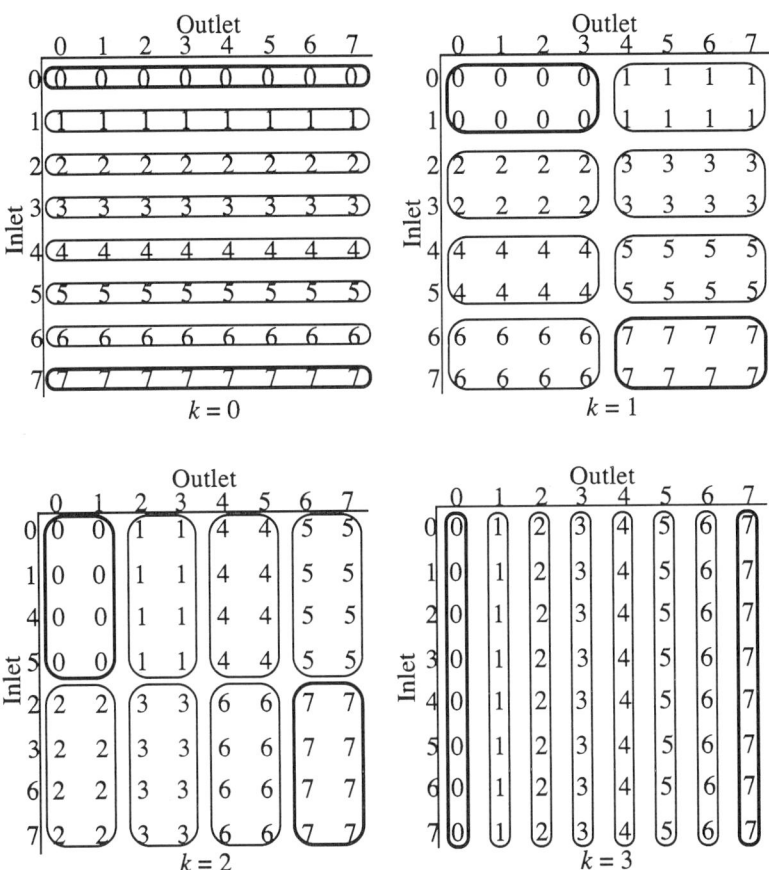

Figure 8 The Constraints Induced on the 8 × 8 Input Matrix by Each Stage of Links in a Banyan. The numbers in the matrix indicate which link a packet from a given inlet to outlet uses at stage k.

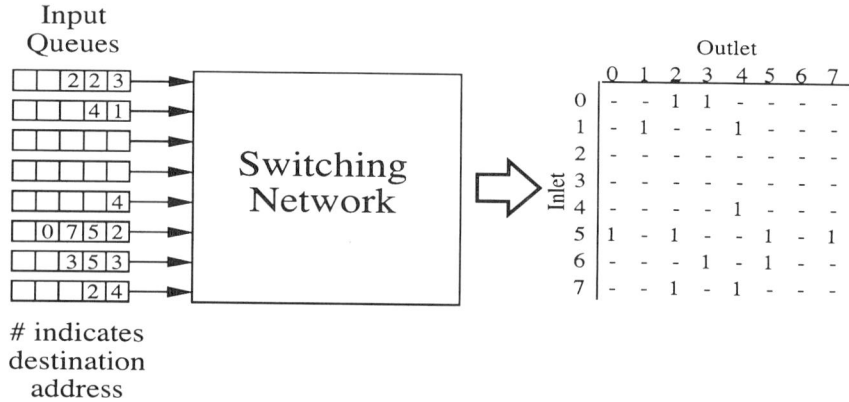

Figure 9 The Mapping of the State of the Queues to a Packet Matrix. A 1 is placed in (i,j) of the matrix if a packet is waiting at input queue i to be sent to output address j.

the outputs, this approach has the maximal throughput even if the controller only randomly chooses a P' from all possible non-blocking sets without regard for P (see [11, 25] for more detail), and thus any more sophisticated controller that takes into account of P will achieve this rate.

Neural Approach

This section designs a neural network that chooses a set of non-blocking set of packets from the queue. The complexity of the design is calculated, and compared with other proposed neural approaches.

For our problem, we use N^2 neurons arranged in an $N \times N$ matrix that corresponds to the packet matrix of Figure 9. This is our neuron set Θ. The input to the problem is a particular packet matrix, that is, neuron (I, D) receives a positive (say 0.5) input if a 1 is in entry (I, D) of the packet matrix, negative (say -1) otherwise. If at the end of the computation, neuron (I, D) is on, then send the packet (I, D). If more than one packet has the same (I, D), then send the first (I, D) that arrived in the queue.

The neural network design follows directly from the constraint sets, S_i, defined in the previous section. Using Θ and $\{S_i\}$ from Section 3.3, we design a multiple overlapping winner-take-all circuit according to (6). By definition, this has the property that no more than one neuron will be on per constraint set. Therefore, the set of neurons that are on at the end of the computation will always correspond to a non-blocking set of packets.

To measure the complexity of this circuit, we count the number of interconnections. Using the construction method of Section 3.5, the number of interconnections in the network is $O(\sum_i |S_i|)$ connections. In the case of the Banyan network, there are $O(N)$ connections for each equivalence class, S_j^k. There are $(n+1)N$ of these classes, so the number of connections in the network is just $O(N^2 log_2 N)$. For the non-blocking switch, there are $2N$ equivalence classes. The number of connections in this case is just $O(N^2)$. We can assume that the minimum circuit in either case would have at least N^2 connections, one for each neuron in the circuit. Thus, we have a circuit that is defined for any N, provably works, and is within a factor of $log_2 N$ of the smallest possible network. We contrast this with the resulting networks from the energy function approach to optimization [8]. Using that approach on the problem of choosing a non-blocking set of packets for a non-blocking switch [23], it is seen that the resulting network has $O(N^3)$ connections instead of $O(N^2)$, a factor of N more connections, and further, the strength of the connections must be experimentally determined for every N.

The MOWTA does not guarantee finding the maximal overlap set P'. In fact it could be argued that it is only performing a simple greedy search. The performance of this network under realistic traffic conditions was analyzed extensively in [25, 11] and shown to have performance significantly better than a pure greedy search and for all but the highest loads, nearly optimal.

Extension for Optimal Non-Blocking Arbitration

Reference [24] describes an architecture that produces optimal results for a non-blocking switch. It uses the fact that in the non-blocking case the packet arbitration problem is equivalent to a maximum flow problem that has an interesting and precise analog solution. The structure is shown in Figure 10. For each waiting packet, the corresponding crosspoint is closed. The current sources at the edges provide one unit of current. A computation consists of setting the crosspoints and letting the circuit settle, and then reading the currents at the crosspoints.

It can be shown that this circuit will always reduce to zero or unit currents at each crosspoint and result in a maximal overlap solution except in two cases: (1) more than one optimal solution exists; (2) the maximum network flow is less than N (a reduced rank solution). In both cases it will result in partial

32   CHAPTER 2

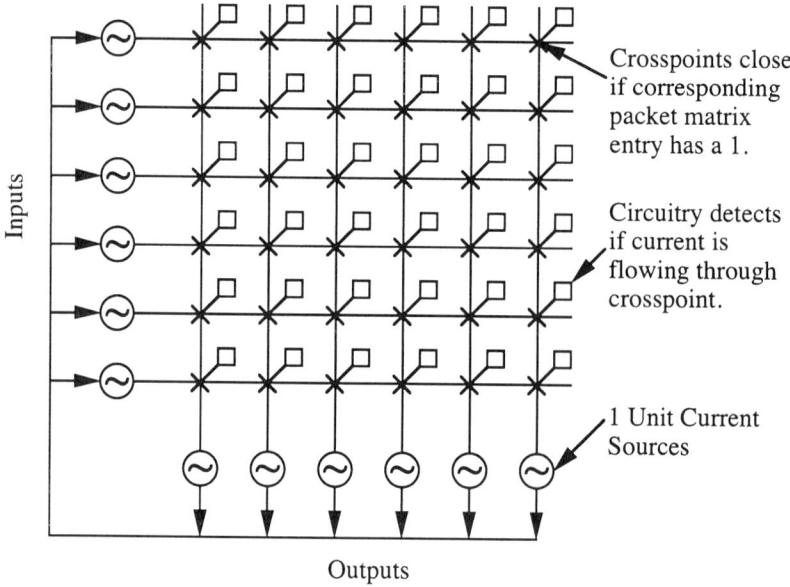

Figure 10 An Analog Computational Solution to the Maximum Flow Problem

currents on crosspoints that are part of the optimal solution. In these cases the trick is to choose a solution from this subspace of chosen nodes. This can be done simply by using the output of the crosspoints from the analog max-flow computation as an input to an $N \times N$ neural network designed for a non-blocking switch as described in the previous section.

In this way, the max-flow computation will always find a maximal solution when possible, and when not, the neural network can choose a solution from the subspace of packets chosen by the max-flow computation.

5   CONCLUSION

This chapter presented a framework for designing neural network solutions using extensions of the winner-take-all circuit. Unlike the Hopfield energy function approach that requires the researcher to first define the constraints of the problem and then go through an imprecise and obscuring energy function to define the weights, these networks have properties that can be directly defined and controlled. This direct approach allows efficient implementations that are scalable to large sizes and as long as the external inputs are within defined limits, the network will always satisfy the constraints embodied in the winner-take-all circuits.

The winner-take-all circuit served as an efficient means for communicating the many constraints of the problems between the neurons in our solutions. The neural network design, based on simple electrical components, has a direct implementation in hardware. Thus, the neural components can be concentrated where they can excel: as highly interconnected massively parallel feedback elements.

While applicable to a variety of optimization problems, a review of the literature reveals that more than 10 results on neural network switch control (all that could be readily found) can be placed within this framework with the immediate benefits of reduced connectivity by a factor of at least one order off magnitude, the elimination of invalid computation outputs, and for the crossbar packet switch, an improved controller that should surpass the performance of any known high-speed controller.

The solutions described here show that neural networks can be applied to practical problems. We emphasize that although each instance of these problems requires a different neural network, the solution is well defined, even for large-size problems where the parallelism of the neural network is most advantageous.

REFERENCES

[1] Beneš, V.E., *Mathematical Theory of Connecting Networks and Telephone Traffic*. Academic Press, New York, 1965.

[2] Inose, H., *An Introduction to Digital Integrated Communications Systems,* Univ. Tokyo Press, 1979.

[3] Schwartz, M., *Telecommunication Networks: Protocols, Modeling, and Analysis,* Addison-Wesley Pub. Co., Reading, MA., 1987.

[4] Hopfield, J. J., "Neurons with Graded Response Have Collective Computational Properties Like Those of Two-State Neurons," *Proc. Nat. Acad. Sci. USA,* vol. 81, pp. 3088–92, May 1984.

[5] Majani, E., Erlanson, R., Abu-Mostafa, Y., "On the K-Winners-Take-All Network," in *Advances in Neural Information Processing Systems I,* Touretzky, D.S. Ed., Morgan Kaufman, San Mateo, CA, 1989. pp. 634–42.

[6] Eberhardt, S.P., Daud, T., Kerns, D.A., Brown, T.X, Thakoor, A.P., "Competitive Neural Architecture for Hardware Solution to the Assignment Problem," *Neural Networks*, vol. 4, pp. 431-42, 1991.

[7] Lazzarro, J., Ryckebusch, S., Mahawold, M.A., Mead, C.A., Winner-Take-All Networks of $O(N)$ Complexity, in *Advances in Neural Information Processing Systems I,* Touretzky, D.S., Ed., Morgan Kaufman, San Mateo, CA, 1989, pp. 703–11.

[8] Hopfield, J.J., Tank, D.W., "Neural Computation of Decisions in Optimization Problems," *Biological Cybernetics,* vol. 52, pp. 141–52, 1985.

[9] Joppe, A., Cardon, H.R.A., Bioch, J.C., "A Neural Network for Solving the Traveling Salesman Problem on the Basis of City Adjacency in the Tour," in *Proceedings of the International Joint Conference on Neural Networks (IJCNN '90),* vol. 3, 1990, p. 961–4.

[10] Brown, T.X, "Neural Networks for Switching," *IEEE Communications Magazine,* vol. 27, no. 11, pp. 72–81, Nov. 1989.

[11] Brown, T.X, *Neural Network Design for Switching Network Control,* Ph.D. Thesis, California Institute of Technology, Pasadena, CA, 1991.

[12] Melsa, P.J.W., Kenney, J.B., Rohrs, C.E., "A Neural Solution for Call Routing with Preferential Call Placement," in *IEEE Global Telecommunications Conference and Expo Proceedings (GLOBECOM '90),* vol. 2, 1990, pp. 1377–81.

[13] Goudreau, M.W., Giles, C.L., "Neural Networking Routing for Random Multistage Interconnection Networks," in *Advances in Neural Information Processing Systems 4,* Touretzky, D.S., Ed., Morgan Kaufman, San Mateo, CA, 1992, pp. 722–9.

[14] Ma, J., Rahko, K., "A Neural Network Controller for the Three-Stage Clos' Packet Switch," in *XIV International Switching Symposium (ISS '92),* vol. 1, 1992, pp. 234–8.

[15] Hakim, N.Z., Meadows, H.E., "A Neural Network Approach to Set Up the Benes Switch," in *The Conference on Computer Communications Proceedings of the 10th Annual Joint Conference of the IEEE Computer*

and *Communications Society (INFOCOM '90),* vol. 2, IEEE Computer Society Press, Los Alamitos, CA, 1990, pp. 397–402.

[16] Ghosh, J., Hukkoo, A., Varma, A., "Neural Networks for Fast Arbitration and Switching Noise Reduction in large Crossbars," *IEEE Transactions on Circuits and Systems,* vol. 38, no. 8, Aug. 1991, pp. 895–904.

[17] Karol, M.J., Hluchyj, M.G., Morgan, S.P., "Input vs. Output Queueing on a Space-Division packet Switch," *IEEE Transactions on Communications*, vol. 35, pp. 1347–56, Dec. 1987.

[18] Hluchyj, M.G., Karol, M.J., "Queueing in High-Performance Packet Switching," *IEEE Journal of Selected Areas in Communications* vol. 6, no. 9, pp. 1587–97, Dec. 1988.

[19] Giacopelli, J.N., Hickey, J.J., Marcus, W.S., Sincoskie, W.D., Littlewood, M., "Sunshine: A High Performance Self-Routing Broadband Packet Switch Architecture," *IEEE Journal on Selected Areas in Communications,*, vol. 9 no. 8, pp. 1289–98, Oct. 1991.

[20] Kuwahara, H., Endo, N., Ogino, M., Kozaki, T., Sakurai, Y., Gohara, S., "A Shared Buffer Memory Switch for an ATM Exchange," in *Proceedings ICC*, 1989, pp. 118–22.

[21] Ali, M.M., Nguyan, H.T., "A Neural Network Implementation of an Input Access Scheme in a High-Speed Packet Switch," in *IEEE Global Telecommunications Conference and Expo Proceedings (GLOBECOM '89),* vol. 2, 1989, pp. 1192–6.

[22] Sun, K.T., Fu, H.C., "A Neural Networking Algorithm for Solving the Traffic Control Problem in Multistage Interconnection Networks," in *Proceedings of the International Joint Conference on Neural Networks (IJCNN '91),* vol. 2, 1991, p. 1136–41.

[23] Troudet, T.P., Walters, S.M., "Neural Network Architecture for Crossbar Switch Control," *IEEE Transactions on Circuits and Systems,* vol. 38, no. 1, Jan. 1991, pp. 42–56.

[24] Morris, R.J.T., Samadi, B., "Neural Networks in Communications: Admission Control and Switch Control," in *International Conference on Communications Conference Record (ICC 91),* vol. 2, 1991, pp. 648–54.

[25] Brown, T.X., Liu, K.H., "Neural Network Design of a Banyan Network Controller," *IEEE Journal of Selected Areas in Communications*, vol. 8, no. 8, pp. 1428–38, Oct. 1990.

[26] Wu, C.L., Feng, T.Y., "On a class of Multistage Interconnection Networks," *IEEE Transactions on Computer*, vol. 29, no. 8, pp. 694–702, Aug., 1980.

_3_

ROUTING IN RANDOM MULTISTAGE INTERCONNECTION NETWORKS

Mark W. Goudreau*
and C. Lee Giles*†

This chapter is a shortened version of [9]. Reprinted with permission.

*NEC Research Institute, Inc.
† UMIACS, University of Maryland

1 INTRODUCTION

The design and use of interconnection networks is a topic that has generated considerable interest for many years. The reason for this interest is the wide applicability of the field. Interconnection networks have been widely studied and used for telecommunications, parallel processing, and distributed computing.

Neural networks have been used to solve certain interconnection network problems, such as finding legal routes [3, 7, 8, 9, 10, 17], finding preferential legal routes [16, 19, 22], and increasing the throughput of an interconnection network [4, 5, 6, 15, 18, 20, 21].

A random multistage interconnection network (RMIN) is defined as a multistage interconnection network that has "random" connections between stages. A RMIN will be constructed out of complete or incomplete crossbar switches. This chapter is concerned with RMINs that have structures that *do not* allow simple routing algorithms to be used. Depending on the requirements of the communication system, exhaustive search routing or greedy routing would be likely candidates to establish routes in such a RMIN. The purpose of this work is to discover whether the neural network router that is described in Section 4 can compete with exhaustive search routing and greedy routing for some RMINs.

To this end, three separate RMINs are used to test the *ability* of each of the three routing schemes to solve the routing problem. These interconnection networks have enough structure that they may be classified as RMINs, but they have insufficient structure to have a simple routing scheme (as is available for multibutterflies, for example). The communication system operates in a cyclic fashion. A distributed computing, synchronous model is assumed, as opposed to a telecommunication, asynchronous model. At the beginning of each message cycle, each input port may desire to communicate with one of the output ports of the RMIN. It is the job of the router to establish as many of the point-to-point connections as possible. For each of these RMINs, all three routing methodologies were simulated for differing message set sizes. Obviously, exhaustively searching for routes through a RMIN will, in general, be a prohibitively difficult approach. However, in terms of the *ability* of the RMIN to establish the maximum number of routes (i.e., if neither *speed* nor *resource utilization* are criteria), the exhaustive search approach is optimal. For the rest of this chapter, the term *optimal router* will be used for a router that has the *ability* to always find the best solutions for routing problems.

Greedy routing performs less well in terms of ability, while it has clear advantages when it comes to speed. The same could be said for neural network routing. The simulation results show that although the exhaustive search routing does indeed solve the routing problem best for the three sample RMINs, the performance is comparable to that of both greedy routing and neural network routing. This would make greedy routing or neural network routing the most viable choices for routing in small RMINs. There is no clear winner: the best approach depends on the needs of the communication system. However, the neural network router is a competitive alternative that may be appropriate for many communication applications involving small RMINs that are not self-routing.

2 THE COMMUNICATION SYSTEM

Assume the system that is being modeled is a shared memory, distributed computing system. Figure 1 contains a diagram of such a system. The purpose of the interconnection network is to allow the processors (on the left) to communicate with the memory modules (on the right). This is a circuit switching problem, as opposed to a message passing problem. The switches in the interconnection network have no buffers. An input/output connection must therefore be established in its entirety for communication to occur (this will be called *establishing connections* or *routing messages* interchangeably). Also, no broadcasting will be allowed. Each processor can only request to access one memory module in each message cycle.

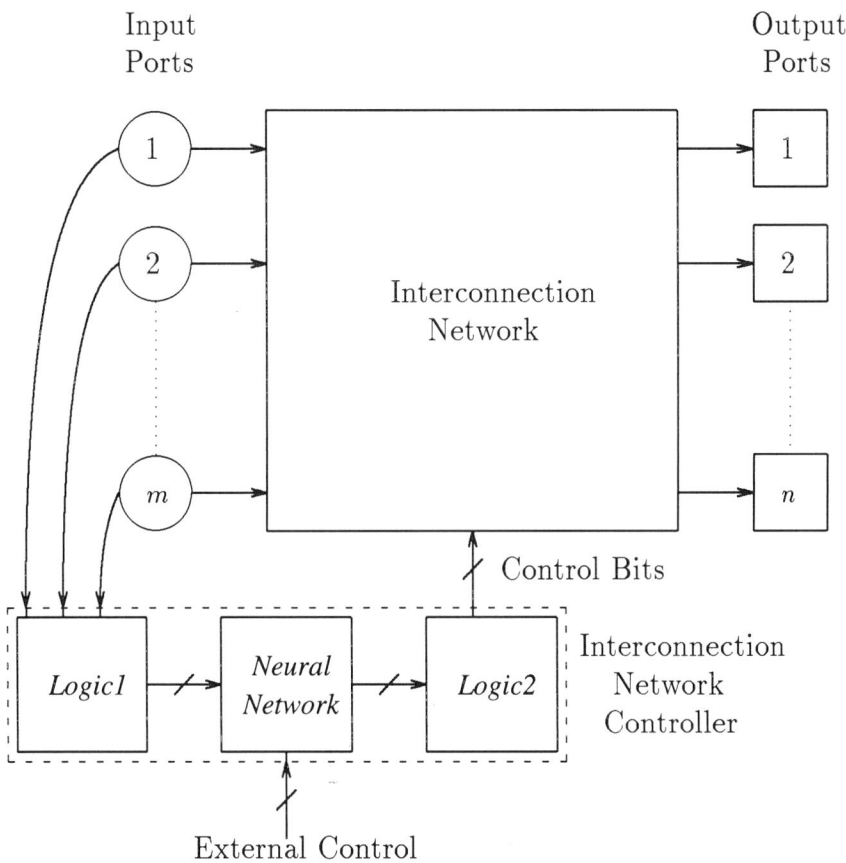

Figure 1 The communication system with a neural network router.

The interconnection networks that are to be examined in this chapter are RMINs that have no self-routing capability. Hence, a separate controller (or router) is required to generate control bits for the interconnection network.

This communication system will operate in a cyclic fashion. At the beginning of each *message cycle*, a processor may request access to a memory module. It is up to the interconnection network controller to find a route if possible. The processor that desires communication with a memory module and the memory module that it wishes to communicate with make up a *message pair*, or simply a *message*. The processors may be indexed by number (from 1 to m), as can the memory modules (from 1 to n). Thus, if processor 3 wants to access memory module 5, we say that message (3,5) is desired. Each message may have zero, one, or more than one possible *routes*. A route is the path that the message takes. A route is defined by the processor number of the message and the number of the output port that the message uses on *each stage* of the interconnection network. Note that the output port of the last stage of the interconnection network will be the same as the memory module address of the message.

Naturally, at the beginning of each message cycle, more than one desired message may exist. The entire set of messages that is desired at the beginning of some message cycle is termed the *message set*. The size of the message set will henceforth be denoted as M. Since each processor can try to communicate with no more than one memory module, $M \leq m$. In general, it may not be possible to establish all M of the desired connections in a single message cycle. This may be due to the limitations of the interconnection network and/or the limitations of the interconnection network controller. If a processor finds that its attempt to communicate with a certain memory module has failed, it may try to communicate again in the next message cycle.

It is the responsibility of the interconnection network controller to route as many messages from the message set as possible. When considering this criterion, an exhaustive search approach will provide an optimal interconnection network router. Clearly, if one examines every combination of routes for a message set, the maximum number of messages may be routed. Any connections that can not be established using an exhaustive search router will be due to the limitations of the interconnection network itself. The greedy router, however, is not optimal. Nor is the neural network router. The limitations of a greedy router or a neural network router may prevent some connections from being established. While an exhaustive search router may be impractical for the vast majority of applications, it provides a convenient upper bound against which to measure the performances of the other routing schemes.

3 EXHAUSTIVE SEARCH AND GREEDY ROUTING

A router that uses an exhaustive search approach will not be practical for most routing problems. However, in terms of the ability of a router to route the maximum number of messages in each message cycle, an exhaustive search router will be optimal. Thus, an exhaustive search router may serve as an optimal router for the purposes of comparison. The exhaustive search implementation that is used for the purposes of this chapter is a brute-force, sequential approach.

The exhaustive search router works in the following manner. An interconnection network is chosen. Every route from each input to each output is stored in a database. (The RMINs that are used as test cases in this chapter always have at least one route from each processor to each memory module.) When a new message cycle begins and a new message set is presented to the router, the router searches through the database for a combination of routes for the message set that has no conflicts. A conflict is said to occur if more than one route in the set of routes uses a single bus in the interconnection network. In the case where every combination of routes for the message set has a conflict, the router finds a combination of routes that will establish the largest possible number of desired connections.

It should be noted that there are many other ways to implement an optimal router for a RMIN. However, the details of the operation of the exhaustive search router are not really important. What is important is that all of the methods that give an optimal solution to this problem will take a great deal of time and/or resources. These methods are therefore impractical. In fact, the exhaustive search implementation that was used for this chapter takes more than exponential time (with respect to the size of the message set) in the worst case.

Greedy routing is a far more practical approach for routing in RMINs than the exhaustive search approach is. However, it is not optimal. A greedy router will not always route the maximum number of messages in a message set.

The greedy router that is used for this chapter is in many ways similar to the exhaustive search router. It uses a mixture of offline and online techniques to arrive at a solution. The calculation of all of the routes is performed offline, and these routes are stored in memory. The choice of which routes to use, given a message set, occurs online. This is also true of the exhaustive search router. Both routers use the same amount of memory resources. It is the online generation of route combinations that differs.

The greedy routing scheme works in the following manner. In a given message cycle, some set of processors wants to communicate with some set of memory modules. First, the router chooses one desired message and looks at the first route on that message's list of routes. The router then establishes that route. Next, the router examines a second message (assuming a second desired message was requested) and sees if one of the routes in the second message's route list can be established without conflicting with the already established first message. If such a route does exist, the router establishes that route and moves on to the next desired message.

In general, the greedy router gives sub-optimal solutions. Fortunately, this is not a critical flaw for the type of communication problem that is described in this chapter, since stymied processors can always try to communicate again during subsequent message cycles. Furthermore, the greedy router is so much faster than the exhaustive search router that it would be the preferred solution for most practical problems. Calculating solutions using this greedy scheme and a given RMIN takes quadratic time (with respect to the size of the message set).

4 THE NEURAL NETWORK ROUTER

In this section the neural network routing scheme that was first proposed in [7] will be explained. This routing scheme in effect calculates the routes for the message set in parallel on numerous small processing elements (neurons). The idea is that this should give the neural network routing method a big advantage over other routing methods in terms of speed. Unfortunately, the neural network routing scheme has not actually been implemented in its parallel form. Rather, its performance was simulated. For this reason, the actual speed advantage (if any) of a neural network router is not known.

What is known is that the neural network router performs sub-optimally and that its performance gets worse as the size of the RMIN increases. This degeneration is especially apparent when the RMIN has many stages. However, for RMINs that are "not too large," the neural network router performs quite well.

There are some additional points that should be noted. First, the neural network router calculates the routes online. Unlike the exhaustive search and greedy routing methods described previously, the neural network router does not need a database of all of the possible routes. In a sense, the neural network router simultaneously finds routes and makes sure the routes that are found are not conflicting. Second, the neural network router is a nondeterministic routing scheme. That is to say, given a RMIN and some message set, the neural network router will not always give the same answer. This is because there is a certain amount of randomness involved in the neural network solution. Because of this randomness, measuring the ability of a neural network router to solve the routing problem is inherently probabilistic.

A description of the routing representation that is used by the neural network can be found in Section 4.1. In Section 4.2, the basics of neural networks will be presented. The specific structure of the routing neural network is given in Section 4.3.

4.1 The Routing Representation

Consider the RMIN in Figure 2. Suppose that the messages $\alpha = (1, 4)$ and $\beta = (2, 2)$ need to be established. One possible route for message α uses the second output port of the first stage of the RMIN, and the third output port of the second stage of the RMIN. This route will can be called $\alpha_1 = (1, 2, 3, 4)$. Similarly, one may use route $\beta_1 = (2, 1, 1, 2)$ for message β. We will now construct a routing representation for these two routes. This representation will be called the routing array. In Section 4.3, a neural network structure will be described. The neural network will be in a state of minimal energy when the neuron outputs directly represent a legal routing array.

Each message route will have a corresponding routing matrix. For example, the routing matrix for route α_1 is shown as the left table in Table 1. The columns of a routing matrix represent the stage of the interconnection network, while the rows represent the output ports for each stage of the interconnection network. The column of processors is represented by column 0 in the tables. If $a_{i,j} = 1$, the message is routed through output port i of stage j. Having $a_{i,j} = 0$ implies that the message is *not* routed through output port i of stage j. The routing matrix for route β_1 is shown as the right table in Table 1.

The routing array for the set of routes is simply constructed by treating each routing matrix as a "slice" and constructing a "loaf". The routing array is

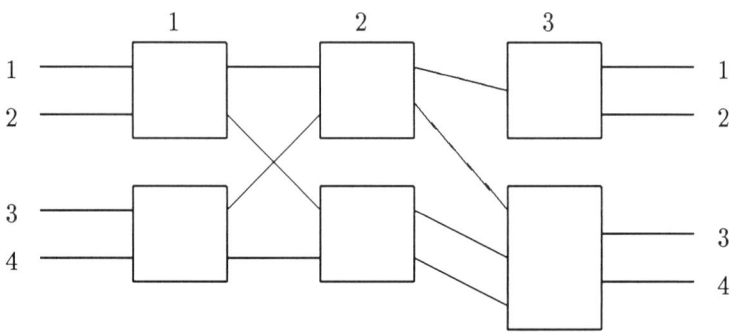

Figure 2 A sample RMIN.

	0	1	2	3
1	1	0	0	0
2	0	1	0	0
3	0	0	1	0
4	0	0	0	1

	0	1	2	3
1	0	1	1	0
2	1	0	0	1
3	0	0	0	0
4	0	0	0	0

Table 1 Matrix representations for routes $\alpha_1 = (1, 2, 3, 4)$, and $\beta_1 = (2, 1, 1, 2)$ for the RMIN shown in Figure 2.

a 3-dimensional representation of a set of routes, and each slice of the array represents a single route. For our example, there are two messages to be routed so the routing array will have two slices. In general, if a system has m input ports, there can be m slices in the routing array.

Each element of the routing array now has three indices. If element $a_{i,j,k}$ is equal to 1 then message i is routed through output port k of stage j. We say $a_{i,j,k}$ and $a_{l,m,n}$ are in the same *row* if $i = l$ and $k = n$. They are in the same *column* if $i = l$ and $j = m$. Finally, they are in the same *rod* if $j = m$ and $k = n$.

A legal routing array will satisfy the following three constraints:

1. one and only one element in each column is equal to 1.

2. the elements in successive columns that are equal to 1 represent output ports that can be connected in the interconnection network.

3. no more than one element in each rod is equal to 1.

The first restriction ensures that each message will be routed through one and only one output port at each stage of the interconnection network. The second restriction ensures that each message will be routed through a legal path in the interconnection network. The third restriction ensures that any resource contention in the interconnection network is resolved. In other words, only one message can use a certain output port at a certain stage in the interconnection network. When all three of these constraints are met, the routing array will provide a legal route for each message in the message set.

The neural network router that is described in Section 4.3 calculates routes in the form of a routing array. Like the routing array, the neural network router will naturally have a 3-dimensional structure. The neural network router, however, does not calculate the first or last columns in each of the routing matrices. Since a processor can only communicate to (at most) one memory module in each message cycle, the first column of each routing matrix is well defined by the message itself. Also, it is assumed here that there is no conflict for memory modules during each message cycle. That is, no two processors will want to communicate with the same memory module in a given message cycle. A message set that has this property will be called a *valid* message set. If the message sets are valid, the last column of each routing matrix is also well defined by the message. It should be noted that the neural network router can be modified in a simple way to accommodate systems for which the valid message set restriction does not hold.

4.2 Neural Networks

The type of neural network that is described here was analyzed by Hopfield [13]. The neural network has N neurons. The input to neuron i is u_i, its input bias current is I_i, and its output is V_i. The input u_i is converted to the output V_i by a sigmoid function, $g(x)$. Neuron i influences neuron j by a connection represented by T_{ji}. Similarly, neuron j affects neuron i through connection T_{ij}. One possible electronic implementation for this type of neural network is given by Hopfield [12, 13].

In order for the Liapunov function (Equation 6) to be constructed, T_{ij} must equal T_{ji}. We further assume that $T_{ii} = 0$. There is also a time constant, denoted by τ.

The network functions in a continuous time mode, and the system response is decided by a set of simultaneous differential equations. The equations which describe the output of a neuron i are:

$$\frac{du_i}{dt} = -\frac{u_i}{\tau} + \sum_{j=1}^{N} T_{ij} V_j + I_i \qquad (1)$$

$$\tau = RC \qquad (2)$$

$$V_j = g(u_j) \qquad (3)$$

For the purposes of this chapter, τ will be set to 1. To make digital simulation possible, Equation 1 is approximated by the formula shown here:

$$\Delta u_i = \Delta t(-\frac{u_i}{\tau} + \sum_{j=1}^{N} T_{ij} V_j + I_i) \qquad (4)$$

The sigmoid function that is used is shown here:

$$g(x) = \frac{1}{1 + e^{-x}} \qquad (5)$$

The equations above force the neural net into stable states that are the local minima of this approximate energy equation:

$$E = -\frac{1}{2} \sum_{i=1}^{N} \sum_{j=1}^{N} T_{ij} V_i V_j - \sum_{i=1}^{N} V_i I_i \qquad (6)$$

For the neural network, the weights (T_{ij}'s) are set, as are the bias currents (I_i's). It is the output voltages (V_i's) that vary to to minimize E.

4.3 The Structure of the Neural Network Router

Figure 1 contains a diagram of the communication system with a neural network router. A neural network of the type described in Section 4.2 is the crucial component of the neural network router. The neural network router has two logic blocks to interface the neural network with the rest of the communication system. The logic block *Logic1* converts the desired message set into a language the neural network can understand: bias currents. The logic block *Logic2* is required to convert the routing array solution of the neural network into a crosspoint form that the interconnection network can understand. The construction of these logic blocks should be a straightforward process.

Neural networks have been used to generate good solutions for difficult optimization problems [12]. These neural networks can also find acceptable solutions to certain constraint satisfaction problems. Solving optimization problems, such as the Traveling-Salesman Problem (TSP), requires minimization of some cost function subject to a set of constraints. Satisfying a set of constraints may also be achieved by minimizing a cost function, and there may be many correct solutions. The problem of routing a set of messages through a general interconnection network may be cast into the form of a constraint satisfaction problem, thereby making a neural network solution possible.

In this section we describe the construction of a neural network in which each neuron directly represents an element in the routing array for an interconnection network and message set. The neural network has a three-dimensional structure just like the routing array. Each $a_{i,j,k}$ of a routing array is represented by the output voltage of a neuron, $V_{i,j,k}$. At the beginning of a message cycle, the neurons have a random output voltage that is bounded by 0 and 1. The neural network is forced into stable states that are the local minima of the Liapunov function, which is the energy function (Equation 6). If the neural network settles in one of the global minima, the problem will have been solved.

We now construct an energy function such that the neural network will be in a global minimum when the output values of the neurons directly represent a legal routing array. The energy function has four components that are shown here:

$$E_1 = \frac{A}{2} \sum_{m=1}^{M} \sum_{s=1}^{S-1} \sum_{p=1}^{P} V_{m,s,p}(-V_{m,s,p} + \sum_{i=1}^{P} V_{m,s,i}) \qquad (7)$$

$$E_2 = \frac{B}{2} \sum_{s=1}^{S-1} \sum_{p=1}^{P} \sum_{m=1}^{M} V_{m,s,p}(-V_{m,s,p} + \sum_{i=1}^{M} V_{i,s,p}) \qquad (8)$$

$$E_3 = \frac{C}{2} \sum_{m=1}^{M} \sum_{s=1}^{S-1} \sum_{p=1}^{P} (-2V_{m,s,p} + V_{m,s,p}(-V_{m,s,p} + \sum_{i=1}^{P} V_{m,s,i})) \qquad (9)$$

$$E_4 = D \sum_{m=1}^{M} \left[\sum_{s=2}^{S-1} \sum_{p=1}^{P} \sum_{i=1}^{P} d(s,p,i) V_{m,s-1,p} V_{m,s,i} \right. \tag{10}$$
$$\left. + \sum_{j=1}^{P} (d(1, \alpha_m, j) V_{m,1,j} + d(S, j, \beta_m) V_{m,S-1,j}) \right]$$

The parameter S represents the number of stages in the RMIN, while the parameter P represents the number of outputs ports per each stage of the RMIN. Note that P may actually be a function of the stage. A, B, C, and D are arbitrary positive constants. E_1 and E_3 handle the first constraint in the routing array (from Section 4.1). E_4 deals with the second constraint. E_2 ensures the third. From the equation for E_4, the function $d(s1, p1, p2)$ represents the "distance" between output port $p1$ from stage $s1 - 1$ and output port $p2$ from stage $s1$. If $p1$ can connect to $p2$ through stage $s1$, then this distance may be set to zero. If $p1$ and $p2$ are not connected through stage $s1$, then the distance may be set to one. Also, α_m is the source address of message m, while β_m is the destination address of message m.

The entire energy function is:

$$E = E_1 + E_2 + E_3 + E_4 \tag{11}$$

Solving for the connection and bias current values, as shown in Equation 6, results in the following equations:

$$\begin{aligned} T_{(m1,s1,p1),(m2,s2,p2)} = & -(A+C)\delta_{m1,m2}\delta_{s1,s2}(1-\delta_{p1,p2}) \\ & -B\delta_{s1,s2}\delta_{p1,p2}(1-\delta_{m1,m2}) \\ & -D\delta_{m1,m2}[\delta_{s1+1,s2}d(s2,p1,p2) \\ & +\delta_{s1,s2+1}d(s1,p2,p1)] \end{aligned} \tag{12}$$

$$I_{m,s,p} = C - D[\delta_{s,1}d(1,\alpha_m,p) + \delta_{s,S-1}d(S,p,\beta_m)] \tag{13}$$

$\delta_{i,j}$ is a Kronecker delta ($\delta_{i,j} = 1$ when $i = j$, and 0 otherwise). The connection values from Equation 12 are well defined for a given interconnection network. That is, once the interconnection network is designed, the neural network and all of its inter-neuron connections can be calculated. When different groups of input-output ports need to be connected, it is only the input bias currents of boundary neurons that are affected. Equation 13 quantifies that change.

If Hopfield's circuit implementation of the neural network from [12, 13] is utilized, changing to a new set of desired messages corresponds to reducing the input bias currents to illegal nodes in the first and last stages of the neural network. The connectivity matrix is defined totally by the interconnection network that is chosen, and so the conductances need not change.

If the user has the ability to make the output of a rod of neurons equal to zero or give a rod of neurons a large negative input bias current, then the neural network can provide a *fault-tolerant* routing scheme. For example, if an output port in some stage of the interconnection network is faulty, the user could set the rod of neurons that represents that output port to zero. This may be accomplished through the use of bias currents that can be applied externally. This is shown as the External Control lines in Figure 1. The rest of the neural network can operate exactly as it did before. Neither the structure nor the weights need to be changed. Similarly, this routing methodology could tolerate faulty input ports and broken buses. However, the user must know if a fault exists and where it exists in the interconnection network.

The neural network router is generally incapable of generating complete solutions to a routing problem, even when such solutions exist. However, even when a complete solution for a routing problem is not calculated, the neural network router will often find a partial solution. For example, if 6 connections need to be established through an interconnection network in a given message cycle, it may be that only 5 of them will be established after the neural network converges to its solution. Again, it is assumed here that if a desired connection is not established in a given message cycle, the processor that requested that connection may try again during the next message cycle. Thus, the neural network router is not an optimal router, but it is felt that a fast implementation of the neural network may allow the neural network router it to outperform other routing methods in terms of throughput.

5 PERFORMANCE MEASURES FOR ROUTERS

There are two measures that are used in this chapter to demonstrate the routing ability of the three routing approaches. Both of these measures are functions of the following three variables:

1. The routing methodology used.
2. The interconnection network.

3. The size of the message set, M.

Thus, in order to compare the relative effectiveness of two routing methodologies, a certain interconnection network and a certain message size M must be chosen.

The first measure is the $CS\%$, or *complete success percentage*. The $CS\%$ is the percentage of message sets that can be routed successfully in a single message cycle (given a routing methodology, an interconnection network, and a message set size).

The second measure is the $SM\%$, or the *successful message percentage*. The $SM\%$ is the expected percentage of messages that can be routed successfully in a single message cycle (given a routing methodology, an interconnection network, and a message set size).

These two measures were calculated for several scenarios for the exhaustive search router ($CS_{es}\%$ and $SM_{es}\%$), for the greedy router ($CS_{gr}\%$ and $SM_{gr}\%$), and the neural network router ($CS_{nn}\%$ and $SM_{nn}\%$). However, the measures were calculated in different ways for the routing schemes. For the exhaustive search router and the greedy router, which have deterministic routing schemes, the measures were evaluated by trying to route *every message set of size M*. The neural network routing scheme, on the other hand, is not deterministic. Even if the exact same message set is to be routed in two different message cycles, the neural network router may come up with a different solution for each message cycle. Thus, while exact values for $CS_{es}\%$, $SM_{es}\%$, $CS_{gr}\%$, and $SM_{gr}\%$ may be obtained, $CS_{nn}\%$ and $SM_{nn}\%$ are probabilistic. Furthermore, simulating the neural network router takes a large amount of computer time since one uses a sequential, digital simulation for a parallel, analog neural network. For these reasons, calculation of $CS_{nn}\%$ and $SM_{nn}\%$ was achieved by trying to route *1,000 random message sets of size M*.

It is important to note that the measures that were calculated are valid for the routing of a random message set, which is not an exact modeling of the way a real system performs. For example, in a real system, if a processor fails in its attempt to communicate with a memory module, it will try to establish communication with the same memory module during the next message cycle. For the measures that were calculated for this chapter, this was not the case. For each message cycle, the messages that were chosen were not dependent on what was or was not routed in the previous message cycle.

As an example of how the measures are calculated, suppose only one test message cycle were run for an interconnection network and this test cycle required $M = 5$ connections to be established. If the routing methodology under examination only established 4 of the 5 desired connections, then the $CS\% = 0.0$ while the $SM\% = 80.0$. Now suppose another test were run where $M = 5$, but this time all 5 of the connections were established. The cumulative effect of both tests now goes into calculating the percentages, so now the $CS\% = 50.0$ while the $SM\% = 90.0$. Note that for a string of tests, $CS\% \leq SM\%$.

6 SIMULATION RESULTS

Three RMINs are considered. They are shown in Figures 3. These RMINs are called RMIN1, RMIN2, and RMIN3. These RMINs are not random in any formal sense. Rather, they were simply constructed from the minds of the authors. They are not chosen because of any particular properties that may make them amenable to routing solutions calculated by any of the routing methodologies. However, these RMINs are constricted in one sense: each input port (processor) can communicate with each output port (memory module). This restriction is made because it is felt that this is a minimal requirement for most reasonable communication systems.

Recall that each message set will be considered only for calculating the measures for the exhaustive search router and the greedy router. A statistical approximation of the measures for any routing methodology may be achieved by trying to route a large number of random message sets. This was the case for calculating the measures for the neural network router, for which 1,000 test random message sets were considered. The parameters of the neural neural network used for these simulations were $A = C = D = 3.0$, while $B = 6.0$. These were chosen by simulating the neural network for many values, and choosing the ones that gave the best results.

Consider RMIN1, as shown in Figure 3. This is a three stage RMIN that has no known effective routing algorithm. For RMIN1, $m = n = 8$. The neural network router for RMIN1 has 88 neurons. Note that there are only 5 buses connecting the first and second stages of RMIN1. This means that no more than 5 messages can be routed successfully in each message cycle. Obviously, this RMIN has very limited capabilities, but it may be suitable if the communication requirements of the system are not too stringent.

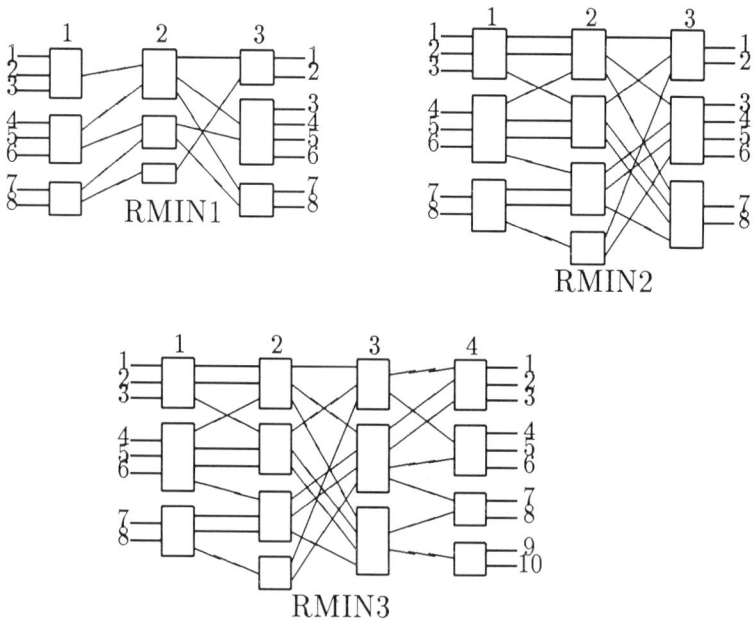

Figure 3 Three random multistage interconnection networks.

The routing results for exhaustive search routing, greedy routing, and neural network routing for RMIN1 are shown in Table 2. The M column contains the message length. Tests were run for $M = 1, 2, \ldots, m$. The $|S_M|$ column contains the size of the class of all valid message sets of size M for RMIN1. The other columns show the results of the routing trials.

In general, the SM_{es} value provides an upper bound for routing methodologies because using exhaustive search one can always route the maximum number of messages. However, in some cases, SM_{nn} is actually greater than SM_{es}. This implies that the neural network router somehow outperformed the exhaustive search router. This is due to the fact that the exhaustive search was tested for every message set while the neural network router was tested for 1,000 randomly chosen message sets. For the neural network router to appear to have done better than the exhaustive search router, the 1,000 test message sets that were randomly generated for the neural network router must have been slightly easier to route than the average message set. Thus, when $SM_{nn} > SM_{es}$, it is reasonable to conclude that the neural network router has performed just as well as the exhaustive search router. If the neural network router had been

Routing in Random Multistage Interconnection Networks

M	$|S_M|$	$CS_{es}\%$	$CS_{gr}\%$	$CS_{nn}\%$	$SM_{es}\%$	$SM_{gr}\%$	$SM_{nn}\%$
1	64	100.0	100.0	100.0	100.0	100.0	100.0
2	1,568	85.7	83.4	86.5	92.9	91.7	93.3
3	18,816	54.3	50.3	53.8	84.2	82.6	83.7
4	117,600	20.8	17.9	21.7	77.1	74.6	74.4
5	376,320	3.5	2.9	2.2	70.6	67.5	64.8
6	564,480	0.0	0.0	0.0	64.9	61.2	57.5
7	322,560	0.0	0.0	0.0	59.4	55.8	52.3
8	40,320	0.0	0.0	0.0	54.1	51.2	47.3

Table 2 Routing results for RMIN1, shown in Figure 3.

tested for every message set, then its performance measures would have been worse than or equal to those of the exhaustive search router. Note that in no case does the greedy router outperform the exhaustive search router. It is only the probabilistic calculation of the neural network router's routing measures that makes it seem that the neural network router is outperforming exhaustive search routing in some cases.

The results for RMIN1, exhibited in Table 2, show that the performance of the three routers is virtually identical for $M \leq 4$, but that the greedy router and neural network router performances start to degrade for larger M. Furthermore, the greedy router executes slightly better than the neural network router for the larger values of M. Note that it is for these larger M that an exhaustive search router will begin to use an excessive amount of time. Since it is impossible for more than 5 messages to be routed in a single message cycle, $CS_{es}\%$, $CS_{gr}\%$, and $CS_{nn}\%$ are all equal to 0.0 for $M \geq 6$.

Suppose the system engineer feels at this point that RMIN1 had insufficient communication capabilities, especially for larger M. A decision may be made to expand RMIN1 into RMIN2, which is shown in Figure 3. As in RMIN1, RMIN2 has $m = n = 8$. The neural network router for RMIN2 has 168 neurons. RMIN2 is indeed an expansion of RMIN1. All of the routes that exist for RMIN1 are still present, but the crossbar switches are expanded to increase the connectivity, and in the second stage a new crossbar switch is added. There are now 10 buses between stages 1 and 2, and 11 buses between stages 2 and 3. Thus, it may be possible to route entire message sets of size $M = 8$.

M	$\lvert S_M\rvert$	$CS_{es}\%$	$CS_{gr}\%$	$CS_{nn}\%$	$SM_{es}\%$	$SM_{gr}\%$	$SM_{nn}\%$
1	64	100.0	100.0	100.0	100.0	100.0	100.0
2	1,568	97.4	97.4	97.5	98.7	98.7	98.8
3	18,816	91.3	90.8	92.7	97.1	96.9	97.5
4	117,600	81.3	79.8	80.5	95.1	94.7	94.9
5	376,320	67.9	65.1	64.8	92.9	92.3	92.1
6	564,480	52.1	48.5	47.9	90.6	89.9	89.4
7	322,560	35.4	31.5	31.2	88.1	87.5	87.5
8	40,320	17.1	14.3	11.4	85.7	85.3	85.0

Table 3 Routing results for RMIN2, shown in Figure 3.

The routing results for RMIN2 are shown in Table 3. The exhaustive search numbers still show that the interconnection network is highly limited (e.g., only 17.1% of the message sets of size $M = 8$ can be routed in a single message cycle). However, the numbers are still much higher than were those for RMIN1. For RMIN2, all three of the routers perform almost equally well. Thus, for RMIN2, the exhaustive search router is even less appealing than it was for RMIN1. This suggests that both greedy routers and neural network routers may be especially useful for RMINs that have a high degree of connectivity.

Now suppose that the system engineer finds that an expansion of the communication system is required so that 10 memory modules can be accessed. RMIN3 is identical to RMIN2 for the first 3 stages, and then stage 4 is added to accommodate the two new memory modules. For RMIN3, $m = 8$ and $n = 10$, and its neural network router has 232 neurons. The routing results for RMIN3 are shown in Table 4. Calculating the results for the exhaustive search router became computationally prohibitive for $M = 5, 6, 7, 8$. The results here show that the neural network router's results are slightly worse than the exhaustive search router's results, and that the neural network router's performance degrades as M increases. The greedy router generally functions worse than the neural network router for RMIN3. It appears that increasing the number of stages in the RMIN worsens the performance of the neural network router with respect to the exhaustive search router.

Overall, the results for the three RMINs show that the routing ability of the neural network router is quite good in comparison to the optimal exhaustive

Routing in Random Multistage Interconnection Networks 55

M	$|S_M|$	$CS_{es}\%$	$CS_{gr}\%$	$CS_{nn}\%$	$SM_{es}\%$	$SM_{gr}\%$	$SM_{nn}\%$
1	80	100.0	100.0	100.0	100.0	100.0	100.0
2	2,520	98.5	88.3	94.4	99.3	94.2	97.2
3	40,320	98.4	72.0	86.5	99.5	90.4	95.5
4	352,800	94.6	53.9	73.1	98.6	87.3	92.9
5	1,693,440	*	36.1	57.3	*	84.4	90.7
6	4,233,600	*	20.9	37.7	*	81.6	87.2
7	4,838,400	*	9.9	16.2	*	78.8	82.9
8	1,814,400	*	3.3	1.7	*	76.2	75.8

Table 4 Routing results for RMIN3, shown in Figure 3. Calculation of the * entries was computationally prohibitive.

search router for RMINs of a certain size. The relationship between the greedy router and the neural network router is unclear. For RMIN1, the greedy router works better than the neural network router. For RMIN3, the reverse is true. What is clear is that the performances of both the greedy router and the neural network router are highly dependent on the structure of the RMIN. Preliminary results show that the neural network router performance degenerates as the number of stages in the RMIN increases.

This section dealt with the abilities of the three routing methodologies. In Section 7, certain other criteria will be discussed: the speed and the resource utilization of the routing schemes.

7 ANALYSIS

We begin with the assumption that we have found a RMIN for which the abilities of the three routing schemes to solve the routing problem are essentially equal. This could be said for the three RMINs that were presented in Section 6. The question now is: how many resources do the routing schemes use and what are the speeds of the routing schemes?

First, the resources will be investigated. Both the exhaustive search router and the greedy router base their decisions on a *comparison* of routes. We will assume that each route takes up only one piece of memory, even though each route will have a value for each stage of the RMIN. The number of processors is m and the number of memory modules is n. Thus, the total number of distinct messages is mn. Let the number of distinct routes from processor i to memory module j be $k_{i,j}$. Then the total number of distinct point-to-point routes in the RMIN is $r = \sum_{i=1}^{m} \sum_{j=1}^{n} k_{i,j}$. Now let the value k represent the average number of routes for each message. That is, $k = \frac{r}{mn}$. For both the exhaustive search router and the greedy router, mnk memory locations are required to store routes, so a memory of size $\Theta(mnk)$ is needed.

Both the exhaustive search router and the greedy router are implemented sequentially using a single processing element.

The neural network router, on the other hand, does not store the routes, but calculates them online. However, if each neuron is considered to be a processing element, then many processing elements are required. Let q_i be the number of output ports for stage i of the RMIN. S is the number of stages. Then $m \sum_{i=1}^{S-1} q_i$ will be the largest number of processing elements required. In fact, it is often the case that fewer processing elements will be required, because some of the neurons might never be able to be equal to 1 due to the structure of the RMIN. For many common interconnection networks, the average value of q_i will be m. Under these conditions, the number of processing elements will be $O(m^2 S)$.

Now the speeds of the algorithms will be analyzed. For a message set of size M, and assuming each message pair has k routes, the exhaustive search router may have to examine, in the worst case, k^M combinations of routes. Once these routes are chosen, each combination of routes is compared for compatibility. Thus, once a combination of M routes is chosen to be examined, $C_2^M = \frac{M^2 - M}{2}$ comparisons must be made. We assume that each comparison can take place in constant time. Once these comparisons are made, a variation of the maximal clique size problem is performed. An exact analysis of that algorithm is difficult, but clearly it is not a constant time algorithm. However, given what we already know it can be said that, in the worst case, the time complexity of the exhaustive search router is $\Omega(k^M M^2)$. The worst case (when every combination of routes must be examined) often occurs since the algorithm will only terminate early when a complete solution is found. Clearly, this approach is impractical for large M (or k). Note, however, that the exhaustive search implementation described here is quite naïve.

For the greedy router, the first route for the message set is chosen in constant time. It is the first route from the top of the list for the message of highest priority. After that, in the worst case, assuming each message pair has exactly k routes, k routes may need to be examined for each remaining message. And each route must be compared to each route that has already been established. Therefore, $1 + k + 2k + 3k + \cdots + (M - 1)k = \Theta(kM^2)$ is the time complexity of the greedy router in the worst case.

It has been shown empirically that certain variations of the type of neural network that is used here can converge to a solution in essentially constant time. For example, this claim is made for the neural network described in [20], which is a slight variation of the model used here. Hence, we assume the time complexity of the neural network router is $O(1)$.

It should be noted that although asymptotic analysis is the preferred method for comparing algorithms in the computer science community, a direct comparison of the resource utilization and time complexities of these routing methods is not valid since the solutions that these methods give differ in quality. Depending on the size of the RMIN, the greedy routing and neural network solutions may be quite poor in comparison to the exhaustive search solutions. Hence, this type of asymptotic analysis is not a very valid approach for comparison.

In addition, it should be noted that there are many ways other than the ones described here to create routers that use exhaustive search and greedy principles. The algorithms that are analyzed in this section, however, are thought to be reasonable approaches.

In any case, it should be clear that the exhaustive search router will be impractical for most problems. Since the asymptotic analysis is of very limited significance, the best way to compare the running times of the greedy router and the neural network router is to build actual implementations and compare them directly.

8 CONCLUSIONS AND FUTURE WORK

A comparison between exhaustive search routing, greedy routing, and neural network routing was performed for three arbitrary RMINs. It was shown that for several moderately sized RMINs, the greedy router and the neural network router performed nearly as well as the optimal exhaustive search router in terms of establishing the maximum number of connections possible for a given message set.

These results should not be extrapolated to all kinds of RMINs. Specifically, preliminary simulations have shown that the neural network router performance degenerates as the number of stages in an interconnection network increase. Furthermore, the complexity of a neural network router may become too great for very large interconnection networks. However, the routing ability of the neural network router appears to be satisfactory for RMINs that are not too large.

Further analysis of the time complexities and the resource utilization of the three routing algorithms revealed that exhaustive search routing was not a practical approach. Greedy routing and neural network routing would appear to be valid approaches for RMINs of moderate size. However, asymptotic analysis of the time complexities of these approaches is invalid since the quality of the solutions that these methods arrive at is not the same. The best way to compare the speeds of these two routing schemes would be to build actual implementations.

However, since the neural network router essentially calculates the routes *in parallel*, it can reasonably be hoped that a fast, analog implementation for the neural network router may find solutions faster than the exhaustive search router and even the greedy router. Recent VLSI designs show the potential that exists for implementing complex neural network circuits. For RMINs of a certain size the neural network router generates solutions that are of relatively high quality. For this type of RMIN, the neural network router may be a viable alternative.

One variation of the neural network router that may improve its performance would be to implement a *Boltzmann machine* [1], rather than the standard neural network that is described in Section 4. A Boltzmann machine tends to avoid local minima through the use of *simulated annealing* [14]. Therefore, Boltzmann machines may generate higher quality solutions. The neurons in the Boltzmann machine have a probabilistic decision rule that may be approximated by simply adding a Gaussian noise input to each of the neurons [2, 11]. For the application described in this chapter, speed is definitely an issue, and Boltzmann machines are slow to converge to a solution due to the temperature changes that accompany the annealing procedure. However, VLSI implementations of Boltzmann machines appear to be promising. The tradeoffs involved with using a Boltzmann machine must be examined: will the higher quality solution make up for the slower convergence speed?

REFERENCES

[1] Ackley, D., Hinton, G., and Sejnowski, T., "A learning algorithm for Boltzmann machines," *Cognitive Science*, vol. 9, pp. 147–169, 1985.

[2] Alspector, J., Gannett, J., Haber, S., Parker, M., and Chu, R., "A VLSI-efficient technique for generating multiple uncorrelated noise sources and its application to stochastic neural networks," *IEEE Transactions on Circuits and Systems*, vol. 38, no. 1, pp. 109–123, January 1991.

[3] Brown, T., "Neural networks for switching," *IEEE Communications Magazine*, vol. 27, no. 11, pp. 72–81, November 1989.

[4] Brown, T., and Liu, K.-H., "Neural network design of a Banyan network controller," *IEEE Journal on Selected Areas of Communication*, vol. 8, no. 8, pp. 1428–1438, October 1990.

[5] Funabiki, N., Takefuji, Y., and Lee, K., "A neural network model for traffic controls in multistage interconnection networks," in *Proceedings of the International Joint Conference on Neural Networks 1991*, p. A898, July 1991.

[6] Funabiki, N., Takefuji, Y., and Lee, K., "Comparisons of seven neural network models on traffic control problems in multistage interconnection networks," *IEEE Transactions on Computers*, vol. 42, no. 4, pp. 497–501, April 1993.

[7] Goudreau, M., and Giles, C., "Neural network routing for multiple stage interconnection networks," in *Proceedings of the International Joint Conference on Neural Networks 1991*, vol. 2, p. A885, July 1991.

[8] Goudreau, M., and Giles, C., "Neural network routing for random multistage interconnection networks," in *Advances in Neural Information Processing Systems 4* (J. Moody, S. Hanson, and R. Lippmann, eds.), (San Mateo, CA), pp. 722–729, Morgan Kaufmann Publishers, 1992.

[9] Goudreau, M., and Giles, C., "Routing in random multistage interconnection networks: Comparing exhaustive search, greedy and neural network approaches," *International Journal of Neural Systems*, vol. 3, no. 2, pp. 125–142, 1992.

[10] Hakim, N., and Meadows, H., "A neural network approach to the setup of the Benes switch," in *Infocom 90*, pp. 397–402, 1990.

[11] Hinton, G., Sejnowski, T., and Ackley, D., "Boltzmann machines: Constraint satisfaction networks that learn," Tech. Rep. CMU-CS-84-119, Department of Computer Science, Carnegie-Mellon University, Pittsburgh, PA, May 1984.

[12] Hopfield, J., "Neural computation on decisions in optimization problems," *Biological Cybernetics*, vol. 52, pp. 141–152, 1985.

[13] Hopfield, J., "Neurons with graded response have collective computational properties like those of two-state neurons," *Proceedings of the National Academy of Science USA*, vol. 81, pp. 3088–3092, May 1984.

[14] Kirkpatrick, S., and Gelatt, C., Jr. "Optimization by simulated annealing," *Science*, vol. 220, pp. 671–680, May 1983.

[15] Marrakchi, A., and Troudet, T., "A neural net arbitrator for large crossbar packet-switches," *IEEE Transactions on Circuits and Systems*, vol. 36, no. 7, pp. 1039–1041, July 1989.

[16] Melsa, P., Kenney, J., and Rohrs, C., "A neural network solution for call routing with preferential call placement," in *Proceedings of the 1990 Global Telecommunications Conference*, pp. 1377–1382, December 1990.

[17] Melsa, P., Kenney, J., and Rohrs, C., "A neural network solution for routing in three stage interconnection networks," in *Proceedings of the 1990 International Symposium on Circuits and Systems*, pp. 483–486, May 1990.

[18] Mittler, M., and Tran-Gia, P., "Performance of a neural net scheduler used in packet switching interconnection networks," in *1993 IEEE International Conference on Neural Networks*, vol. II, pp. 695–700, 1993.

[19] Rauch, H., and Winarske, T., "Neural networks for routing communication traffic," *IEEE Control Systems Magazine*, vol. 8, no. 2, pp. 26–31, April 1988.

[20] Takefuji, Y., and Lee, K., "An artificial hysteresis binary neuron: A model suppressing the oscillatory behavior of neural dynamics," *Biological Cybernetics*, vol. 64, pp. 353–356, 1991.

[21] Troudet, T., and Walters, S., "Neural network architecture for crossbar switch control," *IEEE Transactions on Circuits and Systems*, vol. 38, no. 1, pp. 42–56, January 1991.

[22] Zhang, L., and Thomopoulos, S., "Neural network implementation of the shortest path algorithm for traffic routing in communication networks," in *Proceedings of the International Joint Conference on Neural Networks 1989*, p. 591, June 1989.

4

ATM TRAFFIC CONTROL USING NEURAL NETWORKS

Atsushi Hiramatsu

NTT Communication Switching Laboratories

1 INTRODUCTION

Multimedia information services including voice, video, and data signals are growing rapidly, increasing the demand for the broadband integrated services digital network (B-ISDN) that will provide high-speed communication channels suitable for transporting multimedia signals efficiently [1]. The asynchronous transfer mode (ATM) is a key B-ISDN technology [2, 3] and many communication research laboratories and vendors have been actively pursuing the research and development of ATM systems since the early 1980's. The ATM interface and protocol have been standardized at ITU-TS (CCITT) [7] and products based on ATM technology have already been released [4]. Applications of ATM communication channels operating at hundreds of megabits per second are being investigated [6], and the deployment of ATM technology is expected to change the traditional architecture of public and private networks [5].

The ATM is based on the statistical multiplexing of short fixed-size cells (that is, on cell multiplexing) and can support communication at a wide variety of arbitrary bit-rates and easily merge various communication media into one service. Maintaining the quality of service (QOS), however, will require flexible ATM traffic control [10, 13]. There are QOS parameters particular to the ATM network, such as cell transmission delay and cell loss rate, and the requirements for these parameters differ between users and services. The network controller has to satisfy all of these requirements while utilizing the network efficiently.

Although various ATM traffic control methods for handling these requirements have already been proposed [11, 12], the traditional approaches based on thorough analyses of offered traffic characteristics and service quality are not suitable for handling the wide variety of ATM services and the diversity of their combinations. Such approaches are also not flexible enough to 0 new service installations or unexpected changes in traffic characteristics.

An adaptive ATM traffic control method using neural networks has therefore been proposed [17]. Because neural networks can learn the relation between the offered traffic characteristics and the resulting service quality by on-line training, this approach is expected to yield a simple and flexible network controller that can adapt to almost any situation. Adaptive methods for controlling communication networks, such as routing control with learning automata [22], have already been investigated, but these methods are not suitable for ATM networks because they can handle only a few input parameters and control variables. A neural network, however, is suitable for ATM network control because by using a learning algorithm such as back-propagation [20] it can easily learn a nonlinear function with many inputs and outputs and because recent neural network VLSI technology is expected enable very fast processing.

The remaining sections of this chapter are organized as follows. Section 2 briefly introduces readers unfamiliar with ATM technology to the ATM network architecture and the requirements for ATM QOS control. Section 3 overviews possible neural network applications in the ATM traffic control and describes a strategy for integrating the various adaptive control functions distributed over the ATM network [19]. Sections 4 and 5 give examples of neural network application in call admission control [17] and link capacity control [18]. And finally Section 6 briefly addresses the future directions of research in this field.

2 ATM TRAFFIC CONTROL

2.1 ATM architecture

Cell multiplexing technology

The main purpose of the B-ISDN is to provide users with a universal platform integrating all kinds of multimedia services, from voice services at

ATM Traffic Control Using Neural Networks

some tens of kilobits per second to 140-Mbps high-definition TV and also including the transfer of huge data files. Conventional circuit-switching and packet-switching technologies are not suitable for this purpose because circuit switching would require a different interface for each multimedia service and because the delay in packet switching would be too long for delay-sensitive communications like voice and video.

Proposed as a means of solving these problems, ATM switching is a kind of high-speed packet switching that includes the advantages of circuit switching [Fig. 1(a)]. All information from user terminals is transmitted in a series of 53-byte cells, each consisting of a 5-byte header and a 48-byte data field. The header contains control information such as the ATM channel identifier and the service class, and the data field contains the user data. At an ATM switching node, all the cells received from many users are statistically multiplexed into 150-Mbps or 600-Mbps ATM links. These links are partitioned into multiple time-slots and the cells received from users are transferred, in the first available time-slots, in the order of their receipt at the switching node. In time-division-multiplexing (TDM), time slots are periodically assigned to a specific communication channel, but in the ATM they are used "on demand": this is the meaning of "asynchronous." At the other end of the ATM link, each cell is routed to its destination according to its header information.

ATM node structure and network architecture

A typical ATM switching node consists of traffic monitors for each input link, a self-routing switch, and output buffers for each output link [Fig. 1(b)]. A traffic monitor counts the cells sent from user terminals within a certain period and, when necessary, controls the cell flow (usage parameter control). A self-routing switch distributes cells by hardware, according to their header information, to the output buffer connected to the ATM link leading to the destination terminal. Cells in an output buffer are sent to the link one-by-one in the FIFO manner. "Self-routing" means that the hardware works as if each cell itself selected the route in the hardware switch network. Various kinds of ATM self-routing switches have been proposed, and their details are described in Refs. [8, 9].

The ATM cell transport pathway is divided into a virtual path (VP) and a virtual channel (VC) [Fig. 1(c)]. A VC connects two user terminals, whereas a virtual path basically connects two ATM nodes. Many VC's are multiplexed into a VP. An ATM cross-connect divides the capacity of a physical transmission line into several VP's with virtual bit-rates. Like the ATM switching node,

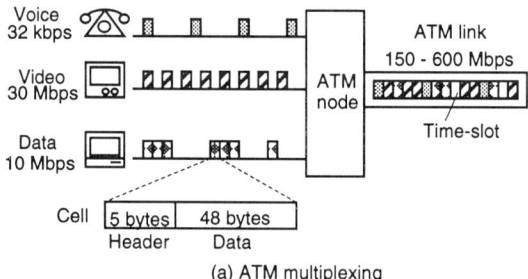

(a) ATM multiplexing

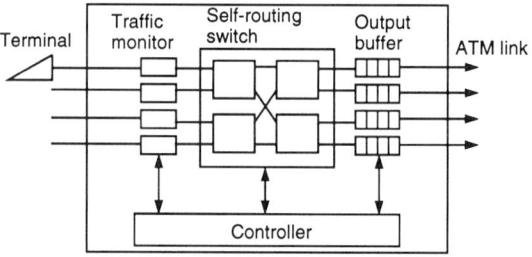

(b) ATM switching node structure

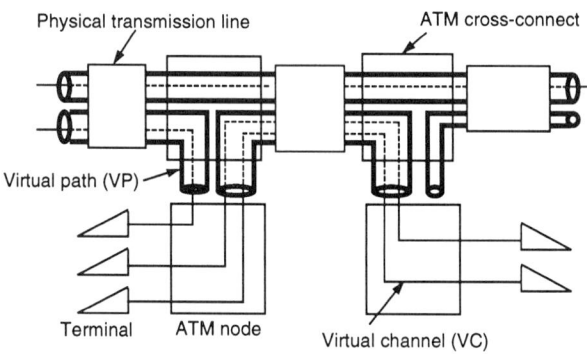

(c) Virtual channels and virtual paths

Figure 1 ATM network architecture.

an ATM cross-connect uses cell multiplexing techniques, and it can change VP bit-rates continuously and flexibly. This enables the traffic controller in an ATM network to change the path capacity more frequently than it can be changed in conventional networks, where the link capacity can take only discrete values and a long time is required for changing the connection pattern in a TDM switch. In the following sections, the conventional terms call, link, and trunk, are respectively used for VC, VP, and transmission line.

ATM network features

- ATM networks can support delay-sensitive communication, such as voice and video, because a cell is small enough to make the cell assembling delay short and is transferred through the network quickly by a self-routing switch.

- To speed cell transmission, slow flow-control mechanisms like cell retransmission are not used in the ATM layer. Bit errors and lost cells are expected to be properly recovered by the upper-layer end-to-end protocols.

- Users can select any types of communications, with any bit-rates, simply by adjusting the cell transmission interval. And an ATM link can be used efficiently because it is occupied only when a user is sending cells.

- Users can divide an ATM link into virtual channels by using the header information to identify each channel. Also multipoint connections can be implemented easily.

2.2 ATM QOS control

QOS parameters

The cell delay and cell loss rate are the essential QOS parameters for ATM network users. Since users send cells at their convenience, many cells sometimes arrive simultaneously at an output buffer in an ATM node. The buffer might then become full and some cells might be lost. Or, at least, cell delay would be increased. Like the conventional networks, the ATM network also has QOS parameters like call loss rate.

The requirements for the QOS parameters of multimedia services differ between services and users, but all QOS parameters should be kept at levels satisfying the users' requirements. For the delay-sensitive services, for example, like telephone, videophone, and TV conference services, the cell delay is crucial but a cell loss rate as high as 10^{-3} may be acceptable. For data communication, on the other hand, cell delay is not critical but cell loss is not acceptable (although the lost cells may be recovered by the upper-level end-to-end protocols). Moreover, users may have the option to select, considering quality and cost, one of several network-provided QOS levels.

To handle this wide variety of QOS requirements, the ATM is expected to support several types of service classes, each of which has different QOS requirements for different traffic characteristics [7]. These different QOS classes can be supported by using priority control at the output buffers.

Cell traffic characterization and burstiness

When considering ATM traffic control, we should understand the variety of the traffic characteristics of multimedia communication. In the ATM network, the effective bit-rate of a call is defined by the number of cells actually sent during a unit period, and in many services it varies with time. Calls are therefore categorized according to their cell generation patterns into two types: constant bit-rate (CBR) and variable bit-rate (VBR). The bit-rate of a CBR call is almost constant, whereas that of a VBR call changes widely with time. VBR calls are also called "bursty" calls because the cells are transmitted intermittently or "in burst." Various traffic parameters such as peak bit-rate, average bit-rate, and average duration of peak bit-rate transmission have been proposed for characterizing a cell generation pattern, but it is hard to accurately characterize bursty calls because these parameters differ with communication service and coding method and even with user and time. And the complexity of each call's traffic characteristics makes the ATM QOS control a very complicated task.

QOS control hierarchy

The ATM network control function is divided into the following three control levels [Fig. 2]: cell transfer control, call control, and network control. The cell-transfer-control level provides functions regulating the flow of cells, such as priority control at the output buffers and the policing function to keep the number of input cells under the declared traffic parameters. The call-control level provides functions controlling the flow of calls, such as the admission control and path selection for call set-up requests. And the network-control

ATM Traffic Control Using Neural Networks

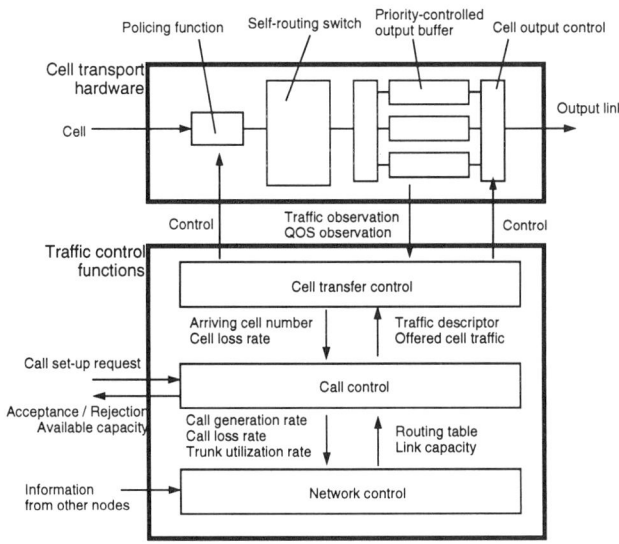

Figure 2 ATM QOS control hierarchy.

level consists of network-wide controls, such as routing control, link capacity assignment, and call congestion control.

Since the functions in each of these control levels are closely related to these in other levels, the cooperation of all of these three levels is indispensable: if the performance characteristics of a lower-level control are changed, the upper-level control must be modified accordingly. And similarly, the lower-level control requires the correct control of the upper level. ATM network control is thus effective only when the methods and parameters used in each level of control are harmonized with those of the other control levels.

3 NEURAL NETWORK APPLICATIONS IN THE ATM NETWORK

Traditional approach vs. neural network approach

In the traditional approach to the ATM QOS control, the determination of control algorithms and of which parameters should be monitored requires detailed network analyses in all possible situations. This approach to the design of the efficient and flexible ATM traffic control functions obviously has some disadvantages. It is hard, for example, to derive good models for analyzing multimedia traffic, and it is of course impossible to enumerate all possible situations. This approach cannot adapt to traffic characteristic changed by unexpected changes in installed services or by the installation of new services. And because each function is designed independently, it is also difficult to optimally integrate the control functions in various control levels.

Neural networks can approximate the complicated input-output relations by autonomously selecting significant inputs and deriving feature parameters from input data. This learning capability can be used to automatically derive a function to estimate QOS from the observed traffic and makes it possible to build an ATM traffic control system that adapts to various types of communication services and to changing situations. Another capability of neural networks is that they can quickly find the near-optimal solutions to combinatorial problems with a number of variables and complicated constraints. This optimization capability can also be used in various control functions in the ATM traffic control system. Figure 3 shows an overview of neural network applications to ATM traffic control. The remaining sections in this chapter mainly discuss applications and features of the learning capability.

Features of neural-network-based ATM traffic control

(i) Exact information about the traffic characteristics is not necessary in the controller design phase: the controller learns this information from the operating network. This kind of traffic control is thus free from the problems related to the difference between the analysis model and the actual traffic.

(ii) The extraction of significant and accurate information from data observed is automatic. If, for example, a measure of burstiness is necessary for

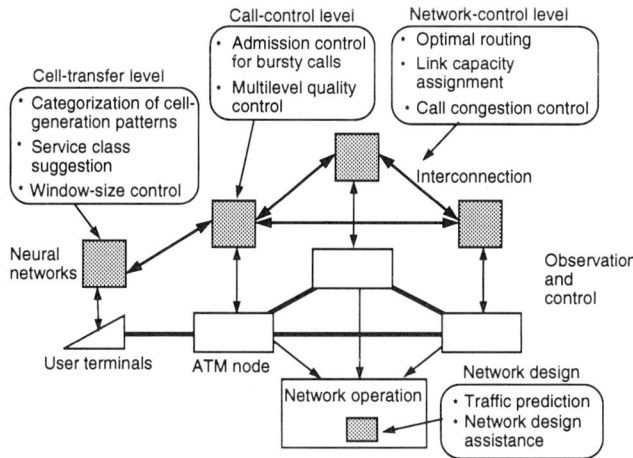

Figure 3 ATM traffic control using neural networks.

control, the neural network may automatically determine it from the history of observed bit-rates.

(iii) The control accuracy improves with experience. When the neural network faces a situation with which it has no experience, it takes a rather long time to adapt to the situation. After it experiences the situation many times, however, it is able to adapt quickly and the period of miscontrol becomes shorter.

(iv) The controller can adapt to changing situations, requirements, and performance characteristics. It is also flexible enough to be suitable for a variety of network configurations.

Integration of distributed neural networks

An neural network's learning capability can be used to optimally integrate various control functions in different control levels. Because it is difficult to train the large control system all at once, these control functions are integrated in a three-step training method. First, to build the control function for each level, each neural network is trained separately to learn system characteristics. Then to optimize control between the different control levels, all the neural networks are trained together. In this step, the controller can adapt to any changes in the mechanism and characteristics of any control level and can optimize the performance of the control system. Finally, for network-wide control by distributed neural networks, the interconnected neural networks are trained together and the information to be exchanged through the interconnections is selected by the neural networks. The neural networks automatically handle the delays and errors in this information exchange.

4 ADAPTIVE CALL ADMISSION CONTROL

4.1 Admission control for the ATM network

Admission control

The basic function of call admission control is to maintain the requested QOS for already-connected calls by rejecting some of the call set-up requests. In this control scheme, each user has to send a call set-up request to an ATM node before staring a call and has to obtain permission to establish a connection between a pair of terminals. A call set-up request usually contains information such as the destination terminal address, the required QOS, the communication service type, and the traffic descriptor (a set of traffic parameters that characterizes the cell generation of the new call).

When the ATM node receives a call set-up request, it selects the requested link according to the call destination and then estimates what the QOS of the requested link would be after the requested call is admitted. If the estimated QOS satisfies the requirements of all already-connected calls and the new call, the node accepts the call set-up request; otherwise, it rejects the request. The QOS during the call is thus guaranteed by the network — assuming that all users send cells conforming to their traffic descriptors. This control is suitable for real-time communication services like voice and video, for which slow flow-control methods can not be used. The key point of this control scheme is the accuracy of the estimated QOS after acceptance of the new call set-up request.

Admission control based on call categorization

ATM Traffic Control Using Neural Networks

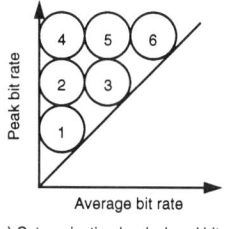

Figure 4 ATM call categorization methods.

Various implementations of the call admission control are based on different traffic descriptors and different QOS estimation methods. See, for example, Refs. [14, 15, 16]. Here a neural network is applied to the "admission control based on call categorization," which has been proposed for handling various kinds of communication services simply.

In this method, calls are categorized into several call-types according to their traffic characteristics or QOS requirements. There are many ways to define call-types, such as by service-type (like 32-kbps voice, 2-Mbps compressed video, and 30-Mbps video), by traffic descriptor values (like peak bit-rate and average bit-rate), or by QOS requirements (like maximum acceptable delay) [Fig. 4]. Because calls of the same call-type have almost the same cell generation characteristics, the cells generated from many calls multiplexed into an ATM link are characterized only by the number of connecting calls. Thus in this scheme, call admission control is implemented simply by counting the numbers of connecting calls for each call-type and estimating the QOS from these numbers.

Figure 5 illustrates the function of the call admission for two call-types. Here only the average cell loss rate is considered as the QOS parameter to control. In this figure, n_1 and n_2 denote the numbers of calls of each type (1 and 2) multiplexed into an ATM link through the same output buffer. If we let $l(n_1, n_2)$ denote the cell loss rate for the combination of n_1 and n_2, by comparing $l(n_1, n_2)$ with the required cell loss rate l_0 (target loss rate) we can divide the n_1-n_2 map into two regions: $l(n_1, n_2) < l_0$ and $l(n_1, n_2) > l_0$. The boundary between these two regions is called the "call admission boundary." When the combination of the numbers of connected calls lies above this boundary, the ATM node should reject new call set-up requests.

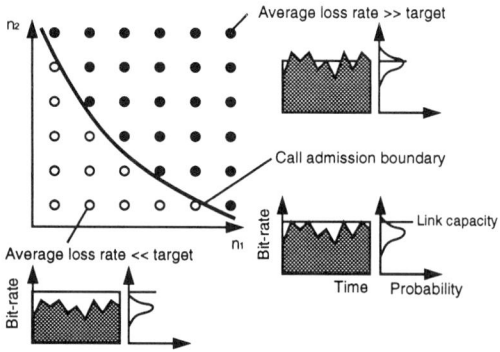

Figure 5 Call admission boundary.

Example of multiplexing two call-types

A simple two-state call model is widely used for analyzing the multiplexing performance of ATM bursty calls [Fig. 6(a)]. Each call has two states, high (H) and low (L), and the state H and L have different cell generation rates v_H or v_L (probability of generating a cell at each cell cycle). The state of a call transits from one to the other at each control cycle randomly with probabilities Δ/t_H (H to L) and Δ/t_L (L to H), where Δ is the cell cycle time. t_H and t_L are the average holding time of the states, and the peak bit-rate v_p is $v_H S$ and the average bit-rate v_a is $(v_H t_H + v_L t_L)S/(t_H + t_L)$, where S is the cell length.

Figure 6(b) shows the simulation model for multiplexing two types of two-state calls on a 150-Mbps ATM link ($\Delta = 2.7 \mu sec$). Call-type 1 is a low-speed call (peak bit-rate = 1.5 Mbps) and call-type 2 is a high-speed call (peak bit-rate = 15 Mbps). All of the calls are multiplexed into a 150-Mbps ATM link with a buffer for 100 cells. Figure 6(c) shows the contours of the cell loss rates observed in simulations of 10^8-cell-cycle for various combinations of n_1 and n_2. Since a cell loss rate smaller than 10^{-7} was not observed, the interpolation algorithm derived zigzag contours for low loss rates. When the target cell loss rate is 10^{-5}, about 140 low-speed calls but only 10 high-speed calls can be multiplexed into the ATM link. This means the best link utilization for low-speed calls is 70% but only 50% for high-speed calls.

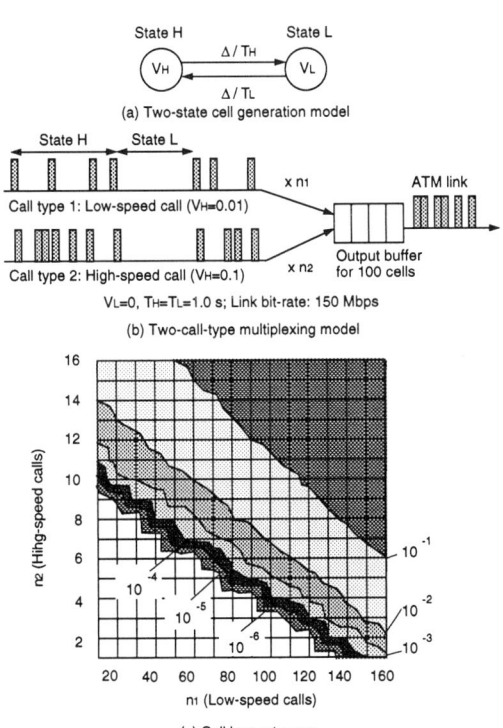

Figure 6 Example of multiplexing two call-types.

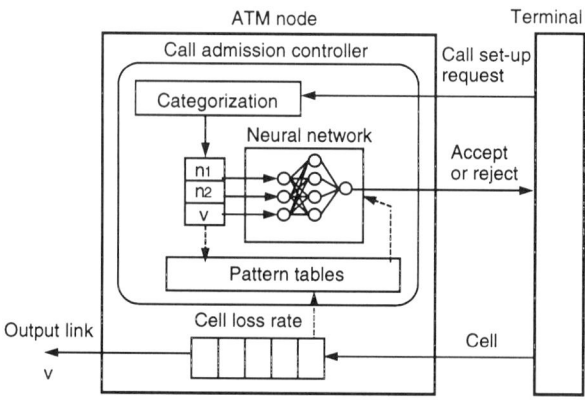

Figure 7 Admission control using a neural network.

4.2 Admission Control using a Neural Network

Features of adaptive call admission

The basic structure of a call admission controller using a neural network is shown in Fig. 7. The neural network learns the call admission boundary from the data gathered at the running node. Since the admission boundary is determined not by modeling-and-analyzing but by learning, this control method is tolerant of errors in the declared traffic characteristics. The exact definition of call-type is not required, and call-types categorized by communication-service type [Fig. 4(b)] can be used. The boundary is expected to be modified appropriately when the characteristics of the multiplexed calls change, and the addition of a new call-type can be accommodated simply by adding input units to the neural network.

Neural network operation

At the node shown in Fig. 7, the numbers of connected calls of each call-type n_i for $i = 1,..k$ are always counted (k is the number of call-types). The cell loss rate l is observed at the output buffer, and the observed data is stored in pattern tables for training. The functions of the pattern tables are explained later in this subsection. The link capacity v of the output ATM link may be changed by an upper-level control, such as link capacity control.

ATM Traffic Control Using Neural Networks

As described in Sec. 4.1, when a new call set-up request arrives at the node, it is first categorized into one of k call-types. Now suppose the new call set-up request for call-type j arrived at time t. The node status at time t is represented by a vector of the numbers of connected calls of each type and the link capacity at that time. If the admission controller accepts the request, the node status becomes

$$(n_1(t), n_2(t), .. n_j(t) + 1, .. n_k(t); v(t)).$$

This vector is the input to the neural network. The neural network's output $d(t)$ for this input vector is a decision value, ranging from 0 to 1, used to determine whether or not the input combination of numbers of connected calls lies above the call admission boundary. The request is accepted if $d(t)$ is smaller than 0.5; otherwise it is rejected.

The neural network is trained with the data observed at the output buffers. The target output, or teacher signal, is generated by comparing the observed QOS with the target QOS value. Suppose, for example, that the only QOS parameter is cell loss rate and that the target value is l_0. When the observed cell loss rate at time t is $l(t)$, then the target output $a(t)$ for the observed status at time t is trained as specified by the following equations:

$$a(t) = \begin{cases} 0 & if \ l(t) < l_0 \\ 1 & if \ l(t) > l_0 \end{cases}$$

Pattern tables

To train the neural network by back-propagation, training data sampled over a wide input domain must be used in a random sequence: if the training input pattern is distributed in only a small part of the domain, the error for the other parts of the domain does not improve. In the on-line training, however, the distribution of the observed data is restricted by the node status at that time. Actually, the numbers of connecting calls of each call-type change very slowly with time. Moreover, because high-loss-rate data is very rarely observed in the well-controlled situation, the neural network learns mostly the low-loss-rate patterns and may therefore not learn the true admission boundary.

To determine the location of the admission boundary accurately, the pattern table method is used [Fig. 8]. The observed data (the node status and the cell loss rate) are categorized into low-loss-rate events and high-loss-rate events and stored in two pattern tables. These categories correspond to the target outputs 0 and 1 described above. Even if the numbers of connected calls changes slowly. if the tables arc largc cnough, aftcr a period of observation,

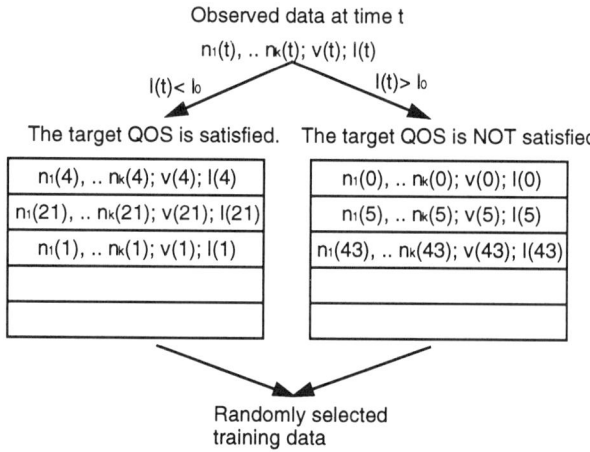

Figure 8 Pattern table method.

they will each contain a number of patterns distributed over a wide area of the input pattern domain. A training pattern is randomly chosen from these tables, so the neural network is trained with various input patterns selected in random sequence from a wide area of the input domain. The admission boundary derived by the neural network can easily be made safer by weighting the selection rate of the high-loss-rate table so that the training rate for the more important high-loss-rate data is more heavily weighted. When the traffic characteristics change rapidly, an optional mechanism discarding old data from the pattern tables is useful because the neural network can then be trained with only the newer data, which more accurately reflect the current environment.

Initial weight values and off-line training

The off-line training is practically important for ensuring that the neural network's performance at the initial stage is adequate. A call admission boundary derived by analytical methods or simulations can easily be implemented by a neural network using off-line training. The neural network can then quickly decide, without complicated mathematical calculation, whether to accept or reject a call set-up request. When the actual traffic differs from the analytical model, the neural network of course adjusts the admission boundary appropriately if the on-line training is used.

4.3 Simulation results

Simulation model

Neural network performance was demonstrated in a simulated call admission control for two bursty call-types. The peak bit-rates for call-types 1 and 2 were respectively set to 0.002 and 0.02 (relative to the link capacity of 1.0), and the average bit-rates were 0.001 and 0.01. To simplify the simulation model, the output buffer was not modeled and the number of cells arriving during each control cycle was approximated by using a normal distribution of the total bit-rate. The average offered traffic load was the same for each call-type.

Constant link capacity

Figure 9(a) shows the call admission boundary (for a constant link capacity $v = 1.0$) obtained by starting with random initial weight values and training the neural network for 10 000 s (10 trainings per second). This admission boundary kept the controlled cell loss rate below 10^{-6}.

Changing link capacity

Then the link capacity v was slowly varied, according to a cosine function in 10 000-s cycles from 0.5 to 1.0, so that the neural network learned call admission boundaries for various link capacities. Figure 9(b) shows call admission boundaries that the trained neural network derived for various link capacities. The analytical call admission boundaries shown in this figure are the points for which the sum of the average and four standard deviations of the total bit-rate is equal to the link capacity. This figure shows that the neural network can adapt to very slow changes in link capacity and can learn the correct call admission boundaries over a wide range of link capacities. The controlled cell loss rate with these boundaries was also less than 10^{-6}.

5 ADAPTIVE LINK CAPACITY CONTROL

5.1 System overview

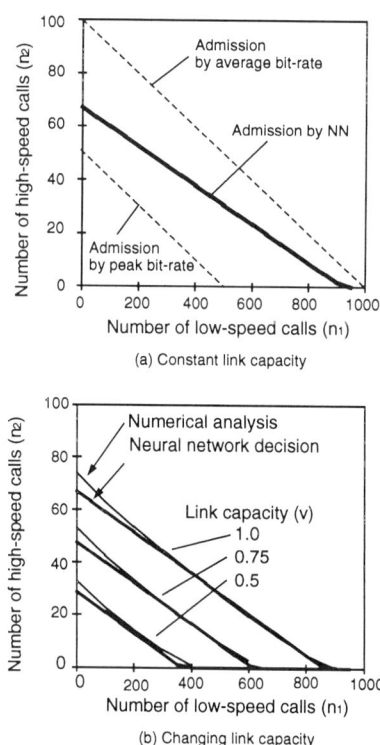

Figure 9 Call admission boundaries derived by a neural network with 20 hidden units.

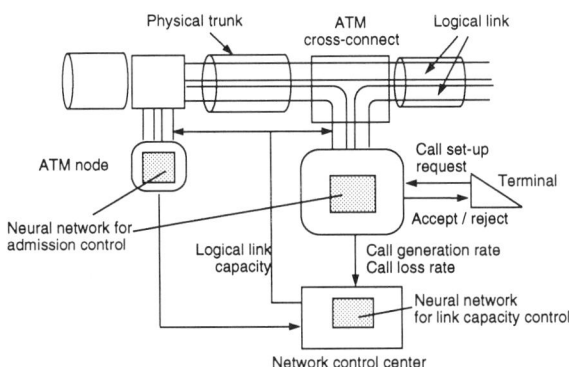

Figure 10 Admission control and link capacity control.

This section first presents the estimation of call loss rate as another example of neural network applications [18]. Then, as an example of the integration of neural-network-based ATM traffic control functions, it shows the integration of adaptive call admission control and adaptive link capacity control. The relation between call admission and link capacity assignment is shown in Fig. 10. The link capacity control regulates the call-level QOS; that is, the call loss rate for a link, by assigning a certain capacity according to the offered traffic. The admission control determines the maximum number of calls that can be multiplexed into the link with that capacity. Neural networks for call admission control learn the call admission boundary as described in Sec. 4.2. The neural network for link capacity control is used to estimate the call loss rate for a link. If the admission boundary moves because of learning, the maximum number of calls carried by the link changes and the link capacity required for achieving a certain call loss rate also changes. Thus the call loss rate estimation must be adaptive. The neural networks in the two levels of control are strongly related to each other and all the neural networks have to cooperate to improve overall network performance.

5.2 Call loss rate estimation by a neural network

Call loss rate estimation in the ATM network

In the conventional circuit-switching network handling calls with the same characteristics, the call loss rate can be usually estimated from the call generation rate (the average number of call set-up requests arriving within 1 s), the average holding time, and the maximum number of calls multiplexed into the link. In the ATM network, however, the maximum number of calls multiplexed into a link depends on the call admission control, and for the emerging multimedia services the distribution of call holding time is not yet certain. This section shows how a neural network can be used to estimate the call loss rate from the link capacity and the call generation rate. Holding time is not used as an explicit parameter, and this method can adapt to changes in the call admission boundary and to the holding times of future multimedia services.

Neural network operation

Figure 11(a) shows the structure of neural network for estimating call loss rate. The inputs to the neural network are link capacity and the observed call generation rate, both normalized within ranges of 0 to 1. An ATM node counts the numbers of generated calls and lost calls for each link and stores them in a pattern table. After a period of on-line training with data randomly selected from the pattern table, the neural network can estimate call loss rate accurately. The target output is amplified by q because the significant call loss rates for link capacity control is much smaller (10^{-4} to 10^{-1}) than the neural network output range, 0 to 1.

Simulation results

To show the accuracy of using a neural network to estimate call loss rate, a simple call multiplexing model with a constant admission boundary was simulated and used to generate data for training the network. Only one CBR call-type (with a bit-rate of 0.01 and an average holding time of 100 s) was simulated. The link capacity v and the call generation rate a were randomly selected within the range 0 to 1, and the corresponding value of the call loss rate ranged between 10^{-1} and 10^{-6}. The inputs to the neural network were v and the number of calls generated in 500 s. Figure 11(b) shows the maximum offered traffic derived by the trained neural network with 10 hidden units with $q = 5$. The errors are small for call loss rates from 10^{-1} to 10^{-5}.

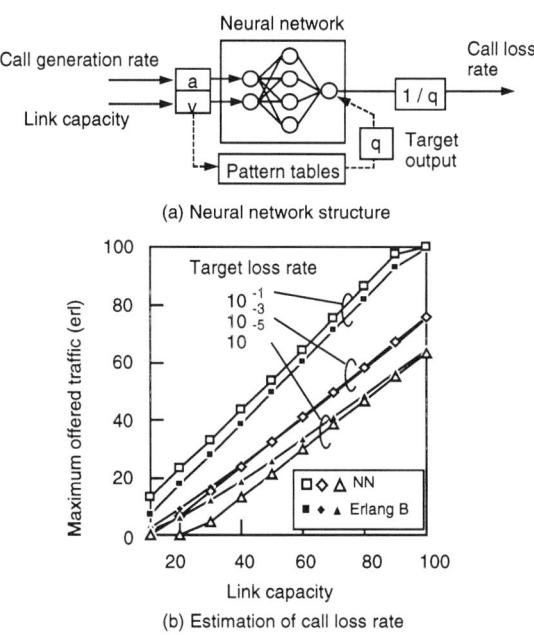

Figure 11 Call loss rate estimation.

5.3 Link capacity control using neural networks

Link capacity control

A link capacity control regulates network-level QOS parameters like call loss rate and link utilization by assigning proper capacity to each logical link or virtual path in the ATM network. As described in Sec. 2.1, switching nodes are connected to each other through a logical link, and the ATM cross-connect can flexibly change the link capacities within the transmission line capacities. When the offered traffic for each link is given, the required link capacity is determined by the target call loss rate. Thus the problem is to determine the logical link path and the capacity of the physical transmission line network. this problem can be formalized into an optimization problem minimizing the objective function.

Objective function for link capacity assignment

Here an actual objective function for one call-type is described. The route of each link on the physical network is assumed to be fixed and only the capacities of the logical links are considered as variables. The goal is to minimize the maximum call loss rate in the whole network:

$$max\{l_1, l_2, ... l_N\} \longrightarrow min,$$

where l_i is the call loss rate for logical link i ($i = 1,...N$) and N is the total number of logical links. As described in Sec. 5.2, the value of l_i is determined by a neural network from the capacity v_i and offered traffic a_i for each link. The constraint on link capacity assignment is that the sum of the capacities of all links passing through a physical transmission line must not exceed the capacity of the transmission line:

$$\sum_{i=1}^{N} p_{ij} v_i \leq c_j \; for \; j = 1,...M,$$

where p_{ij} is 1 if link i passes through the transmission line j and 0 otherwise, c_j is the capacity of the transmission line j, and M is the total number of transmission lines. Thus the objective function E is represented as

$$E(\vec{v}) = max\{l_1, l_2, ...l_N\} + C \sum_{j=1}^{M} H(\sum_{i=1}^{N} p_{ij} v_i - c_j) \longrightarrow min.$$

Here $\vec{v}$ denotes $(v_1, ..v_N)$, C is a sufficiently large positive number to weight the constraint term, and H is the following function:

$$H(x) = \begin{cases} 0 & if \; x < 0 \\ x & if \; x \geq 0 \end{cases}$$

Random optimization method

The Matyas algorithm, which is the simplest random optimization method [21], was used to optimize the objective function. The algorithm steps are as follows:

(i) Select a random vector $\vec{v}(0)$ within the region satisfying the link capacity constraint. Set $\vec{v_b}$ to $\vec{v}(0)$, E_b to $E(\vec{v}(0))$, and i to 1.

ATM Traffic Control Using Neural Networks

(ii) Set $\vec{v}(i)$ to $\vec{v_b} + \vec{dv}$. Each component of the vector $\vec{dv}$ is generated by a random value from the normal distribution $N(0, \sigma)$.

(iii) If $E(\vec{v}(i)) < E_b$, set v_b to $\vec{v}(i)$ and E_b to $E(\vec{v}(i))$.

(iv) If $i < R$, add 1 to i and return to step (ii).

Here R is the number of iterations, σ determines the distribution of test points, and $\vec{v_b}$ is the best link capacity assignment found during the search.

Simulation results

The effect of integrating link capacity control and call admission control was demonstrated by simulating the adaptive link capacity control. The network model used was the 4-node full-mesh shown in Fig. 12(a). Each node had a neural network for call admission control, and the network control center had a neural network estimating the call loss rate of all links. Each neural network had 2 input neurons, 10 hidden neurons, and 1 output neuron.

In this simulation, only one burst call-type was considered (the peak bit-rate was twice the average bit-rate, and the holding time was 100 s), and its average bit-rate was used as the unit for measuring link capacities. The trunk capacities were designed by assigning the link capacity between each pair of nodes to 150, so that each link could multiplex up to 75 calls by peak-bit-rate-based admission control and the call loss rate for 60 erl would be kept under 0.01 (when the holding time is 100 s, 1 erl corresponds to call generation rate of 0.01).

The neural networks for call admission control were initially trained to accept call set-up requests only when the total peak bit-rates of connecting calls did not exceed the corresponding link capacity. The neural networks estimating call loss rate were trained by using the call loss rate derived from the Erlang B equation based on this admission control. Then in on-line training, the neural networks for call admission improved the call admission boundary, and the neural network for estimating call loss adapted to the change of this boundary. The optimal link capacity assignment was repeatedly calculated from the call generation rate observed over 500-s intervals.

Figure 12(b) shows the change in offered traffic for each link and the resulting maximum call loss rates. When the on-line training started, the adaptability of the call admission control first increased the maximum number of connectable

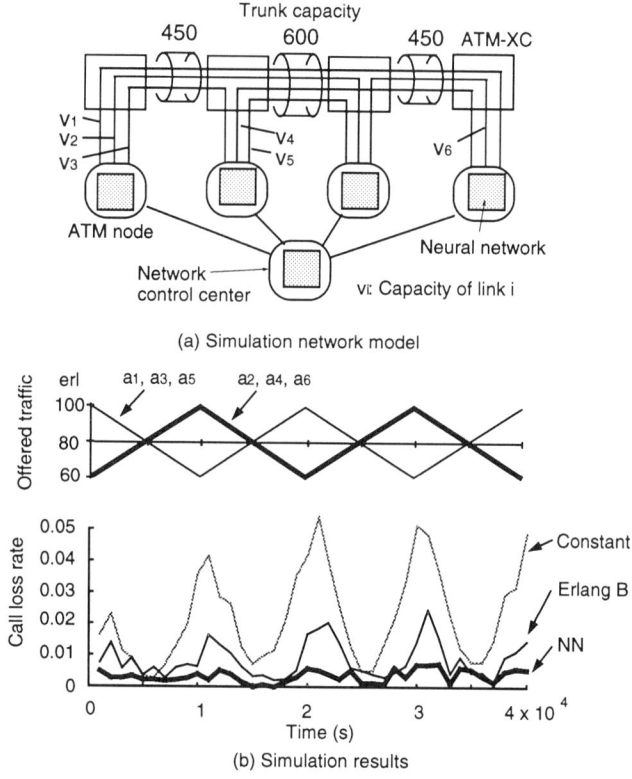

Figure 12 Simulation of link capacity control.

calls under the cell loss rate requirement of 10^{-6}. As the result of this admission boundary change, a link with capacity of 150 carries about 80 erl with the call loss rate of 0.01 even though it had been designed to carry only 60 erl. The curve labeled "Constant" in Fig. 12(b) shows that the constant link capacity assignment results in a call loss rate that changes widely when the offered traffic balance changes. The curve labeled "Erlang B" shows the result of controlling the link capacity according to the call loss rate estimated by the initial neural network trained with the Erlang B equation; that is, without the on-line training. The call loss rate changes much less than it does with constant link capacity assignment but still increases with changes in the traffic balance. The curve labeled "NN" represents the result obtained by the neural network with on-line training: the call loss rate is much lower and nearly constant for all traffic conditions. This implies that adaptive link capacity control adapted accurately to changes in the call admission control, thereby allocating link capacities efficiently.

6 FURTHER RESEARCH DIRECTIONS

This chapter describes the potential effects of using neural networks in ATM traffic control. The neural network capabilities of adaptability and optimization can be used in various control functions at all ATM traffic control levels. The adaptability of these networks can make the traffic controller efficient and flexible: on-line training will enable the ATM network controller to adapt to the various traffic characteristics of multimedia communication services and to changes in traffic characteristics. The recent rapid growth of the ATM LAN system will promote the use of neural networks in private ATM networks (for such applications as data traffic control) as well as in the public ATM networks. Despite the great potential of neural networks and their suitability for ATM traffic control, several problems must be overcome if these networks are to be installed in commercial systems:

(i) Because the data observed from a running system is often widely distributed and has an unknown probability distribution, a stable and accurate method for training with "noisy" data is required.

(ii) Mechanisms assuring safe control in the on-line training are needed. In the call admission control, for example, the QOS is more important than high link utilization. The period during which cell loss rate is high should thus be very short even during training.

(iii) For each application, guidelines for determining the training time, the amount of data, and the number of hidden units required to make the neural network accurate enough for practical control should be established.

Acknowledgments

I would like to thank Tatsuro Takahashi of NTT Communication Switching Laboratories for his directions and valuable suggestions.

REFERENCES

[1] Armbrüster, H., and K. Wimmer, "Broadband Multimedia Applications Using ATM Networks: High-Performance Computing, High-Capacity Storage, and High-Speed Communication," IEEE J. Sel. Areas in Commun., 10, December 1992, pp. 1382–1396.

[2] Kulzer, J.J., and W. A. Montgomery, "Statistical switching architectures for future services," Int'l Switching Symposium, 1984, pp. 43A1.1.1–6.

[3] Turner, J. S., "New direction in communications," Int'l Zürich Seminar, 1986, pp. A3.1–8.

[4] Biagioni, E., E. Coopre, and R. Sansom, "Designing a Practical ATM LAN," IEEE Network, March 1993, pp. 32–39.

[5] Newman, P., "ATM Technology for Corporate Networks," IEEE Commun. Mag., 30, April 1992, pp. 90–101.

[6] Wright, D. J., M. Wright, W. Verbiest, N. Shimasaki, and M. De Prycker, Eds., "B-ISDN Applications and Economics," IEEE J. Sel. Areas in Commun., 10, 9, December 1992.

[7] Kawarasaki, M., and B. Jabbari, "B-ISDN Architecture and Protocol," IEEE J. Sel. Areas in Commun., 9, 9, December 1991, pp. 1405–1415.

[8] Zegura, E. W., "Architecture for ATM Switching Systems," IEEE Commun. Mag., February 1993, pp. 28–37.

[9] Stephens, W. E., M. DePrycker, F. A. Tobagi, and T. Yamaguchi, Eds., "Large-scale ATM Switching Systems for B-ISDN," IEEE J. Sel. Areas in Commun., 9, 8, October 1991.

[10] Eckberg, A. E., "B-ISDN/ATM Traffic and Congestion Control," IEEE Network, September 1992, pp. 28–37.

[11] Shoraby, K., L. Fratta, I. S. Gopal, and A. A. Lazar, Eds., "Congestion Control in High-speed Packet Switched Networks," IEEE J. Sel. Areas in Commun., 9, 7, September 1991.

[12] Yazid, S., and H. T. Mouftah, "Congestion Control Methods for BISDN," IEEE Commun. Mag., July 1992, pp. 42–47.

[13] Wernik M., O. Aboul-Magd, and H. Gilbert, "Traffic Management for B-ISDN Services," IEEE Network, September 1992, pp. 10–19.

[14] Guérin, R., H. Ahmadi, and M. Naghshineh, "Equivalent Capacity and Its Applications to Bandwidth Allocation in High-Speed Networks," IEEE J. Sel. Areas in Commun., 9, 7, September 1991, pp. 968–981.

[15] Saito, H., and Shiomoto, K., "Dynamic Call Admission Control in ATM Networks," IEEE J. Sel. Areas in Commun., 9, 7, September 1991, pp. 982–989.

[16] Murase, T., H. Suzuki, S. Sato, and T. Takeuchi, "A Call Admission Control Scheme for ATM Networks Using a Simple Quality Estimate," IEEE J. Sel. Areas in Commun., 9, 9, December 1991, pp. 1461–1470.

[17] Hiramatsu, A., "ATM Communications Network Control by Neural Networks," IEEE Transactions on Neural Networks, 1, March 1990, pp. 122–130.

[18] Hiramatsu, A., "Integration of ATM Call Admission Control and Link Capacity Control by Distributed Neural Networks," IEEE J. Sel. Areas in Commun., 9, 9, September 1991, pp. 1131–1138.

[19] Takahashi, T., and A. Hiramatsu, "Integrated ATM Traffic Control by Distributed Neural Networks," Int'l Switching Symposium '90, Vol. III, May 1990, Sweden, pp. 59–65.

[20] Rumelhart, D. E., G. E. Hinton, and R. J. Williams, "Learning internal representations by error propagation," in Ed. D. E. Rumelhart, J. L. McClelland, and the PDP Research Group, "Parallel distributed processing," vol. I, Cambridge, MA: MIT Press, 1986.

[21] Matyas, J., "Random optimization," Automation and Remote Control, 26, 1965, pp. 246–253.

[22] Narendra, K., and P. Mars, "A study of telephone traffic routing using learning algorithms," IEEE Int'l Conf. of Commun., 1981, pp. 55.4.1–5.

5

LEARNING FROM RARE EVENTS : DYNAMIC CELL SCHEDULING FOR ATM NETWORKS

Daniel B. Schwartz

*Phoenix Corporate Research Laboratories,
Motorola, Inc.*

1 INTRODUCTION

The large gap between the systems analyzed by classical queueing theory and ATM networks cries out for solutions based on learning. Even those few problems amenable to classical analysis techniques have a component that defies explicit analysis – the intrinsic uncertainties associated with describing and controlling the flows in high performance, mixed service networks. Unlike the stationary, uncorrelated sources characteristic of queueing theory, the pattern of arrivals in these systems are highly correlated in time. Exploiting these correlations, either directly, with a predictive model, or indirectly, as in reinforcement learning, can result in significant performance gains. Queuing systems have another property requiring the application of non-traditional control techniques; the rewards, or more often penalties, associated with control actions are delayed and non-linear. For example, a control action can allow a buffer to fill up without incurring a short term performance penalty but may result in a large delayed penalty if a subsequent burst of activity causes an overflow. Control problems with delayed rewards require techniques that find optimal sequences or trajectories instead of responding to just the instantaneous state of the system.

The enormous state spaces, characteristic of all but the simplest queueing systems, suggest the use of a continuous representation of the state variables because of both the resulting decrease in complexity and the generalization that follow from defining a metric on the state space. However, the large bias in prior distribution of observed states dooms the direct application of most connectionist approximation functions to failure. As a simple example, consider the M/M/1 queue. In this system, customers enter the queue randomly at rate λ and wait until they reach the head of the queue, at which time they occupy the server for an exponentially distributed interval of average duration τ. Under these circumstances, the probability of finding n customers waiting in the queue at any time is simply [1]

$$p(n) = (1 - \lambda\tau)(\lambda\tau)^n.$$

The goal in controlling most queueing system is to avoid, subject to suitable constraints, having an excess of customers in the queue at any instant. Thus 'congested' states, such as those with high queue occupancies n, are the most interesting from the viewpoint of the control system but also the least frequent. This is a serious problem for most connectionist networks since the most frequent training examples tend to obliterate any memory of the rare ones, especially when trained on-line as required here.

2 ATM NETWORKING

Flexible dynamic control mechanisms are essential for the efficient sharing of network resources by diverse applications. A ubiquitous feature of ATM networks is the partitioning of traffic into classes based upon their *quality of service* requirements. In this work, a traffic class s is defined by its maximum allowable queueing delay τ_{max}^s and maximum cell loss probability ϵ_{max}^s, where cells can be lost either by exceeding the maximum delay or by being discarded as a congestion control measure. On arrival, cells are sorted into m fifo queues, one for each traffic class. The output controller must decide in what sequence to service the queues in order to maximize the number of cells transmitted subject to the quality of service constraints. In anticipation of or reaction to congestion the output controller must also decide when to discard cells. In this work we address the cell scheduling problem. As an additional congestion control mechanism, the time stamp of each cell is examined just prior to transmission and the late ones are discarded.

The most common service discipline for a system with multiple traffic classes is static priorities. In a system with two service classes and worst case queueing delays $\tau_{max}^I < \tau_{max}^{II}$ static priorities service class I until its queue is empty and only then service class II. Static priorities are inefficient because they do not account for the size of the load on each queue; when class II is heavily loaded and class I lightly loaded, a better strategy is to delay cells in class I to reduce the queueing delays for class II. A simple example of this is the Earliest Deadline First algorithm. In EDF the time stamp of the cell at the head of each queue is examined and the cell closest to being late is transmitted. When the objective is to minimize the total cell loss regardless of class, and late cells are discarded, the EDF algorithm is optimal. When the objective is to satisfy separate quality of service constraints for each service class a family of optimal algorithms have been proposed [4]. However, these algorithms are primarily of theoretical interest because they assume the course of an entire busy period is known in advance and require a sort whose computational cost is quadratic in the length of the busy period. Practical algorithms can use traffic predictors to produce causal algorithms but also require heuristics to reduce the computational cost. For example, the MARS algorithm [5] combines a first order traffic predictor with a heuristic to limit the cost of a computationally expensive sorting operation.

3   ON-LINE DYNAMIC PROGRAMMING

In this work, we use reinforcement learning to generate the service discipline without recourse to a model as required by traditional dynamic programming. The resulting algorithm is a member of a class of stochastic service disciplines described by weights α^s where each traffic class s is serviced with a probability

$$P^s = \frac{\alpha^s h(n^s)}{\sum_s \alpha^s h(n^s)}$$

that is proportional to its weight. Here, h is a step function,

$$h(x) = \begin{cases} 1 & x > 0 \\ 0 & x \leq 0 \end{cases}$$

and n^s is the occupancy or number of cells waiting transmission in queue s. This family of service disciplines can approximate those described earlier to any degree of accuracy. For example, if $\alpha^1 \gg \alpha^2 .... \gg \alpha^{m-1} \gg \alpha^m$ the result is static priorities. To approximate EDF, let $\alpha^s = (\tau^s/\tau^s_{max})^\mu$, where τ^s is the elapsed time the cell at the head of the queue for traffic class s has been waiting for transmission. When $\mu = 1$, each queue is serviced with a probability proportional to the age of its oldest cell whereas when μ is large, the probability of the oldest cell being serviced approaches 1.

To reduce the computational cost of this algorithm, the weights α^s are re-evaluated once every T time steps where time is measured in units of the transmission time of a single cell. The cost incurred in servicing traffic class s during a cycle is a function of the queueing delays incurred by that class; it can be readily described in terms of the queueing delay τ_t^s which we define as:

- The waiting time of the cell at the head of the queue for traffic class s if it was serviced at time t.
- τ_{max}^s if the cell at the head of queue for class s was discarded at time t.
- 0 if class s was not serviced at time t.

The cost u_i^s incurred over a cycle is then

$$u_i^s = \frac{\sum_{t=iT}^{(i+1)T-1}(\tau_t^s/\tau_{max}^s)^{\mu^s}}{\sum_{t=iT}^{(i+1)T-1} h(\tau_t^s/\tau_{max}^s)}.$$

where h is step function as defined above and the denominator is the total number of cells transmitted or discarded from class s over the cycle. When the exponent $\mu^s = 1$, u_i^s is the mean queueing delay for class s normalized to the maximum tolerable queueing delay τ_{max}^s. As $\mu^s \to \infty$ the cost u_i^s approaches the probability of a cell from traffic class s being discarded during the cycle. The delayed rewards characteristic of this problem suggests basing the service discipline directly upon the infinite horizon discounted estimate of u_i^s, a quantity we call the urgency U^s [2, 3]

$$U^s(\vec{x}) = <\sum_{i=t}^{\infty} \gamma^{i-t} u_i^s > . \tag{1}$$

where $\vec{x}$ is the state vector of the system, as will be defined shortly and the expectation $< \ldots >$ is computed over all possible sequences of states that start at $\vec{x}$. For reasons of convergence, the discount factor $0 < \gamma < 1$. The discount parameter determines how far into the future the urgency attempts to predict. The temporal indices in this expression and those that follow are measured in units of the cycle time T. The policy or service discipline is computed directly from the urgency, $\alpha^s = U^s(\vec{x}_t)$.

To understand how the controller works, it is informative to consider what happens when μ^s is large. In this case, the controller strongly favors the class closest to its transmission deadline and thus most likely to lose cells. By reducing the exponent, the risk of cell loss tolerated by the controller for a given traffic class can be increased, allowing the loss characteristics of each class to be tailored independently of its delay requirement and the loss requirements of the other traffic classes.

The natural choice of state variables for the controller are the queue occupancies n^s, especially since this is one of the few statistics readily available in real switches. However, in early experiments we found the performance resulting from this choice of state vector to be lacking for obvious reasons – the queueing occupancy alone does not distinguish between the case where cells are just starting to build up from the start of a burst and that where the queue is coming out of congestion. In the former case, the cells can be delayed whereas in the latter they are likely to be close to their deadlines. This problem can be surmounted by adding the waiting time of the cell at the head of each queue at the start of the cycle to the state vector. The resulting state vector $\vec{x}_t = (\vec{n}_t, \vec{\tau}_t)$ is of dimension twice the number of traffic classes in the system. Even with this expanded state vector, the system still has hidden variables associated with the unobservable internal states of the sources.

3.1 Computing the Urgency

The classical approach to computing evaluation functions like the urgency $U^s(\vec{x})$ in sequential decision problems is dynamic programming, a general optimization technique most applicable to sequential decision problems when complete information about the dynamics of the system is available (for example, Ross [6]). For stochastic Markov systems, the dynamics are uniquely specified by the probability $P_{\vec{x}\vec{x}'}$ of making a state transition $\vec{x} \to \vec{x}'$. Given the transition matrix and a cost function $u(\vec{x})$, dynamic programming provides a collection of closely related techniques to efficiently compute the cumulative cost of a sequence of state transitions. For example, the expected cost $U^{(2)}(\vec{x})$ of a sequence of two transitions starting at state $\vec{x}$ can easily be computed,

$$U^{(2)}(\vec{x}) = u(\vec{x}) + \gamma \sum_{\vec{x}'} P_{\vec{x}\vec{x}'} u(\vec{x}').$$

By recursion, the estimated cost of a sequence of n+1 transitions can easily be computed from that for n transitions,

$$U^{(n+1)}(\vec{x}) = u(\vec{x}) + \gamma \sum_{\vec{x}'} P_{\vec{x}\vec{x}'} U^{(n)}(\vec{x}').$$

When $n = \infty$ the recursion relation becomes a self consistency condition for the infinite horizon discounted cost $U \equiv U^{(\infty)}$,

$$U(\vec{x}) = u(\vec{x}) + \gamma \sum_{\vec{x}'} P_{\vec{x}\vec{x}'} U(\vec{x}') \qquad (2)$$

which can in principle be solved directly to obtain the evaluation function U.

Several factors conspire to make computation of the urgency U^s by the direct application of dynamic programming intractable; the most obvious is the size of the state space. In this work, we discuss an experimental example with three traffic classes where both the queue occupancies and waiting times have ranges of at least 1000, resulting in a state space with at least 10^{18} reachable states. Since the dynamics of the system are dominated by the behavior of the sources and hence unknown, the transition matrix must be inferred from sequences of measurements on the system. And finally, although computation of the urgency is superficially a Markov estimation problem, in practice it is a control problem since the urgency directly determines policy through the weights α^s. As a result, the cost $u^s(\vec{x})$, the urgency $U^s(\vec{x})$ and transition matrix are all interrelated, necessitating interleaving estimation of the cost and transition matrix with the solving of the self-consistency condition for the urgency, equation 2.

Two key algorithms make the computation tractable :

- The urgency is computed with an on-line learning rule based on the method of temporal differences [7, 8]. Temporal differences learning generates the evaluation function or urgency directly from a sequence of observations without estimation of either the cost or transition matrix.
- The components of the state vector $\vec{x}$ are treated as continuous parameters and a smooth, local representation is used for the urgency. Thus the properties of a state that has never been visited can be inferred from its neighbors.

3.2 The Learning Procedure

When compared to the computational cost of full dynamic programming, the method of Temporal Differences as used here is strikingly simple. At time $(t+1)T$ the change in the urgency is

$$\Delta U^s(\vec{x}_t) = \eta(u_t^s + \gamma \hat{U}^s(\vec{x}_{t+1}) - \hat{U}^s(\vec{x}_t)) \tag{3}$$

where η is a learning rate parameter. When no cells are transmitted from class s during a cycle its urgency is not updated. Note that the update for $U^s(\vec{x}_t)$ is performed at time $(t+1)T$ using $\vec{x}_{t+1}$ and the cost incurred by the transition from $\vec{x}_t$ to $\vec{x}_{t+1}$ as this is characteristic of TD learning. A direct comparison between the TD learning rule, equation 3, and the self-consistency condition for the evaluation function, equation 2, suggests that TD learning can be understood as Monte Carlo estimation of the evaluation function. The tradeoff involved in using a learning rule of this type as opposed to full dynamic programming is between computation and efficiency. Because full dynamic programming keeps detailed information about past observations in the transition matrix, it can squeeze the most information out of the current observation by combining it with earlier observations of its predecessor states. Learning procedures that lie between the extremes of having no system model and building a complete model are an area of active research [9, 10]. As described later in this chapter, we bridge this gap by keeping a limited record of past state transitions and make strategic use of them to 'back up' rare states.

An important feature of dynamic programming based control techniques is that under restricted conditions, the resulting controllers are optimal. However, as posed, urgency scheduling is an optimization problem only in a loose sense. The objective is to transmit as many cells as possible subject to the quality of service constraints but we never explicitly estimate the link utilization as a part of the control strategy. Instead, following the strategy assumed by the earliest deadline first algorithm and its derivatives, we attempt to maximize link utilization by delaying the transmission of each cell for as long as possible. Because we place strong restrictions on the choice of policy the optimality of the result is unclear. However, we can rely on theoretical results [11, 12] on the convergence of optimal control algorithms for insight. In particular, we know the controller must occasionally take actions that are suboptimal according to the current estimate $\hat{U}^s(\vec{x})$ of the evaluation function. Otherwise, the controller can become trapped by a policy that is locally optimal with respect to a restricted trajectory through state space but globally suboptimal. We follow such a program crudely by weakening the policy,

$$\alpha^s(\vec{x}) = (\hat{U}^s(\vec{x}))^\beta,$$

where $0 \leq \beta < 1$. This is similar in spirit to weighing the selection of actions with Boltzman factors (for example [13] and references therein) since for small values of β, all actions are equally likely and for large values, the queue with the highest urgency is always served. By default, we allow β to vary linearly with time, starting at zero and reaching one at the end of training. A more sophisticated learning procedure might adjust the the amount of experimentation to match the reliability of the current estimate of the urgency.

3.3 Representing the Urgency

Development of an on-line representation of the urgency is made difficult by the interaction between the service discipline and current estimate of the urgency. A CMAC [14] failed at this task due to a pair of interrelated factors – first, instabilities resulting from interference between new and stored observations and second, the enormous skew in the prior distribution of observations as described in the introduction of this chapter. Global approximation functions like multi-layered perceptrons would be, of course, even worse than semi-local approximation functions such as a CMAC. The challenge is to develop a functional representation and learning algorithm that efficiently utilize rare observations without being overwhelmed by the commonplace.

A powerful approach to function approximation problems of this type is to simply store every data point, generating estimates of the function by retrieving the n data points nearest to the desired point and applying some type of regression to them [15, 16]. In its simplest form, a piecewise constant approximation can be obtained by using the value of the data point nearest the query point as the function estimate. If the data is stored in a tree structure, like a kd tree [17, 18], the retrieval cost is asymptotically proportional to $\log N$ where N is the number of data points and thus manageable. When the data is uniformly distributed across a high dimensional space, the constant of proportionality is large, but when the data can be embedded onto a lower dimensional manifold, the constant of proportionality is substantially reduced.

The resulting learning rule is to simply store

$$U^s(\vec{x}_t) = u_t^s + \gamma \hat{U}^s(\vec{x}_{t+1})$$

for every observation. The rate of convergence of the controller is dominated by the frequency of rare events; in the simulations described here typical experiments were 10^8 time steps long. With a cycle length of 64 this corresponds to approximately 1.5×10^6 observations. At 50 bytes per record, this is a substantial amount of storage even by contemporary standards. We circumvent this difficulty by averaging together points that are closer together than a threshold distance. When a new observation $\vec{x}$ is within d_{min} of an old data point $\vec{x}_i$ the two are coalesced,

$$y' = \frac{y + \omega_i y_i}{1 + \omega_i}$$
$$\vec{x}' = \frac{\vec{x} + \omega_i \vec{x}_i}{1 + \omega_i}$$
$$\omega' = 1 + \omega_i.$$

The weight ω is initialized to one when a point is created. Even with sixteen nearest neighbors in the six dimensional space of this work, the simplex formed by the retrieved data points frequently collapses, leaving the query point on the outside. In this case, simple linear regression is unreliable so we use kernel regression to estimate the urgency,

$$\hat{U}^s(\vec{n}) = \frac{\sum_{i=1}^{N} K(d_i/d_N) y_i \omega_i}{\sum_{i=1}^{N} K(d_i/d_N) \omega_i}$$

where d_i is the distance to $\vec{x}$'s i'th nearest neighbor and $K(x) = (1 - x^2)^3$. To de-emphasize data points acquired during the early stage of training when the level of experimentation is high and reliability of the estimate of the urgency low, each data point is tagged with its acquisition time, t_{acq}, which is then used to compute an exponential weight, $\omega \propto exp(-\alpha(t - t_{acq}))$.

The accuracy of locally weighed regression is sensitive to the choice of metric used to define the neighborhood of the query point. We restricted our attention to metrics of the form

$$\|\vec{x}\|^2 = \sum_i x_i^2 / \sigma_i^2$$

computed using two different methods. The simpler of the two is to let σ_i^2 be the variance of the i'th coordinate of the data used to estimate the urgency. A more complicated procedure with superior performance is to allow the controller to run with the discount factor $\gamma = 0$ in which case the urgency reduces to a simple estimate of the cost u_i^s. We then used the simplex method [19] of gradient descent to minimize the prediction error ε_p averaged over all those data points whose cost u^s were greater than a threshold value,

$$\varepsilon_p = \sum_{y_i > y_{min}} (y_i - \hat{U}(\vec{x}_i))^2$$

where $\hat{U}$ is the estimated value of the undiscounted urgency at $\vec{x}_i$.

3.4 Backing Up Rare States

The rate of convergence of the urgency is dominated by the frequency of occurrence of rare events with non-trivial cost. Motivated by the success of prioritized sweeping [9], we have developed a technique that allows rare states to be 'backed up' without attempting to model the dynamics of the system. In effect, we try to maximize the information gained from rare events by computing the effect this information would have had on the urgency if we had it at an earlier time. With conventional dynamic programming, whenever a large change in the current estimate of the evaluation function is observed for a state $\vec{x}_{novel}$, the system model can be used to update the evaluation function of states that could have preceded $\vec{x}_{novel}$ by iterating equation 2, the self consistency condition. Instead of building a model, we record the history of the system and use the history to update the evaluation function of possible predecessors of the novel state $\vec{x}_{novel}$. The history of the past behavior of the system is kept by storing in a kd tree, without clustering, every 'interesting' observation of the system. Each record in the history tree contains

- The observed cost u_i^s of leaving the state.
- Pointers to the records corresponding to each observation predecessor and successor states.

States are added to the history tree only if one or more components of the cost u_i^s exceeds a threshold, RecordThreshold, and states are backed up only if their costs exceed a second, higher threshold, BackUpThreshold. Because of the close correlation between the probability for the occurrence of a state and its cost, these thresholds effectively control the size of the history tree.

When the cost of a new state exceeds BackUpThreshold, the first step is to update the urgency as usual and then back up those points affected by this local change in the urgency. To simplify the notation, it is helpful to define some pseudo-code in which each record is a structure whose fields are

- $\vec{x}$, the point in state space corresponding to the record.
- u^s, the cost incurred in leaving this state.
- $pred$, the predecessor of the current state.
- $succ$, the successor of the current state.

In this notation if r is a record then r.$\vec{x}$ is the state vector by which it is indexed and r.pred.$\vec{x}$ is the state vector of its predecessor. If N is the number of points used to estimate the urgency we back up a state $\vec{x}$ by finding its N nearest neighbors, R[1..N], in the history tree. We then generate an update for the predecessor of each of these states by storing

$$U^s(R[i].pred.\vec{x}) = R[i].pred.u^s + \gamma \hat{U}^s(R[i].\vec{x})$$
$$\text{for } i = 1 \text{ to } N \text{ and } R[i].pred.u^s > BackUpThreshold.$$

The effect is as if the new information on the urgency was available when the old transitions occurred. In practice, the addition of the backing up of rare states to the learning algorithm is to speed up the convergence of the urgency by a factor of up to five.

4 EXPERIMENTAL EVALUATION

To provide a realistic environment in which to evaluate scheduling with urgency prediction we have developed a model system with three traffic classes would co-exist in an ATM switch with 155 Mbit ports serving as a hub connecting multiple ATM LANs. Traffic class I carries compressed video as found in a high performance video phone, class II supports high performance graphic applications as generated by CAD tools or visualization software and class III carries TCP-IP traffic. These sources are all bursty, reflecting the type of traffic expected in real ATM networks.

The Video Source

Traffic class I is a video source based on measurements [20] of the output of a video codec when the face of a human engaged in conversation is used as input. The characteristic feature of this source is the periodic emission of bursts, thirty times a second second, in sync with the refresh rate of an NTSC video signal. The length of the bursts varies slowly as a function of time as determined by the presence or absence of motion in the subject. The maximum queueing delay for the video source was $\tau_{max}^{I} = 1000$ and the average link utilization 0.045.

Unlike the quasi–periodic video sources, both the interactive graphic application and the TCP-IP traffic were modeled by interrupted Poisson processes. Such a model has two states, **on** and **off** and three parameters, ρ, P_{off} and P_{on}. When the source is on, it transmits a cell at each time step with probability ρ. The transition from **off** to **on** occurs with probability P_{on} and from **on** to **off** with probability P_{off} resulting in an average data rate of

$$\nu_{IPP} = \rho \frac{P_{on}}{P_{off} + P_{on}}.$$

Interactive Graphics

In the absence of measurements of an interactive graphics application the parameters for sources in traffic class II were determined from typical specifications for a high performance color workstation, resulting in an average burst size of 2000 cells (approximately 100 Kbytes) with an average spacing between bursts of 100 milliseconds and a link utilization of 0.034. The maximum queueing delay was set to $\tau_{max}^{II} = 2000$.

TCP-IP Sources

The parameters for the TCP-IP traffic as modeled by traffic class III were obtained from experimental measurements of the average packet length on a campus internet [21]. The experimental measurements can be roughly modeled by randomly interspersing bursts with an average length of ten cells or 500 bytes. The worst case delay for this traffic class was $\tau_{max}^{III} = 5000$ and the average link utilization 0.09.

5 SIMULATIONS

The primary goal of this work was to produce a robust controller capable of good performance when presented with a broad spectrum of traffic mixes without relying on adaptation to respond to short term fluctuations in the characteristics of the traffic. To provide the variety of traffic scenarios needed to achieve this performance the controller was trained with sources constantly being created and destroyed, just as caused by the diurnal variations in loading of real networks. The lifetimes of individual sources were exponentially distributed with a mean of 250,000 cell times and the source creation events were Poisson distributed with the same mean. If we imagine a three dimensional space whose axes are the number of active sources of each type, the point describing the current source configuration executes a slow random walk which will eventually fill in the space. To prevent gross congestion the creation of new sources was blocked if they would have caused the link utilization to exceed 0.85. Typical training sessions were 10^8 time steps long at the end of which the urgency for each traffic class was dumped to disk where it could be retrieved for performance testing.

To test the controller, the performance was measured for different combinations of fixed sources and the result compared to that obtained by applying the earliest deadline first service discipline to *the identical stream of traffic*. The use of identical streams of traffic substantially reduces the variance of the ratio of the schedulable volumes when comparing the two service disciplines. The results of a single performance measurement is an ordered pair of vectors $(\vec{c}, \vec{\epsilon})$ where the components of $\vec{c}$ are the number of each type of source used in the simulation and those of $\vec{\epsilon}$ are the fraction of the cells lost from the corresponding traffic class. Overall performance is most easily compared by counting the total number of points $\vec{c}$ for which the quality of service constraints can be met, a quantity we refer to as the *schedulable volume*, following the concept of the schedulable region introduced in [5]. Typical performance measurements consisted of 10^7 time steps per source configuration.

5.1 Controlling the Total Cell Loss Rate

In these simulations the quality of service constraint was somewhat arbitrarily chosen to be a total loss rate $\epsilon_{max}^{total} < 10^{-3}$. For this quality of service constraint, earliest deadline first is optimal. Based on a series of experiments, the exponent μ in the definition of the urgency (equation 1) was set to eight for all three traffic classes. Experimental result are shown in table 1. The significance of these

Service Discipline	Schedulable Volume
Earliest Deadline First	74
Urgency Scheduling	72
Naive Scheduling	49
Static Priorities	42

Table 1 Experimental results where the objective is minimizing the total cell loss rate. The 'naive' service discipline is obtained by using the queueing delay of the cell at the head of each queue exponentiated by μ as a crude estimate of the urgency to demonstrate that the urgency has significant predictive power.

results is primarily to show that urgency scheduling can achieve near optimal performance, at least when the quality of service constraint is minimizing the total cell loss. The results for urgency scheduling were only weakly dependent on the value of the discount factor γ, an unsurprising result since the optimal algorithm requires knowledge of only the cell at the head of each queue.

5.2 Heterogeneous Service Objectives

A more interesting application of urgency scheduling occurs when the most delay sensitive traffic class, in this case video, can tolerate higher losses than a delay tolerant traffic class like TCP-IP traffic. This is not an unrealistic scenario – video codecs that can tolerate high cell losses have already been demonstrated. For these simulations the loss thresholds ϵ^s and exponents μ^s were set to $(10^{-2}, 8)$, $(10^{-3}, 4)$ and $(10^{-4}, 2)$ for video traffic, graphics traffic and TCP-IP traffic, respectively. As in the previous example, the values of the exponent were adjusted by experiment, building on the assumption of $\mu^I > \mu^{II} > \mu^{III}$ to reflect the desired loss tolerance. Some experimental results measured as a function of the discount parameter are shown below in figure 1 Urgency based scheduling clearly outperforms the earliest deadline

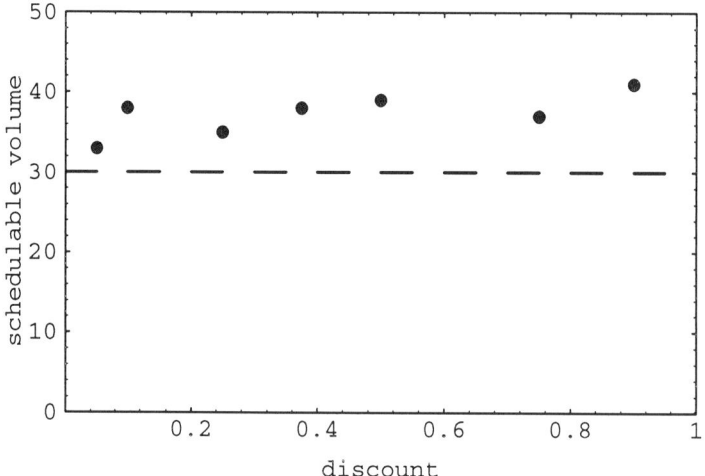

Figure 1 Performance of urgency scheduling with heterogeneous service objectives as a function of the discount parameter. The vertical axis is the size of a subset of the schedulable volume and the dotted line denotes the performance obtained from EDF. In these experiments the cell loss probability was measured over 10^6 time step windows. Failure to meet the quality of service constraints in any particular window caused that source configuration to be removed from the schedulable region.

first algorithm for all values of the urgency, achieving on average a 25% improvement for $\gamma = 0.9$. In a set of experiments explicitly designed to look for worst case behavior urgency scheduling was found to always equal or better the performance of EDF with an average improvement of 25% as shown in the data.

The observed performance advantage of urgency scheduling, although modest, is of practical interest since for many configurations of sources it corresponds to a real increase in available bandwidth. Furthermore, in ATM switching equipment urgency based scheduling has some subtle performance advantages. The controller can easily be modularized hierarchically to reduce its computational cost and simplify the required hardware. At the bottom and most time critical level a module would be dedicated to executing the schedule from the weights α^s, acting on a cell-by-cell basis. At the next level in the hierarchy, the memory based function approximator used to approximate the urgency can be replaced with a more parsimonious representation like a multi-layered perceptron or a very sparse tree based approximation function. This replacement can reduce the computational cost of evaluating the urgency to the point where, when amortized over an entire cycle, the computational cost of urgency scheduling can be comparable to or lower than Earliest Deadline First. At the top level of the hierarchy the memory based function approximator can be retained but this representation of the urgency probably needs only be updated in response to rare states and their immediate predecessors. As appropriate, this slowly changing representation of the urgency can be used to update that of the module below it.

6 CONCLUSIONS

We have applied on-line dynamic programming to a realistic ATM network element and demonstrated it to be capable of matching the performance of the Earliest Deadline First algorithm when the quality of service constraints are such that EDF is optimal. For more complex quality of service constraints, we have shown that our algorithm, urgency scheduling, outperforms EDF by a substantial margin. For control in ATM networks, adaptive algorithms like urgency scheduling will be critical to the deployment of complex networks because of the large uncertainties in the quantity and characteristics of the traffic that will flow through them.

The learning algorithm and function approximation techniques used here are less well known than the familiar paradigm of supervised learning combined with neural network approximation functions. However, the domination of ATM control problems by delayed rewards and rare events forces their use. The combination of memory based function approximators and on-line dynamic programming demonstrated here could readily be extended to other control problems in high speed networks. For example, an effective cell discard policy that drops cells at the input to a switch in anticipation of future congestion is sorely needed. Development of an effective admission control policy is also an open problem well suited to these techniques; unlike the scheduling and cell discard problem, specially designed hardware will not be required to implement such an algorithm since call set times can be relatively long. The problem of learning to control a system characterized by delayed rewards and rare events is not limited to ATM networks; indeed, many areas of telecommunication networks have the same difficulties, suggesting that the techniques used to develop urgency scheduling should be of broader interest.

Acknowledgments

The author would like to acknowledge the support of GTE Laboratories where this work was begun and helpful discussion with L. D. Jackel, R. S. Sutton, S. Whitehead and L. J. Norton.

REFERENCES

[1] M. Schwartz. *Telecommunications Networks : Protocols, Modeling and Analysis*. Addison-Wesley, Reading, Massachusetts, 1987.

[2] D. B. Schwartz. ATM Scheduling with Queueing Delay Predictions. Proceedings of the Tenth International Conference on Machine Learning. In *Proc. of the Tenth Int. Conf. on Machine Learning*. Morgan Kaufman Publishers, Inc. pages 306–313, 1993.

[3] D. B. Schwartz. ATM Scheduling with Queueing Delay Predictions. To be published in *Proceedings of SIGCOM'93*.

[4] J. M. Peha and F. A. Tobagi. Evaluating scheduling algorithms for traffic with heterogeneous performance objectives. In *Proc. GLOBECOM '90*, pages 21–27, 1990.

[5] J. M. Hyman, A. A. Lazar, and G. Pacifici. Real-time scheduling with quality of service constraints. *IEEE J. Select. Areas Commun.*, 9:1052–163, 1991.

[6] S. H. Ross. *Introduction to Stochastic Dynamic Programming*. Academic Press, New York, 1983.

[7] R. S. Sutton. Learning to predict by the method of temporal difference. *Machine Learning*, 3, 1988.

[8] A. G. Barto, R. S. Sutton, and C. J. C. H. Watkins. Learning and sequential decision making. In M. Gabriel and J. Moore, editors, *Learning and Computational Neuroscience : Foundations of Adaptive Networks*, pages 539–602. MIT Press, Cambridge, MA, 1990.

[9] A. G. Moore and C. G. Atkeson. Memory-based reinforcement learning : Converging with less data and less real time. Technical report, Massachusetts Institute of Technology, 1992.

[10] R. S. Sutton. Integrated architecture for learning, planning and reacting based on approximating dynamic programmming. In *Proceedings of the 7'th International Conference on Machine Learning*. Morgan Kaufman, 1990.

[11] C. J. C. H. Watkins. *Learning from Delayed Rewards*. PhD thesis, Cambridge University, Cambridge, UK, 1989.

[12] D. P. Bertsekas and J. N. Tsitsiklis. *Parallel and Distributed Computation: Numerical Methods*. Prentice-Hall, Englewood Cliffs, NJ, 1989.

[13] A. G. Barto, S. J. Bradtke, and S. P. Singh. Real-time learning and control using asynchronous dynamic programmming. Technical Report 91-57, University of Massachusetts at Amherst, 1991.

[14] J. S. Albus. A new approach to manipulator control: the cerebellar model articulation controller (CMAC). *J. Dyn. Sys. Meas. Cont.*, 97:220–227, 1972.

[15] C. G. Atkeson. Using local models to control movement. In *Advances in Neural Information Processing Systems*, volume 2, pages 316–323, San Mateo, CA, 1989. Morgan Kaufman Publishers, Inc.

[16] A. G. Moore and C. G. Atkeson. An investigation of memory-based function approximators for learning control. Technical report, Massachusetts Institute of Technology, 1992.

[17] J. L. Bentley. Multidimensional binary search trees used for associative searching. *Comm. of the ACM*, 18:509–516, 1975.

[18] R. F. Sproull. Refinements to nearest-neighbor searching in k-dimensional trees. *Algorithmica*, 6:579–589, 1991.

[19] W. H. Press, B. P. Flannery, S. A. Teukolsky, and W. T. Vetterling. *Numerical Recipes in C : The Art of Scientific Computing*. Cambridge University Press, Cambridge, 1988.

[20] B. Maglaris, D. Anastassiou, P. Sen, G. Karlsson, and J. D. Robbins. Performance Models of Statistical Multiplexing in Packet Video Communications. *IEEE Trans. on Commun.*, 36:834–843, 1988.

[21] A. Schmidt and R. Campbell. Internet Protocol Traffic Analysis with Applications for ATM Switch Design. Technical Report. University of Illinois at Urbana-Champagne, 1992.

6

A NEURAL MODEL FOR ADAPTIVE CONGESTION CONTROL IN ATM NETWORKS

Xiaoqiang Chen

AT&T Bell Laboratories

1 INTRODUCTION

The problem of congestion control in packet-switching networks has been the subject of extensive research over the past two decades [1, 2]. A variety of congestion control strategies have been proposed in the literature and some have been applied, more or less successfully, to conventional data networks. Nevertheless, the behavior of network dynamics in the presence of congestion and the proper ways of handling traffic in order to obtain more reliable and predictable performance is not yet sufficiently understood. The difficulty stems mainly from the uncertainties about traffic patterns and the time-varying nature of network conditions.

Congestion control in future broadband ATM networks is further complicated by the diverse mix of traffic types and service requirements and by the increasing speed of transmission. To effectively use network resources, congestion control in ATM networks must be able to adapt gracefully to the dynamic behavior of traffic loads and network conditions during the network operation. Adaptation capability is particularly important when new services are being continually introduced after network design and installation. With the high-speed transmission, control algorithms should also be effective in terms of taking effect immediately at the onset of congestion, preferably be incorporated into hardware implementation. These ongoing design requirements represent a significant challenge to the next generation of congestion control.

Congestion is bound to take place in ATM networks. This will happen in spite of *preventive* control such as call admission control (CAC) and policing function, as discussed in the proceeding chapters. Unexpected surge of instantaneous traffic rates and network component failures are potential causes of congestion inside the network. Therefore there is a need for additional *reactive* control to dynamically regulate the volume of input traffic based on the information of congestion obtained from cells sent into the network. This is based on the belief that the more we can learn from inside the network, the more effective measures we may be able to find to interact gracefully. To develop a neural network model for adaptive congestion control is the purpose of this chapter. [1]

In the first part of the chapter, we formulate the adaptive congestion control as a general quality-of-service (QOS) control problem. This is followed by a brief review of backpropagation neural networks where an adaptive learning rate algorithm is proposed in order to improve the learning performance. The neural control scheme and learning algorithms for control are then described in detail. Later in the chapter, two examples of dynamic queueing systems are included to illustrate the methodology, and performance results obtained from simulations are given.

2 PROBLEM FORMULATION

The general problem of adaptive congestion control in ATM networks consists of adaptively regulating access of external traffic into the network in order to guarantee the desired performance given in the form of a performance bound. Let $\lambda^*(t)$ define the average traffic arrival rate and let $d^*(t)$ be the required performance bound; $d(t)$ denotes the performance observed from the network. $d(t)$ and $d^*(t)$ can be measured as the loss rate, delay or delay variance of interest. Note that all quantities considered are time-dependent averages (i.e. vary smoothly with time) to capture the dynamics of both the traffic load and the network condition. Control can only be achieved in a statistical sense. As external traffic arrivals are assumed to be independent of network state, the absence of control on traffic load can lead to the severe violation of the given performance bound. The objective of the QOS control is, therefore, to devise an adaptive controller — in this case, a neural network placed at the entrance of the network — that will maximize the input traffic to the network within the specified performance bound. The actual traffic rate admitted to the network by the controller is described as $\lambda(t)$, with the traffic constraint

$$0 < \lambda(t) \leq \lambda^*(t) \tag{1}$$

---

[1] For other applications, refer to the relevant chapters in this book.

A Neural Model for Adaptive Congestion Control

It is convenient to describe the traffic handling rule by the control variable, $\eta(t)$, defining the portion of the offered traffic, $\lambda^*(t)$, that can be admitted to the network; i.e.

$$\lambda(t) = \eta(t)\lambda^*(t) \qquad (2)$$

A diagrammatical representation of the control problem is given in Figure 1. Thus, the control problem in terms of the bounded performance and the traffic

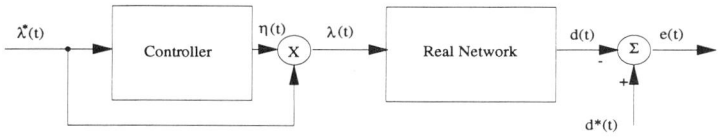

Figure 1 A general QOS control problem

constraint can be qualitatively stated as follows:

$$\begin{aligned} max \quad & \lambda(t) \\ s.t. \quad & d(t) \leq d^*(t) \\ & 0 < \eta(t) \leq 1 \end{aligned}$$

As the underlying dynamics of the real network are unknown beforehand, it is treated as a black box in our model. The real network represents a practical ATM network, such as a switching node, an end-to-end connection, or a simple queue. Previous knowledge concerning the characteristics of traffic is also not required. By continuously monitoring the traffic load and the corresponding performance measure, the neural controller should be able to learn the dynamics of the underlying network and subsequently recognize the changes in the network conditions and adopt a suitable adaptive control strategy. We assume that the real network has a single-input and single-output, but the method developed in this section can be easily extended to multi-input and multi-output networks, where $\lambda(t)$ and $d(t)$ are replaced by variable vectors.

3 ADAPTIVE LEARNING RATE ALGORITHM

A backpropagation neural network is a multi-layered network consisting of an input layer, an output layer and at least one hidden layer of nonlinear processing elements called *neurons*. The neurons between layers are interconnected by variable connections, which are internal parameters referred to as *weights*. Changing these weights will alter the behavior of the whole network. The rules for the adaptation of the weights are referred to as the *learning algorithm*. A backpropagation neural network may be considered as a tool to solve nonlinear function approximation problems [3]. To approximate an unknown function $f(X)$, with X being a vector representing inputs, consider a backpropagation neural network described accordingly by $\hat{f}(X, W)$, with W being a vector representing variable weights. The task is to train $\hat{f}$ to approximate f by sequentially applying input vectors (*training set*) to the neural network, while adjusting network weights according to a predetermined performance objective. Often, the performance objective of weight adaptation is to reduce the error averaged in some way over the training set. The most common error function is the mean-square error (MSE) which defines the error signal $e(n)$ at step n to be the difference between the desired response f and the actual output of the network $\hat{f}$:

$$e(n) = f(X(n)) - \hat{f}(X(n), W) \quad (3)$$

The MSE function is then

$$J = \sum_n e^2(n)$$

The most popular approach to MSE reduction is based upon the method of steepest descent. Adaptation of weights by this method is described as follows:

$$\triangle W = -\frac{1}{2}\mu \frac{\partial e^2(n)}{\partial W}, \quad \text{new } W = W + \triangle W \quad (4)$$

where μ, known as the *learning rate*, is a parameter that controls stability and rate of convergence. The gradient of the MSE function is measured as $\partial e^2(n)/\partial W$. This procedure is repeated, causing the MSE to be successively reduced on average and causing the weight vector to approach an optimal value. Details can be found in the first chapter of this book.

Strictly speaking, the adjustment of weights should be carried out by determining the gradient of J (*epoch learning*); however, the procedure commonly followed is to adjust them at every instant of time based on the error $e(n)$ (*pattern learning*). This is especially useful in a dynamic and non-stationary environment, where a neural network experiences a continuous flow of new training data. In pattern learning, one training sample is repeatedly presented for training, and the adjustment of weights continues until $|e(n)| < \varepsilon$ (a small constant), or a predetermined maximum number of adjustments is reached for that sample. In the original algorithm, the learning rate μ is set to be a small constant (e.g. $\mu = 0.1$), serving to adjust the size of the average of the weight

A Neural Model for Adaptive Congestion Control

changes. If μ is too small, convergence can be very slow; if too large, continuous instability can result. As illustrated in Figure 2, if a large μ is chosen, $e(n)$ oscillates alternatively between $e(n) < -\varepsilon$ and $e(n) > \varepsilon$, and never converges. This may be avoided by choosing a small value of μ, but it will increase the

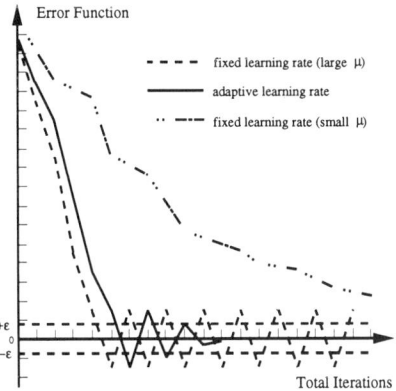

Figure 2 Illustration of fixed learning rate against adaptive learning rate

training time significantly. To overcome this problem, we propose an adaptive learning rate algorithm intended to adjust learning rate automatically as the learning process proceeds. This algorithm starts with a basic learning rate to enhance the learning speed, and each time $e(n)$ changes its sign, the learning rate is reduced according to the iterative formula

$$\text{new } \mu = \tau\mu \qquad (5)$$

where τ is an updating coefficient in the range $(0, 1)$. From Figure 2, there seems to be a definite advantage to the adaptive learning rate algorithm. It is shown that the speed of convergence of the backpropagation algorithm is significantly improved with this easily implemented change, giving rise to a substantial reduction in training time. Further, the adaptive learning rate algorithm brings improvement in the learning performance as μ also determines the accuracy of learning.

4 NEURAL NETWORK CONTROL SCHEME

The aim of the neural controller shown in Figure 1 is to make control decisions over λ based on real-time interaction with the real network. Owing to time-dependent relationship, the control decision λ depends on the history of the observed performance and the traffic input rate. Without loss of generality, the λ that we have control over takes the form

$$\lambda(n) = G(\lambda^*(n), \lambda(n-1), \cdots, \lambda(n-k); d(n), d(n-1), \cdots, d(n-m)) \quad (6)$$

where $\lambda(n-i)$ and $d(n-j)$ are the time-series of controlled traffic input rates and performance observations, respectively. This function indicates that each value of λ and d over time plays a part in determining the network condition. The dimension of the function $(k+m+2)$ depends on the complexity of the real network considered and on the control accuracy required. Substituting Eq. (6) into Eq. (2), we obtain

$$\begin{aligned} \eta(n) &= \lambda(n)/\lambda^*(n) \\ &= g(\lambda^*(n), \lambda(n-1), \cdots, \lambda(n-k); d(n), d(n-1), \cdots, d(n-m)) \end{aligned}$$

where $\lambda^*(n)$ is absorbed into the function g.

Since the dynamics of the real network (i.e. function g) is assumed unknown, a neural network is used to learn the relationship between the control variable $\eta(n)$ and the network dynamics. The neural network for the controller has $(k+m+2)$ inputs that are the delayed values of relevant signals and one output that can be written as

$$\eta(n) = \hat{g}(\lambda^*(n), \lambda(n-1), \cdots, \lambda(n-k); d(n), d(n-1), \cdots, d(n-m), W) \quad (7)$$

where W represents the adjustable weights of the neural controller. Referring to Figure 1, the error signal for the neural controller is

$$e_c(n) = d^*(n) - d(n) \quad (8)$$

It can be seen from Figure 1 that the error backpropagation algorithm cannot be applied directly to the neural controller because of the location of the real network. The distal error $e_c(n)$ cannot be propagated through the real network, and there is no direct learning algorithm available to propagate this error back to the controller.

A second neural network is therefore needed to overcome this difficulty by using the so-called *indirect control model* in control theory [4, 5]. This added

A Neural Model for Adaptive Congestion Control

neural network is trained to behave like the real network itself and is therefore called an *emulator* (or *forward model*). The idea behind this method is to translate the distal error signal to the output error of the controller through the neural emulator. The dynamics of the real network can be described by an unknown function

$$d(n+1) = f(\lambda(n), \lambda(n-1), \cdots, \lambda(n-k); d(n), d(n-1), \cdots, d(n-m)) \quad (9)$$

The overall structure of the control scheme is therefore shown in Figure 3. The

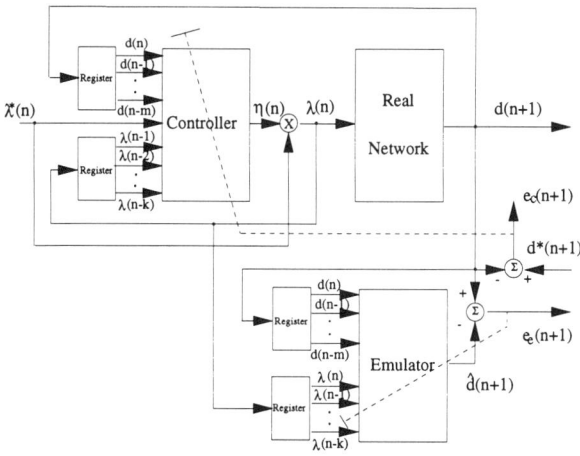

Figure 3 The neural control scheme

output of the emulator is

$$\hat{d}(n+1) = \hat{f}(\lambda(n), \lambda(n-1), \cdots, \lambda(n-k); d(n), d(n-1), \cdots, d(n-m); V) \quad (10)$$

where V represents the weights of the neural emulator. The training of the neural emulator to approximate the function f is a straightforward application of backpropagation neural network, with the following error signal and adaptation of weights:

$$e_e(n) = d(n) - \hat{d}(n), \quad \Delta V = -\frac{1}{2}\mu\frac{\partial e_e^2(n)}{\partial V} \quad (11)$$

---

[2] In the control engineering literature, the process of training a neural emulator is referred to as the *identification* of dynamic systems.

To compute the gradient for the adjustment of weights W in Eq. (8), the chain rule is applied to the error function $e_c^2(n)$, yielding the result

$$\frac{\partial e_c^2(n)}{\partial W} = -2(d^*(n) - d(n))\frac{\partial \hat{d}(n)}{\partial \lambda(n)}\lambda^*(n)\frac{\partial \eta(n)}{\partial W} \qquad (12)$$

where $\partial d(n)/\partial \lambda(n)$ is replaced by $\partial \hat{d}(n)/\partial \lambda(n)$ as $\hat{d}(n)$ tends to approach $d(n)$ asymptotically. This expression describes the propagation of the distal error $e_c(n)$ backward through the emulator and down into the controller where the weights W are updated. The neural emulator in a sense translates the error in the final output of the real network to the equivalent error in the controller output. $\partial \hat{d}(n)/\partial \lambda(n)$ and $\partial \eta(n)/\partial W$ are referred to as *backpropagation-to-input* and *backpropagation-to-weights*, respectively. The quantity $\partial \hat{d}(n)/\partial \lambda(n)$ can be calculated by using techniques similar to the error backpropagation algorithm, although some minor modification is needed.

It can be seen from Figure 3 that there is no feedback loop in the neural emulator as the delayed inputs come from the real network (the so-called *series-parallel model*). In contrast, the computation of $\partial \eta(n)/\partial W$ (i.e. $\partial \lambda(n)/\partial W$) is affected by the past values of $\lambda(n - i)$ ($1 \leq i \leq k$) because of the feedback from the output of the neural controller to its input. Neural networks with feedback from output to input are often called *recurrent* networks [6]. For a recurrent backpropagation network, modification must be made to evaluate the gradient $\partial \eta(n)/\partial W$. It can be verified from variable calculus that $\partial \eta(n)/\partial W$ is the solution of the following difference equation:

$$\frac{\partial \eta(n)}{\partial W} = \frac{\partial \hat{g}}{\partial W} + \sum_{i=1}^{k}\frac{\partial \hat{g}}{\partial \lambda(n-i)}\lambda^*(n-i)\frac{\partial \eta(n-i)}{\partial W} \qquad (13)$$

It should be noticed that $\partial \hat{g}/\partial W$ and $\partial \eta(n)/\partial W$ differ in that the former treats the inputs as constants. Since $\partial \hat{g}/\partial \lambda(n-i)$ and $\partial \eta(n-i)/\partial W$ can be computed on-line at every instant of time, the desired gradient can be generated accordingly. Since the first term of Eq. (13) reflects the previous basic learning algorithm, the recurrent or dynamic algorithm requires much more computation effort.

In general, the learning control algorithm proceeds in two phases. In the first phase, the emulator is trained using the basic backpropagation learning algorithm. In the second phase, the controller is trained across the composed network. During the second phase only the weights of the controller are changed, and the weights of the emulator are held fixed. The error signal is propagated through the trained or partially trained emulator for the adaptation of the controller weights. The controller is then able to control correctly the traffic input rate by using the emulator as a guide. It is very important to notice that this control is constrained by the external traffic $\lambda^*(n)$, as it is impractical to produce a controlled traffic rate which is greater than the external traffic input rate. Specifically, there should be no adaptation of the weights in the controller and consequently no control action imposed on input traffic if the output of the neural network $\eta(n)$ approaches 1 while the measured performance does not exceed the given performance bound. This constraint can be easily incorporated into the above learning control process. In practice, it is not necessary that the training of the neural emulator always precedes the training of the neural controller. Identification and control can be done simultaneously in on-line control. It has been shown that the controller does not require an exact emulator — an approximate emulator often suffices [5]. As the emulator begins to be trained, the control errors decrease and consequently the controller improves.

A principal advantage of the above neural control scheme is the flexibility it offers in establishing various performance objectives (i.e. error signals) that can be properly incorporated into the learning control process. For illustration, we often have two conflicting requirements in the context of the QOS control, to minimize the performance $d(n)$ (delay or loss rate) on one hand and to maximize the throughput or the actual input traffic $\eta(n)$ on the other. Suppose we associate a cost or loss with the increase of $d(n)$, and a profit or reward for the traffic admitted into the network. The performance objective is then to minimize

$$e_c(n) = \alpha d(n) - \beta \eta(n)$$

where α is a weighting constant representing the performance cost and similarly β is a weighting constant reflecting the reward for the admission of traffic. Also, the above function can be measured over any time period of interest by using epoch learning.

Although we confine our attention to the QOS control in which the input traffic to the network is required to be controlled, the basic control structure and the learning algorithms are broadly applicable to other performance-driven controls, in which, for example, bandwidth requirements, number of buffers, routing decisions, or priorities of transmission and queueing can be considered as control variables.

5 SIMULATION RESULTS

In this section, two dynamic queueing systems for the representation of a real network, taken from [7, 8], are used to test the performance of the suggested control scheme. These systems with known dynamics are chosen so that the performance of the neural controller can be easily checked. Only the first-order difference approximation for these systems is assumed, which is aimed at the development of a system that can be conveniently used for evaluating the performance of a neural controller. This is done to simplify the system and to focus on the control mechanism. No attempt is made to validate its accuracy, though the approximation can be made precise by using high-order difference equations.

In all those neural network simulations, initial weights are chosen randomly from a uniform distribution on the interval $(-1, 1)$. The basic learning rate is chosen as $\mu = 0.1$ and the adaptive learning rate algorithm is used with the updating coefficient $\tau = 0.5$. Three-layer backpropagation neural networks are used in all simulations. For ease of expression, we shall denote a neural network by the notation $N(a, b, c, d)$ such that it consists of a input neurons, b first-layer hidden neurons, c second-layer hidden neurons, and d output neurons, where the output of each node in the previous layer is an input to each node in the next layer. No attempt is made to optimize the number of hidden neurons or their connectivity. The input neurons have linear activation functions, and hidden and output neurons have nonlinear functions of the form $s(x) = (1-e^{-2x})/(1+e^{-2x})$. The inputs and desired output of a neural network are scaled to lie between $(-1, 1)$ before being fed into the network.

5.1 Example 1

In the first example, we consider an M/M/1 dynamic queueing system as a "real" network. With first-order approximation, the system is described by the difference equation $d(n + 1) = f(d(n)) + \lambda(n)$, where $d(n)$ and $\lambda(n)$ represent the time-dependent average delay and arrival rate, respectively. (Note that time units are normalized in such a way that the service capacity is equal to unity.) The function f, assumed unknown, has the form $f(d(n)) = d^2(n)/(1 + d(n))$. The objective of control is to regulate the arrival rate $\lambda(n)$ subject to the delay bound $d^*(n)$ specified. It is recognized that the traffic input rate in this system appears separately from the unknown function f, resulting in a simple control structure that is particularly suitable for control using a neural network. In fact, the control variable can be obtained explicitly from the emulator parameters so that the controller does not require a separate neural network implementation.

Since f is not known *a priori*, it is estimated on line as $\hat{f}$ by using a neural network of $N(1, 20, 10, 1)$. The output of the neural network is

$$\hat{d}(n+1) = \hat{f}(d(n), W) + \lambda(n)$$

subject to the following:

$$\begin{cases} \lambda(n) = d^*(n+1) - \hat{f}(d(n), W) \\ 0 < \lambda(n) \leq \lambda^*(n) \end{cases} \quad (14)$$

It follows that the control variable at time step n can be derived directly as

$$\eta(n) = min[1, \frac{d^*(n+1) - \hat{f}(d(n), W)}{\lambda^*(n)}] \quad (15)$$

with the following input rate admitted into the real network:

$$\lambda(n) = \eta(n)\lambda^*(n) \quad (16)$$

It can be seen that the neural network learns to predict the future performance of the real network upon which the control decision is made.

In Figure 4, the response of the real network with control is compared with the response without control. The external input rate is a sinusoid $\lambda^*(n) = 0.6 + 0.3\, sin(\pi n/125)$, and the desired delay bound is specified by

$$d^*(n) = \begin{cases} 4.5 & 0 < n < 700 \\ 8 - n/200 & 700 < n < 1300 \\ 1.5 & n > 1300 \end{cases}$$

It is evident from Figure 4 that the delay performance has been satisfactorily controlled and kept below the given delay bound. Since the identification of the real network dynamics is not complete for small values of n, this result reveals that efficient on-line control can be possible even with a partially trained emulator. It is also apparent that the control performance improves as the neural network tracks the dynamics of the real network. Figure 5 also gives the controlled and uncontrolled input rates to the real network and the corresponding control variable.

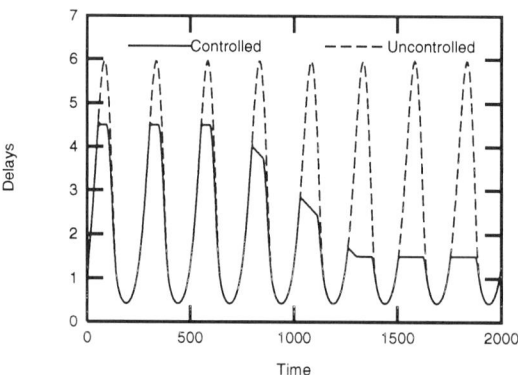

Figure 4 Controlled and uncontrolled delays

5.2 Example 2

In the second example, the real network considered consists of an M/M/K dynamic queueing system without a queue. This system is often used to model a loss system in performance evaluation. Unlike the previous example, the

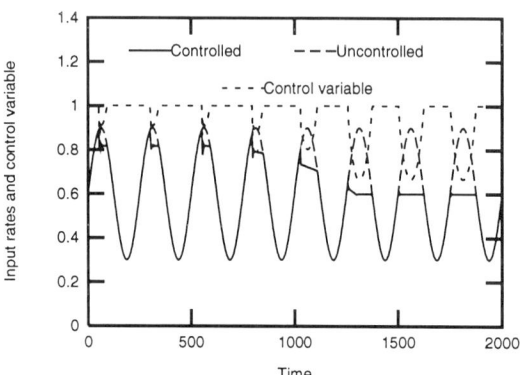

Figure 5 Controlled and uncontrolled input rates and control variable

A Neural Model for Adaptive Congestion Control

performance we are concerned with, in this case, is the time-dependent average loss rate. The system is described by the difference equation

$$x(n+1) = x(n) - \Phi(x(n)) + \lambda(n) \tag{17}$$

together with

$$l(n) = 1 - \frac{x(n)}{\Phi(x(n))} \tag{18}$$

where $x(n)$ and $l(n)$ represent the average number of packets and loss rate in the system, respectively. The function Φ is given by

$$\Phi(x) = \begin{cases} x & x < K/2 \\ K/2 - 7\ln(2(K-x)/K) & K/2 \leq x < K \end{cases} \tag{19}$$

It can be derived that $l(n+1)$ is the function of $l(n)$ and $\lambda(n)$, that is, the system can be described by a first-order difference equation as before; i.e.

$$l(n+1) = f(l(n), \lambda(n)) \tag{20}$$

which is determined jointly by Eq. (17), (18) and (19) but it is not possible to give an explicit expression for this function (a neural network can learn it).

A neural emulator is first used to learn the dynamics of the real network implicitly described by Eq. (20) with the parameter K chosen to be 12. The neural network belongs to the class of $N(2, 10, 20, 1)$ with two inputs. An important difference between static problems and dynamic problems is that it is generally infeasible for a neural network to learn a practical dynamic system by using an arbitrary random process as training input, because of the considerations of stability. For instance, if $\lambda(n)$ in Eq. (20) is allowed to be a random process in the range (0, 10), then the system state $x(n)$ can become negative. This yields data that is of little use for learning this system. On the other hand, the input signals for learning must be sufficiently general to be useful in the subsequent stage of practical control. A solution to this problem is to smooth the input random process while maintaining the stability of the dynamic system under study. Our approach is therefore to generate the input signals from the following recursive equation

$$\lambda(n+1) = \lambda(n) + r$$

with the constraint

$$0 < \lambda(n+1) \leq 10$$

where r is a random variable in the range $(-1, 1)$.

In learning dynamic systems where a neural network faces a continuously changing environment and may not see the same input vector twice, pattern learning may result in temporal instability. Because it updates the weights after each observation, we found that the process of learning the dynamic system given by Eq. (20) never converged and, in fact, it wandered aimlessly or oscillated wildly. This was due to the fact that the weight adjustments of pattern learning disrupted what had already been learned. To solve this problem, we use the method suggested in [9] which involves adding a term, called *momentum*, to the weight adjustment. The momentum is proportional to the amount of the previous weight change. Once an adjustment is made it is remembered and serves to modify all subsequent weight adjustments. The adaptation of weights given by Eq. (4) is now modified into

$$\triangle W = \nu \frac{\partial e^2(n-1)}{\partial W} - \frac{1}{2}\mu \frac{\partial e^2(n)}{\partial W} \qquad (21)$$

where the momentum ν is chosen to be 0.5 in our simulation. This method has proven very effective in learning the dynamic system while retaining the stability.

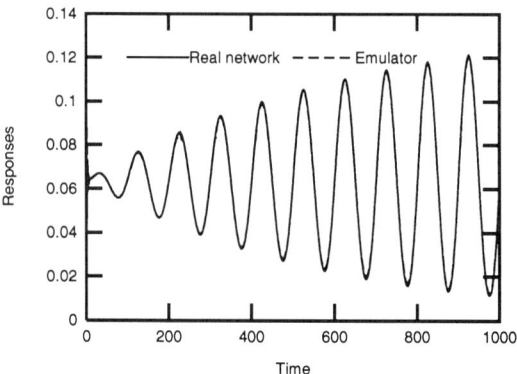

Figure 6 Responses of the real network and the emulator

The training of the neural emulator terminates after $100,000$ time steps, and the performance of the emulator is tested off-line using its output as feedback to its inputs (*parallel model*). Figure 6 is a plot of the responses of the emulator and the corresponding real network when the input rate is given by

A Neural Model for Adaptive Congestion Control

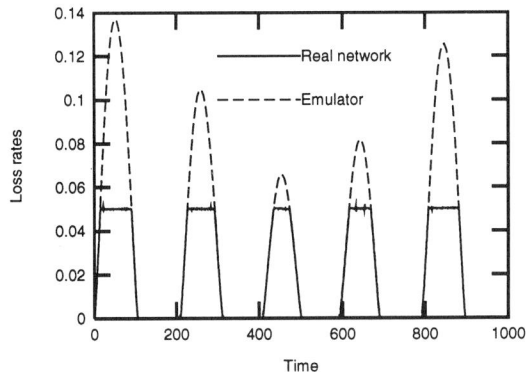

Figure 7 Controlled and uncontrolled loss rates ($l^*(n) = 0.05$)

$\lambda(n) = 8 + 2(1 - exp(-0.0015n)) \, sin(\pi n/50)$. The curves show that a neural network can learn a dynamic system successfully. With the trained emulator, the neural controller using a neural network $N(2, 10, 20, 1)$ is then taught. The outputs of the real network with and without control are shown in Figure 7 with the external input rate $\lambda^*(n) = 6 + 3 \, sin(\pi n/100) + sin(\pi n/125)$ and the desired

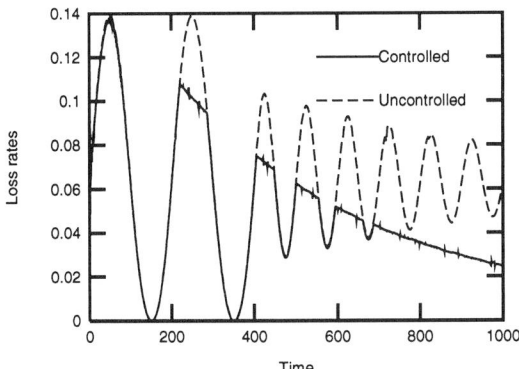

Figure 8 Controlled and uncontrolled loss rates ($l^*(n) = 0.17 \, exp(-0.002n)$)

loss rate bound $l^*(n) = 0.05$. Figure 8 shows another result with the desired loss rate bound $l^*(n) = 0.17\,exp(-0.002n)$ while the external input rate is $\lambda^*(n) = 8 + 2\,sin(\pi n/100)$ and changed to $\lambda^*(n) = 8 + 2\,exp(-0.0015n)\,sin(\pi n/50)$ at time 400.

6 SUMMARY

It appears that the neural network approach is attractive and promises unique advantages over the conventional strategies in dealing with congestion control in ATM networks. These advantages are the adaptivity to changes in traffic loads and/or network conditions, the hardware implementation and the ability to satisfy different performance objectives. The unique feature of this approach is that it makes no assumptions about the detailed knowledge of the underlying network or about the nature of the traffic sources. The approach relies only on learning and observation of performance to adapt to changing network conditions. It is very encouraging that the control scheme described and evaluated here has been found to work well when controlling the real networks which consist of relatively simple dynamic queueing systems. The fact that neural networks are based on the massive parallelism of hardware implementation suggests that the proposed scheme may also work well for multivariable networks with high dimensions. Although we have validated the neural control scheme in terms of bounded performance control, we feel that the proposed scheme is general and can be applied to many other applications of congestion control in future ATM networks that fit within this control framework.

There are some ways the research described here can be extended. Since the performance observations have been assumed available to the controller without delay, a desired extension will be to consider the effect of delayed observations on the control performance. Additional work can also be carried out to study more complex ATM networks, e.g. considering the catenation of simple queues to emulate an end-to-end connection, or replacing the real network by a practical network simulator. It must be emphasised that, however, application of neural control will be rather limited until special-purpose hardware or neurocomputer designed to implement neural network topologies and emulate the operation of learning algorithms becomes readily available. Only with these can the real advantages of adaptive neural control techniques be realized for practical congestion control.

REFERENCES

[1] Gerla, M. and Kleinrock, L., "Flow Control: A Comparative Survey," IEEE Trans. Commun., Vol. COM-28, No. 4, Apr. 1980, pp. 553–574.

[2] Maxemchuk, N. and Zarki, M., "Routing and Flow Control in High-Speed Wide-Area Networks," Proc. IEEE, Vol. 78, No. 1, Jan. 1990, pp. 204–221.

[3] Poggio, T. and Girosi, F., "Networks for Approximation and Learning," Proc. IEEE, Vol. 78, No. 9, Sept. 1990, pp. 1481–1497.

[4] Narendra, K. and Parthasarathy, K., "Identification and Control of Dynamical Systems Using Neural Networks," IEEE Trans. Neural Networks, Vol. 1, No. 1, Mar. 1990, pp. 4-27.

[5] Jordan, M. and Rumelhart, D., "Forward Models: Supervised Learning with a Distal Teacher," Technical Report Occasional Paper #40, MIT Center for Cognitive Science, 1990.

[6] Williams, R. and Zipser, D., "A Learning Algorithm for Continually Running Fully Recurrent Neural Networks," Technical Report ICS-Report-8805, Institute for Cognitive Science, Oct. 1988.

[7] Filipiak, J., Modelling and Control of Dynamic Flows in Communication Networks, Springer-Verlag, Berlin Heidelberg 1988.

[8] Agnew, C., "Dynamic Modeling and Control of Congestion-Prone Systems," Operations Research, Vol. 24, No. 3, May-Jun. 1976, pp. 400–419.

[9] Rumelhart, D., Hinton, G. and Williams, R., "Learning Internal Representations by Error Propagation," in Parallel Distributed Processing: Explorations in the Microstructure of Cognition. Vol. 1. Foundations, pp. 318–362, MIT Press/Bradford Books, Cambridge, MA, 1986.

7

STRUCTURE AND PERFORMANCE OF NEURAL NETS IN BROADBAND SYSTEM ADMISSION CONTROL

Phuoc Tran-Gia and Oliver Gropp

Institute of Computer Science
University of Würzburg

1 INTRODUCTION

This chapter[1] is dedicated to the use of neural networks for the connection admission control (CAC) in Asynchronous Transfer Mode (ATM) networks. An overview of ATM is provided by Hiramatsu in a preceding chapter. The major aim is to present and to compare possible neural net structures which can be applied to CAC and to show the performance of a basic neural net under various stationary and non-stationary load conditions. In Section 2 basic principles of the use of feed-forward neural networks with back-propagation learning in connection admission control are discussed and different alternative neural net structures are compared. A simple neural net is selected as an example in Section 3 and Section 4 to show the acceptance control performance and to discuss numerical aspects of the neural net under consideration.

2 NEURAL NETWORKS FOR CONNECTION ADMISSION CONTROL

Depending on the information available to the CAC function and its location in the communication network, different neural net structures can be developed. In this section we will briefly present these alternatives and discuss in particular the basic function and learning procedure of a back-propagation neural network used as admission controller.

---
[1] This chapter is an extended and updated version of [1].

2.1 Admission control in broadband networks

The connection admission control plays an important role during the resource allocation procedure of an ATM network (cf. [2]). According to CCITT [3, 4] the CAC function is defined as: " ... the set of actions taken by the network at the call set-up phase (or during the call re-negotiation phase) in order to establish whether a (virtual channel or virtual path) connection can be accepted or rejected."

Given that the CAC function is able to estimate the quality of service (QOS) before and after having accepted the requested connection, it can make the acceptance decision, i.e. the request will be rejected if the required QOS cannot be maintained.

In the following, we consider a number M of different connection types to be served by the network.

We distinguish two cases:

i) <u>CAC based on network state</u> : In this case, we assume that the entire information about the number of all connections being multiplexed is available.

   The system state seen by the network is denoted by $X = \{n_1, n_2, .., n_M\}$, where n_i is the number of active connections of type i being in the system. The main CAC function can now be represented by a mapping of the system state X to a decision vector Z defined by $Z = \{z_1, z_2, .., z_M\}$, where $z_i = 1$ denotes the acceptance of a connection establishment request of type i and $z_i = 0$ its rejection. The CAC is thus reduced to the implementation of a mapping $f : X \to Z = f(X)$ according to the predefined quality of service of the network.

   The mapping f can further be simplified by using the state $X^\star = \{n_1, n_2, \ldots, n_i + 1, \ldots, n_M\}$, i.e. the system state just after accepting the connection request of type i. The decision vector is reduced to $Z^\star = \{z_i\}$ and the CAC mapping to $f^\star : X^\star \to Z^\star = f^\star(X^\star)$.

ii) <u>CAC based on bit-rate process</u> : In this case, only the superimposed bit-rate process is available to the connection admission control function. This is the case if an intermediate ATM switching node does not have the whole system information, but only knows the bit-rate processes to be transferred.

Neural Nets in Broadband System Admission Control

Denoting the observed total bit-rate function during the time interval $(t, t+\Delta t)$ by Y, the CAC function can again be represented by the mapping $g : Y \to Z = g(Y)$.

2.2 Neural network as admission controller

As discussed in the previous subsection, the connection admission control function can be interpreted as a mapping of the state vector X into the acceptance decision vector Z. This functional mapping divides the M-dimensional state space into two regions: the acceptance region and the rejection region. In other words, the CAC problem can be formulated like a pattern recognition problem: upon recognition of the load pattern X, a yes/no decision has to be made to accept/reject the connection request. This property in conjunction with the use of a neural net for connection control purposes in ATM systems is thus quite obvious. In this chapter we will use the class of feed-forward neural nets with back-propagation learning algorithm as described in Chapter 1 to solve the CAC problem.

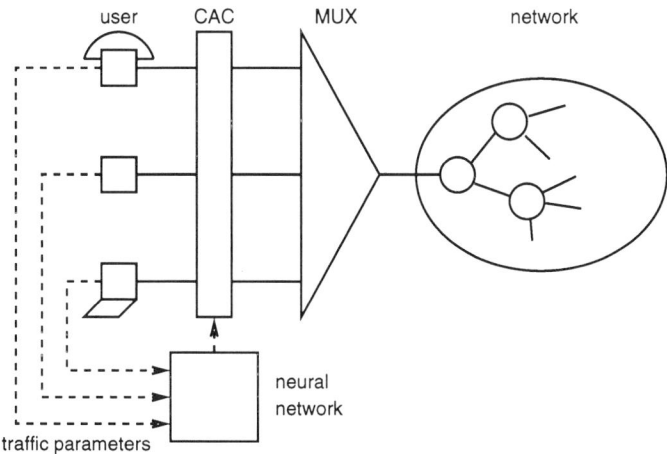

Figure 1 *Neural net for admission control*

The use of a neural network to control connection acceptance is illustrated in Fig. 1. This basic structure has been proposed in [5] and further developed in [6]. Traffic streams offered by different types of sources are multiplexed at the entry node of the high-speed communication network. In the proposed neural net structure in [5] the bit-rate function is used as input to the neural net. As a quality of service indicator, e.g., the cell blocking probability at the multiplexer can be used. During the learning phase of the neural net, the input/output patterns are as follows. Inputs are formed by the bit-rate pattern including the bit-rate process generated by the actual connection request. The resulting QOS will be observed and compared to the target QOS. If the target QOS is still held, the output of the current input/output pair is $Z = 1$, i.e. the connection can be accepted, the bit-rate pattern is a "good" pattern. If the resulting quality of service is lower than the target one the output is then $Z = 0$, i.e. the connection should not be accepted in the current load situation, the bit-rate pattern is a "bad" pattern. After learning input/output pairs have been presented, the neural net can be used in a recall mode to perform the CAC function. One of the disadvantages of this mechanism is the difficulty to generate a significant number of good- and bad-patterns for the neural net to learn. The CAC performance of the neural net is thus strongly dependent on the statistical significancy of load patterns during the learning phase. Therefore we decided to design a modified learning process for the neural net.

In the current study we devote our attention to the neural net structure depicted in Fig. 2. The neural net is designed to perform the mapping depicted in Section 2.1 case i). We consider a number M of different classes of connections, each with different known bit-rate characteristic. The pairs of input/output patterns for the neural net to be learned is computed as indicated in Fig. 2. Starting with a state vector $X = \{n_1, n_2, .., n_M\}$ as the input part of a pattern the multiplexed bit-rate function is determined. Having this bit-rate function as traffic stream, the cell blocking probability can be estimated giving the actual quality of service. Upon a comparison of this measure with the target QOS, the acceptance decision Z can be made. This can be interpreted as the decision to be made to accept/reject a connection request of type i if actual system state is $\{n_1, n_2, .., n_i - 1, .., n_M\}$. The working mode of the neural net during the recall phase is as shown in Fig. 2, where the net will answer with an accept/reject decision Z^* for a connection request of type i when the input vector $X = \{n_1, n_2, .., n_i + 1, .., n_M\}$ is presented.

Thus, after the learning phase, the neural net performs the CAC by separating the M-dimensional input state space in two regions corresponding to a $(M - 1)$-dimensional decision surface. The decision surface, which separates the "accept" region from the "reject" region in the state space, can be thought of as stored in the weight vectors of the neural net.

Neural Nets in Broadband System Admission Control

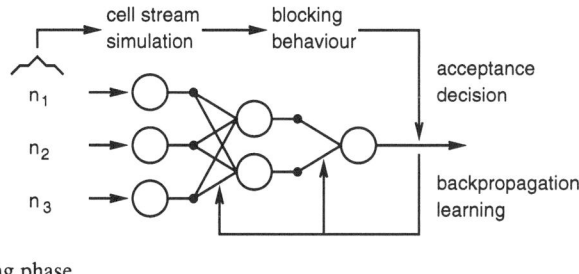

a) Learning phase

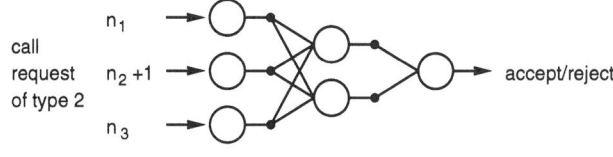

b) Recall phase

Figure 2 *Working modes of a neural net based admission controller*

2.3 Alternative neural net structures for admission control

In the previous subsection we introduced the two cases: i) CAC based on network state, where information about the number of all connection being multiplexed is available, i.e. the detail state vector X of the system is known by the connection admission control function and ii) CAC based on bit-rate process, i.e. only the superimposed bit-rate process is available or measurable to the connection admission control function. According to these two cases different neural net structures can be developed, as shown in Fig. 2 and 3.

i) CAC based on network state

A simple backpropagation neural net with only one output neuron is depicted in Fig. 2 b). The same functionality can be obtained using the neural net structure shown in Fig. 3 a).

ii) CAC based on bit-rate process

Fig. 3 b) depicts a feedforward neural net for the mapping of a bit-rate pattern to an accept/reject decision. This reflects a communication network architecture with less signaling efforts involved, where only the superimposed bit-rate process is available or measurable to the connection admission control function.

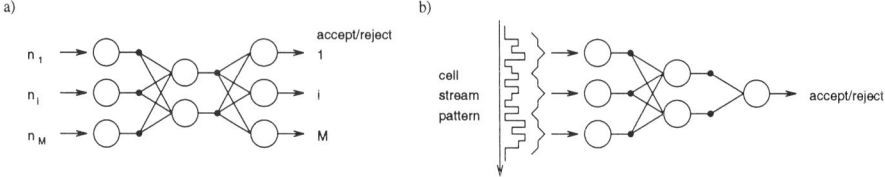

Figure 3 *Alternative neural net structures for admission control*

3 EXAMPLE OF A NEURAL NETWORK FOR ADMISSION CONTROL

In this section we illustrate the performance of the neural net depicted in Fig. 2 b) in connection admission control.

3.1 Traffic assumptions and configuration parameters

The parameters of the ATM multiplexer and the connection types are as follows. The output of the multiplexer has a capacity of 600 Mbps, the buffer space is 0.5 Mb large. To model approximately the VBR (variable bit-rate) sources we consider sources with first-order Markovian bit-rate processes, where the two basic types are used as shown in Fig. 4: a) on/off-sources and b) binomial sources with two parameters: mean bit-rate m and peak bit-rate h. The time axis is discretized by $\Delta t = 100\,msec$. The bit-rate R will be expressed in number of basic units $\Delta B = 1\,Mbps$. In each Δt we assume each source to have an independent bit-rate following the distribution:

a) on/off-sources:
$$p_{ON} = P\{R = \tfrac{h}{\Delta B}\} = \tfrac{m}{h};$$
$$p_{OFF} = P\{R = 0\} = 1 - \tfrac{m}{h};$$

b) binomial sources:
$$p_i = P\{R = i\} = \binom{\tfrac{h}{\Delta B}}{i}\left(\tfrac{m}{h}\right)^i\left(1 - \tfrac{m}{h}\right)^{\tfrac{h}{\Delta B}-i}, \quad i = 0, 1, ..., h.$$

We consider three connection types:

Type 1: on-off, m = 10 Mbps, h = 40 Mbps, c_R = 1.73.

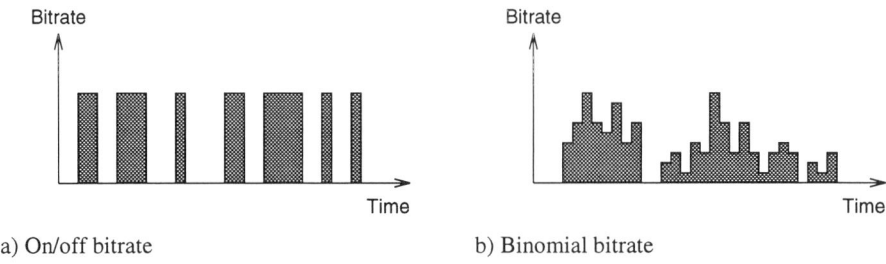

Figure 4 *Basic traffic source models*

Type 2: binomial, m = 5 Mbps, h = 40 Mbps, $c_R = 0.42$.

Type 3: binomial, m = 5 Mbps, h = 80 Mbps, $c_R = 0.19$.

where c_R denotes the coefficient of variation of the bit-rate R.

On connection traffic level, the arrival process of connection requests is assumed to be Poisson with a mean interarrival time chosen according to the simulated load scenario. To obtain patterns for the neural net learning process, the cell stream traffic is simulated. During the simulation time the amount of lost cells is estimated by a fluid flow model (cf. Fig. 5). The connection duration is assumed to be exponentially distributed with mean 20 sec. This mean value is intentionally chosen to be short to enable simulation runs without loosing the qualitative significancy of the results obtained.

We simulate the traffic on burst level and the cell loss depending only on the actual sum of the bit-rates of the sources of active connections, the capacity of the output line and the buffer space of the multiplexer. Fig. 5 shows how the buffer occupancy b) depends on the bit-rate a). Cells are only stored in the buffer if the bit-rate exceeds 600 Mbps. The lost period is shaded dark in Fig. 5 b).

3.2 Alternative CAC methods for performance comparison

Since an agreement on CAC mechanisms for ATM system is not yet available, we will select a few methods proposed in the literature (cf. [7, 8, 9, 10, 11, 12]) to compare with the neural net CAC approach. The parameters taken into account for CAC purpose are the numbers of active sources with given connection types, and for each connection the mean bit-rate m and the peak bit-rate h. The aim

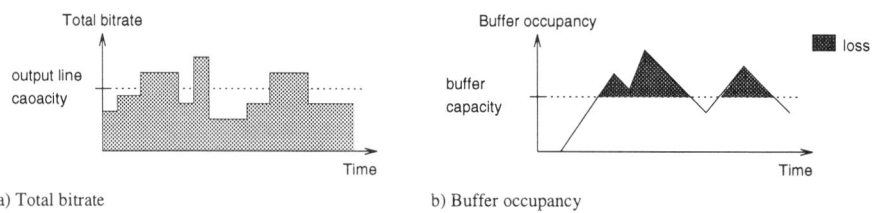

Figure 5 *Estimation of cell blocking probability*

of the CAC is to keep the QOS, i.e. the cell loss rate below a given value, say 10^{-5}.

Peak reservation method (PR)

The most simple and robust method to limit the cell loss probability is to reserve the peak bit-rate for each accepted connection. New connections are only admitted if the sum of the peak bit-rates of the active connections and the new connection is smaller than the capacity of the output line. Thus no loss will ever appear. This method reduces ATM rather to STM (Synchronous Transfer Mode). Obviously, for more bursty bit-rate traffic the output channel is used in an inefficient way and the multiplexer utilization may be intolerably low. This peak bit-rate reservation method is considered here only as a lower bound for admission control methods aiming to high multiplexer utilization. In this section the improvement of utilization achieved by more sophisticated algorithms in comparison to this simple method will be shown.

Equivalent bandwidth method (EB)

The expression "equivalent bandwidth" is introduced in [2]. Each source of type i has its equivalent bandwidth k_i, which depends on its mean bit-rate m_i, its peak bit-rate h_i and the capacity of the multiplexer output line:

$$k_i = C_1 m_i + C_2 \frac{m_i(h_i - m_i)}{c} \qquad (1)$$

The constants C_1 and C_2 depend on the buffer space of the multiplexer and the maximum cell loss rate and have to be determined empirically. If a connection request of type i arrives the following inequality is checked:

$$K + k_i \leq c \qquad (2)$$

where K denotes the sum of the equivalent bandwidths of the actual active connections. If it holds, the new connection is accepted, otherwise rejected.

Weighted Variance method (WV)

The original method proposed in [13] has to be modified in the context of this study due to simulation reasons. The original method only works sufficiently well if the peak bit-rates of the subscribers is less than one percent of the output line capacity (cf. [11]). The modified algorithm works as follows, where m_j represents the mean bit-rate of connection j, h_j its peak bit-rate and c the capacity of the output line. Connection k is the new connection to be admitted, connections 1 to $k-1$ are already admitted. If

$$\sum_{j=1}^{k} h_j \leq c \qquad (3)$$

holds, connection k is accepted. If this inequality does not hold, the following one is employed:

$$\alpha \sqrt{\sum_{j=1}^{k} m_j(h_j - m_j)} + \sum_{j=1}^{k} m_j + \max_{1 \leq j \leq k} h_j \leq c \qquad (4)$$

If this inequality holds, connection k is accepted, otherwise it is rejected.

The term $m_j(h_j - m_j)$ is an estimate of the variance of the bit-rate of connection j. Thus the constant α determines the influence of the variances of the source s on the CAC process. The term α has to be found in advance by simulation.

Neural network CAC (NN)

We use a three layered feed-forward neural net to evaluate the CAC function. The neural net structure is the one depicted in Fig. 2 b).

The input consists of the vector $X^\star$ of the numbers of active sources of each class where the component of the class of the arriving request is incremented by one. The result of the feed-forward computations at the output unit is a real number between 0 and 1. If the output value is less than a threshold (say 0.5) the new connection is accepted, otherwise rejected. The decision of the neural net depends on its internal set of weight matrices which have to be determined in advance during the learning phase as discussed in the previous section.

3.3 Neural net convergence and numerical issues

As mentioned before, the neural net needs a learning process to fix its weight vectors. This process uses a set of patterns to be learned. Each pattern consists of an input vector X^* and the corresponding output value Z^*, which have to be chosen that the network has the capability to work as a mapping function f^* for CAC (cf. section 2.1). We obtain this pattern using a simulation of the multiplexer state process, i.e. we fix the number of active connection of each traffic class at certain values and determine the corresponding loss rate at the multiplexer buffer. If this loss rate is less than a predefined value (in this study 10^{-5}) this set of connection can be accepted, otherwise it should be rejected. We perform this simulation for the vectors

$$X_i^* = \{i_1 k, i_2 k, .., i_M k\} \quad \text{with} \quad 0 \leq i_j \leq \frac{600 Mbps}{mean\ bit - rate\ of\ type\ j} \qquad (5)$$

and step size k. Thus we get an equally spaced M dimensional grid whose nodes are named with A(ccept) or R(eject). This grid can be separated by an $M-1$ dimensional decision surface in an "accept" and a "reject" region. As an example Fig. 6 shows this grid for $M=2$, where n_1 and n_2 denote the numbers of active connections of class 1 and 2. In this case the decision surface is just a line.

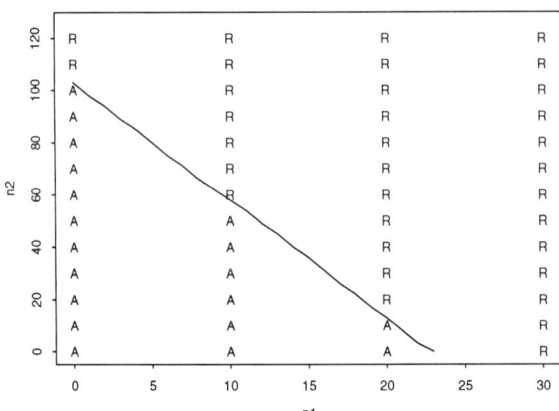

Figure 6 *Learning patterns of neural net*

To complete the pattern each X_i^* gets a $Z_i^* = 0.2$ if the node name is "A" and a $Z_i^* = 0.8$ if the node name is "R." The values 0.2 and 0.8 instead of 0 and 1 are used to obtain a more appropriate learning algorithm.

In the literature a number of different algorithms to adapt the weights of the neural net can be found. In this chapter we use the BFGS algorithm of [14], well known from the theory of unconstrained optimization, aiming for a better convergence speed and a better numerical stability (cf. [15]). First we have to transform the learning process into a function minimization problem. Given the set of learning patterns we define an error function

$$E(W) = \frac{1}{2N} \sum_{i=1}^{N} (Z_i^\star - F(X_i^\star, W))^2 \qquad (6)$$

The term W denotes the vector of weights and $F(X_i^\star, W)$ the output of the neural net upon $X_i^\star$ presented to the input layer. Using the BFGS algorithm $E(W)$ is minimized in the weight space:

1. Initialize W with random values ranging from -0.5 to 0.5.

2. Calculate the search direction in the weight space and perform a line search in this direction to get the next W with a smaller $E(W)$.

3. Check the stop condition ($E(W)$ small enough or local minimum of $E(W)$ is reached). If the condition is not true continue with step 2.

If the final $E(W)$ is small enough, the learning process is terminated. For the recall phase the internal weights of the neural network are now fixed to their final values. The neural network is thought of to has learned the functional mapping $f^\star(X^\star) = Z^\star$ correctly only for the training patterns. Then it is able to perform this mapping also for all other input patterns with the help of the learned decision surface. This property is often referred to as "learning by examples."

3.4 Performance results and discussion

The load control performance of the neural net will be discussed in this subsection, taking into account stationary and non-stationary load conditions.

The neural net operates in the recall mode. Results are obtained by means of simulations with different mixtures of the three connection types described in Section 3.1. For the "Equivalent Bandwidth" and the "Weighted Variance" methods the parameters C_1 and C_2 or respectively, α, had to be determined to guarantee a cell loss rate smaller than the threshold 10^{-5}. Table 1 shows the multiplexer utilization for the CAC methods considered. The column 'Mix' indicates the mixture of the connection types used. Without any admission

Mix	PR	EB	WV	NN
1/2	20.2 %	47.9 %	47.4 %	47.5 %
1/3	17.9 %	45.4 %	43.6 %	48.0 %
2/3	10.3 %	69.1 %	67.0 %	66.7 %
1/2/3	15.8 %	49.5 %	50.6 %	55.4 %

Table 1 Multiplexer utilization

control the utilization of the multiplexer would be about 91 %, without maintaining the desired QOS. As expected, PR is the most restrictive method and has a bad performance, whereas the other methods perform almost on the same level. Only in the case with all the three connection types involved a slight advantage of the NN control can be observed. The reason for this fact is the difference in rejection behavior of the EB and WV method on the one hand and NN on the other hand as shown in Table 2. As shown in Table 2 the two

Type	PR	EB	WV	NN
1	78.2 %	51.2 %	50.9 %	55.0 %
2	78.4 %	36.0 %	33.0 %	22.0 %
3	94.7 %	42.5 %	38.2 %	21.6 %

Table 2 *Call request rejection rates*

methods EB and WV have almost the same connection blocking probabilities. The conclusion of the comparison of their performance with the NN solution is that the NN method rejection decision depends mainly on the mean bit-rate of the connection type while the decision of EB and WV depend on mean bit-rate and variance.

Fig. 7 shows the decision surface of the considered connection admission control methods, which separates the accept and reject regions. The accept region lies on the left hand side of the decision surface. The two methods EB and WV have almost the same decision line, which again indicates the similarity of their performances. The NN decision surface is extremely different. It can be observed that the NN algorithm accepts much more sources with small mean bit-rate (type 2) and less sources with high bit-rate (type 1) than the EB and

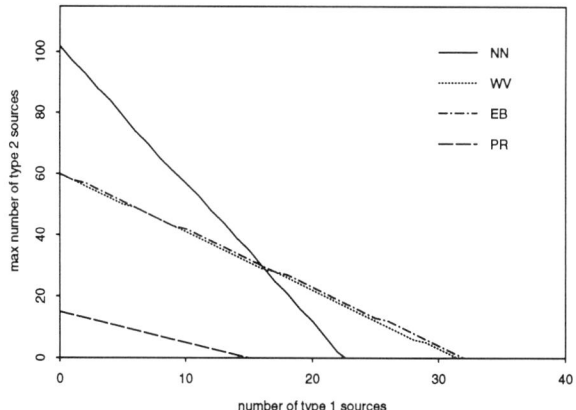

Figure 7 *Decision surface for CAC*

WV methods. From communication network point of view, this results in the same multiplexer utilization (Table 1, row '1/2'), whereas from user viewpoint the differences for the user groups using connection type 1 or 2 are significant.

To show the overload control performance of the CAC methods under consideration, it is necessary to study the CAC response on a non-stationary overload pattern. In the diagrams to follow we use as overload pattern a rectangular overload pulse as illustrated in Fig. 8 and observe the time-dependent CAC reaction in terms of cell and connection blocking probabilities. Clearly, a better CAC mechanism should react to the overload phase with smaller connection blocking probability while keeping the cell loss rate on the same level as under normal load conditions.

Results of the non-stationary load cases are shown in Fig. 8 and Fig. 9. It should be noted that the coefficients of variation of the source bit-rate processes of the three connection types are 1.73, 0.42 and 0.19 respectively.

Fig. 8 shows a comparison of the non-stationary connection blocking probabilities of connection type 3 of the four CAC control methods. In this case it can be seen that the overload performance of the neural net solution is the most efficient.

In Fig. 9 the overload control performance of the neural net is shown where the three connection types are taken as input. As mentioned above, the connection blocking probability is more sensitive to the mean bit-rate than to the variation of the bit-rate process.

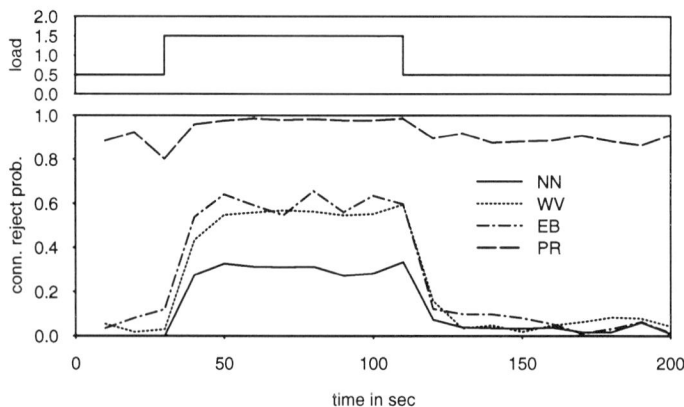

Figure 8 *Comparison of overload performances*

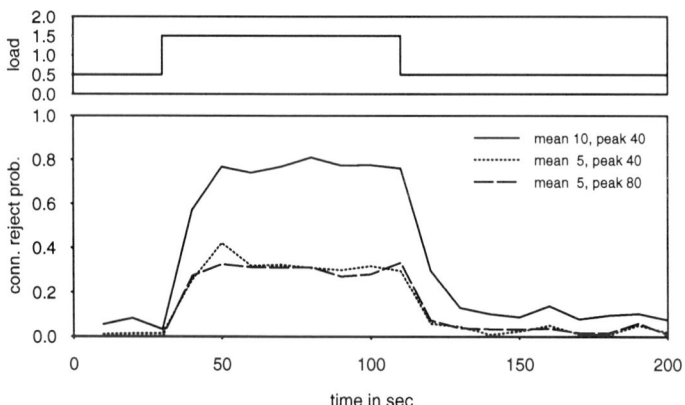

Figure 9 *Impact of traffic variation on overload control performance*

4 CONCLUSIONS

In this chapter different aspects concerning possible use of neural nets to perform connection admission control (CAC) in broadband integrated services networks have been discussed. The formulation of CAC problem as functional mapping and in consequence, the use of learning algorithms to represent the required mapping were shown and architecture alternatives for the CAC neural nets using the class of feed-forward structures in conjunction with back-propagation learning are depicted. In order to discuss performance aspects a basic net example has been investigated. The neural net performance has been compared with other connection admission control mechanisms like the peak bit-rate, the equivalent bandwidth and the weighted variance method. Numerical results for stationary and non-stationary pulse-form overload patterns have been obtained to illustrate the capability of neural nets used as connection admission controller in ATM environments.

In most of load scenarios under consideration the CAC performance of the investigated neural net structure is comparable with and in some cases better than the CAC methods mentioned above, even by using a very small and simple neural net.

To improve the performance of CAC by neural nets, other neural net structures or other input representations can be developed. One promising candidate is a combined solution of an adaptive neural net with learning patterns, which contain more information about the past of the observable load situation.

REFERENCES

[1] P. Tran-Gia and O. Gropp, "Performance of a Neural Net used as Admission Controller in ATM Systems," Proc. Globecom 92, Orlando, FL, pp. 1303-1309.

[2] COST 224 Committee (ed. J. Roberts), COST 224 Final report, Performance evaluation and design of multiservice networks, Paris, October 1991.

[3] CCITT, Draft Recommendation I.311, B-ISDN General network aspects, COM XVIII-R 34-E, Geneva 1990.

[4] CCITT, Temporary Document 43/XVIII, On networking and resource management, Study group XVIII, Matsuyama 1990.

[5] A. Hiramatsu, "ATM communications network control by neural networks," Proc. IEEE Int. Conf. on Neural Networks 1989, Vol. I, pp. 259-266.

[6] A. Hiramatsu, "ATM communications network control by neural networks," IEEE Transactions on Neural Networks, (1990)1, pp. 122-130.

[7] P. Castelli, E. Cavallero, and A. Tonietti, "Policing and call admission problems in ATM networks," Proc. 13th Int. Teletraffic Congress, Copenhagen 1991, pp. 847-858.

[8] A. E. Eckberg, D. T. Luan, and D. M. Lucantoni, "Meeting the challenge: Congestion control strategies for broadband information transport," Proc. Globecom 1989, Dallas, pp. 1769-1773 (49.3).

[9] G. Galassi, G. Rigolio, and L. Verri, "Resource management and dimensioning in ATM networks," IEEE Network Magazine, (1990)5, pp. 8-17.

[10] S. B. Jacobson, K. Moth, and L. Dittmann, "Load control in ATM networks," Proc. XIII Int. Switching Symposium, Stockholm 1990, pp. 131-138.

[11] A. Lombardo, S. Palazzo, and D. Panno, "Admission control over mixed traffic in ATM networks," Int. Journal of Digital and Analog Communication Systems, (1990)3, 155-159.

[12] C. Rasmussen, J. Sørensen, K. S. Kvols, and S. B. Jacobsen, "Source-independent acceptance procedures in ATM networks," Journal on Selected Areas of Communications, JSAC-(1991)4, pp. 351-358.

[13] E. Wallmeier, "A connection acceptance algorithm for ATM networks based on mean and peak bit rates," Int. Journal of Digital and Analog Communication Systems, (1990)3, 143-153.

[14] R. Fletcher, Practical methods of optimization, Wiley 1980.

[15] R. L. Watrous, "Learning algorithms for connectionist networks: Applied gradient methods of nonlinear optimization," Proc. IEEE Int. Conf. on Neural Networks 1987, pp. 619-627.

8
NEURAL NETWORK CHANNEL EQUALIZATION

William R. Kirkland, D. P. Taylor*

CRL, McMaster University, Canada

University of Canterbury, New Zealand

1 INTRODUCTION

The present trend in the communication industry is towards digital transmission of both analog and digital information over analog channels. The continuous growth of the telecommunications market has demanded increased spectrum efficiency on band limited channels. This has been met by the use of higher speed data transmissions which are more sensitive to channel disturbances. In turn, this has created the need for more effective equalization schemes to combat the effects of the channel. Channel disturbances may be of either an additive and/or multiplicative form [1] resulting from background thermal noise, impulse noise, co-channel and adjacent channel interference and fades which manifest themselves in the form of frequency translation and attenuation, nonlinear or harmonic distortion and time dispersion.

Our concern is with the effects of time dispersion resulting from frequency selective fading and the effects of thermal noise. Time dispersion results when the frequency response of the channel deviates from the ideal response of constant amplitude and linear phase[1]. The result of time dispersion is that the effect of a transmitted symbol extends beyond the time interval used to represent that symbol. This is known as intersymbol interference, i.e. ISI. In the telephone channel time dispersion results from the presence of echoes on the telephone line[1]. In this case the channel is unknown but does not change with time. In wireless communication, in particular in line-of-sight digital microwave radio (DMR), [2], the channel disturbance results from the presence of multipath propagation. Multipath propagation may be viewed as transmission through a group of channels with differing relative amplitudes and delays [1]. In this case the equalizer must track time varying channel characteristics.

Conventional equalization schemes have normally used an adaptive linear filter based upon the LMS (least mean squares) or ZF (zero forcing) adaptation algorithms. While these equalizers have demonstrated good performance, in many instances, the equalizer falls considerably short of the matched filter performance bound, obtained by considering the reception of an isolated transmitted pulse [1, 3].

The optimum (minimum error probability) equalization scheme, subject to various constraints, has been shown to be a nonlinear structure either in the form of the maximum-likelihood sequence estimator [4], e.g. the Viterbi algorithm, or as a nonlinear tapped delay line structure [5, 6]. In [6]-[8] Gibson and Cowan *et al.* illustrate through the use of a simple channel model that the optimal (minimum noise enhancement) combining network for a tapped delay line structure is nonlinear in nature. To achieve an adaptable nonlinear equalizer they have made use of feed-forward neural network type structures [6]-[8]. Ungerboeck's[5] motivation for a nonlinear approach is "...that intersymbol interference between signals with quantized pulse amplitudes is of a discrete nature. Exploiting this discreteness, nonlinear methods can reduce intersymbol interference with less noise enhancement than linear methods." Ungerboeck proposes two sub-optimal nonlinear equalization schemes that resemble a feed-forward neural network architecture. Both the work of Gibson and Cowan *et al.* and Ungerboeck suggest that there is merit in investigating the application of neural neural networks to adaptive nonlinear equalization. Neural networks with their ability to form arbitrary complex functional mappings appear to offer a flexible and adaptable equalization structure that is capable of achieving performance close to that of the optimum equalizer.

The outline of this chapter is as follows. The first section provides an introduction to digital transmission systems. This is followed by a section that discusses Bayesian detection and estimation, linear filter theory and the application of neural networks to equalization. This section discusses the performance of neural network equalizers in a simplified transmission system. The next section details the application of neural network equalizers to the more general case of two dimensional signalling through the use of complex baseband channel models along with simulation results. This is followed by a short summary and some conclusions.

2 A BRIEF INTRODUCTION TO DIGITAL TRANSMISSION THEORY

In the following discussion we restrict ourselves to the use of real valued data for reasons of simplicity although the results are easily extended to the more general case of complex valued data.

In a general transmission scheme, the baseband transmitted signal $s(t)$ is given as

$$s(t) = \sum_k s_k \, p(t - kT) \tag{1}$$

where $p(t)$ is the transmitted pulse shape, T is the symbol interval in seconds and s_k is an M-ary (M is even) symbol such that $s_k \in \{\pm 1, \pm 3, \ldots, \pm M - 1\} \equiv \xi$. In order to restrict the maximum amplitude of the signal to ± 1 the values of s_k may be normalized by a factor of $\frac{1}{M-1}$ without any loss in generality. By restricting $p(t)$ to be a unit amplitude rectangular pulse of duration T seconds we form a pulse amplitude modulated (PAM) signal with amplitude values in the set

$$\frac{1}{M-1}[\pm 1, \pm 3, \ldots, \pm M - 1]$$

In many cases the effect of a data pulse extends over many symbol intervals. This is the result of the pulse shaping filters used in the transmitter and receiver to limit the bandwidth of the signal and minimize or control intersymbol interference (ISI). One such pulse shape is the square-root-raised cosine spectral pulse defined as

$$p_{\sqrt{RC}}(t) = 8\beta \frac{\cos[(1/T + 2\beta)\pi t] + \sin[(1/T - 2\beta)\pi t](8\beta t)^{-1}}{(\pi T^{1/2})[(8\beta t)^2 - 1]} \tag{2}$$

where β is the "excess bandwidth" parameter. The combined response at the output of the receiver's pulse shaping filter is the raised cosine spectral pulse

$$p_{RC}(t) = \frac{\cos 2\pi\beta t}{1 - (4\beta t)^2} \left(\frac{\sin \pi t/T}{\pi t/T} \right) \tag{3}$$

Note that both pulses are band-limited to $\beta + 1/2T$ and have an infinite time response but the raised cosine pulse has evenly spaced zero crossings every T seconds. There will be zero intersymbol interference provided the receiver samples the signal every T seconds at the appropriate time instants.

The effects of channel distortion may be easily modelled with the use of a finite impulse response (FIR) filter. In the case of multipath fading the tap delays correspond to the delays associated with the signal paths in the channel and the tap weights represent the attenuation from the signal paths. In general only a small number of coefficients are used to model the channel, i.e. a two or three tap fading model. For simple models, the tap delays may be integer multiples of the unit delay operator z^{-1} which may be taken to be one-symbol interval. In other channel models, such as those used to model the line-of-sight digital microwave radio channels[9] where the relative transmission time differences between paths is quite small, the delays may be a fraction of the symbol interval. If the channel is modelled as a T-spaced N-tap FIR filter (N is small, e.g. in the range of 2 to 5), then only N adjacent data symbols will interfere with each other resulting in the output of the channel having only a finite number of possible states (M^N). If the tap spacing is a fraction of the symbol period then, when a spectrally efficient pulse is used (e.g. the raised cosine pulse) a large number of data symbols will interfere with each other. This may be 20 or more symbols depending upon how fast the overall data pulse decays in time. Clearly the effect of the pulse shaping filters should not be over looked.

The combined effects of the channel and of the transmit and receive filters may be represented by a single filter which is formed by taking the convolution of all the filters in the path of the signal. If its impulse response is denoted as $h(t)$, then upon sampling the received signal at time $kT + t_0$, where t_0 accounts for the channel delay and sampler phase[1], the resulting received signal, $r(kT)$, is

$$r(kT) = s_k h(t_0) + \sum_{j \neq k} s_j h(t_0 + kT - jT) + n(t_0 + kT). \qquad (4)$$

The first term on the right hand side is the desired signal s_k weighted by $h(t_0)$. The last term is the additive noise with variance σ_n^2 while the middle sum is the intersymbol interference from adjacent symbols. Generally, $h(t_0)$ will be handled by the receiver's automatic gain control system leaving the job of ISI cancellation to the adaptive equalizer.

3 SIGNAL DETECTION AND ESTIMATION

For a received signal given by equation (4) the optimum signal detector is a maximum likelihood sequence estimator [4] which at time k detects the entire data sequence $\{s_k, \ldots, s_0\}$ based upon the entire received sequence $\{r_k, \ldots, r_0\}$. This receiver is quite complex as it must search through all the possible combinations of transmitted signals to determine the most likely combination that resulted in the received sequence. Such a receiver will not be detailed here, rather what is desired is a sub-optimum structure whose performance is quite close to optimum. Note that is possible to take multiple samples per symbol of the received signal, however, we shall restrict ourselves to one sample per symbol. The format that we wish to concern ourselves with is symbol by symbol detection (or estimation) based upon a small subset of the entire received sequence. We wish to detect or estimate s_{k-d} given the m-element received signal vector $x_k^T = \{r_k, \ldots, r_{k-d}, \ldots, r_{k-m+1}\}$ where d is a suitably chosen decision delay, i.e. at time k, s_{k-d} is detected, not s_k. The receiver structure is shown in Figure 1. Aside from the processor function the parameters that we are free to chose are the number of received signal samples m, the spacing between the received samples and the decision delay. All of these affect the performance of the system. Once these factors have been set it is up to the processor unit to make the most of the situation. If the channel response is known it is possible to determine the optimum processor function. Generally the channel response is not known so it is desirable to have a network (e.g. neural network) that can be trained to learn the desired processing function. Hence, an error signal e_k has been included in the system so that it is possible to train the processor and to allow the processor to adapt to time varying changes in the channel response.

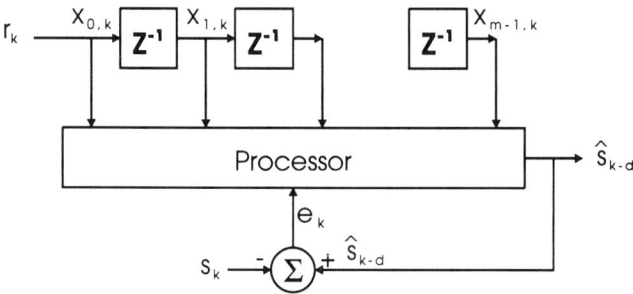

Figure 1 General Equalization Structure

If the output of the processor is continuously valued or soft quantized then the equalizer structure is based upon an estimation type of structure as it attempts to produce an estimate of the transmitted symbol. If the output is hard quantized then the equalizer is based on a detection type of structure, that is, its goal is to detect which one of the m signals out of the m ary alphabet was transmitted. By soft quantization we mean that if the output symbol is represented by binary bits then more than $log_2(m)$ bits are used, whereas in the case of hard quantization only $log_2(m)$ bits or m decision lines are used.

The rational for choosing an estimation or a detection type of structure is arbitrary but is influenced by the architecture following the equalizer. In the case of convolutional coding and trellis coded modulation (TCM) [10] the decoding structure following the equalizer is based upon soft decisions at the output of the equalizer. Hard decisions could be used but this results in a stiff penalty in SNR (signal to noise ratio) performance of about 2-2.5 dB between hard quantization and soft decisions (based upon three bits of quantization per symbol interval).

3.1 Bayesian Detection

To determine the optimum signal detector for a finite impulse response channel filter, $H(z) = \sum_{i=0}^{n} h_i z^{-i}$, we consider the optimum signal detector for the tap input vector x_k where $x_k^T = [r_k, \ldots, r_{k-m+1}]$ and

$$r_k = \sum_{j=0}^{n} s_{k-j} h_j + n_k \qquad (5)$$

This detector is based upon Bayesian detection theory [11][12][8]. To determine the detector all of the possible states of the tap input vector x_k must be considered. Let S_θ denote the set of all possible combinations of the channel input sequence $[s_k, s_{k-1}, \ldots, s_{k-d}, \ldots, s_{k-m+1-n}]$ where the desired signal $\hat{s}_{k-d} = \theta$. Denote the resulting set of states for the received signal vector x_k for the case of noiseless transmission as X_θ and let $X_{j,\theta}$ represent the jth element of this set. The M Bayesian decision variables corresponding to the M possible values for $\hat{s}_{k-d}$ are computed as

Neural Network Channel Equalization

$$d_{\theta, k-d} = \sum_{j=1}^{M^{n+m-1}} \frac{p(X_{j,\theta})}{\sqrt{2\pi\sigma_n^2}} exp\left[\frac{-||x_k - X_{j,\theta}||^2}{2\sigma_n^2}\right] \qquad (6)$$

The detected symbol is chosen such that $\hat{s}_{k-d} = \hat{\theta}$ where $d_{\hat{\theta}, k-d} = max(d_{\theta, k-d})$. The resulting decision boundaries for $m = 2$, $d = 0$ and $s_k \in \{\pm 1\}$ are shown in Figures 2a, b for the two channels, $H_1(z) = 1.0z^0 + 0.5z^{-1}$ and $H_2(z) = 0.5z^0 + 1.0z^{-1}$. These two channels have been used extensively by Gibson and Cowan et al. [6]-[8] to illustrate the nonlinear behavior of the optimum equalizer. In both cases the decision boundary, the line separating the regions where $s_{k-d} = 1$ and $s_{k-d} = -1$, is nonlinear but only in the first case, $H_1(z)$, are the decision regions linearly separable.

$H_1(z)$ and $H_2(z)$ are known as minimum and maximum phase channels respectively. A minimum phase filter in the z-domain has all of its poles and zeros located within a unit circle centered at the origin in the z-domain whereas a maximum phase filter has all of its poles and zeros outside of this unit circle. In the multipath channel, minimum phase indicates that the signal from the principle path arrives first as opposed to non-minimum phase where the signal from one or more of the secondary paths arrives first at the receiver. The principle signal path or ray is usually taken to be that path with the greatest signal strength. The significance in the phase of the channel is in how the decision delay should be chosen. This will be described further in section 3.3.

The main problems with this structure are that it requires knowledge of the channel impulse response to calculate the channel states and its complexity. While the complexity grows linearly with the number of states of x_k the number of states grows exponentially with the number of non-zero ISI terms, n, of equation (4) and the length of the tap input vector x_k, m. The total number of possible states for x_k is M^{m+n}. In the case of pulse shaping, e.g. a raised cosine pulse, the number of states can be quite large. Rather than evaluate equation (6) directly it is desired to train a neural network to approximate the action of the Bayes signal detector. This has been the basis for much of the work of Gibson and Cowan et al. [6]-[8].

In mapping the M possible values for s_{k-d} to the output of a neural network there are two possible choices. The first choice is to use a network with M output nodes, each node representing one of the decision variables. The detected symbol is then chosen as in the case of the Bayesian detector. The network is trained by indicating that the desired node should be at a high level,

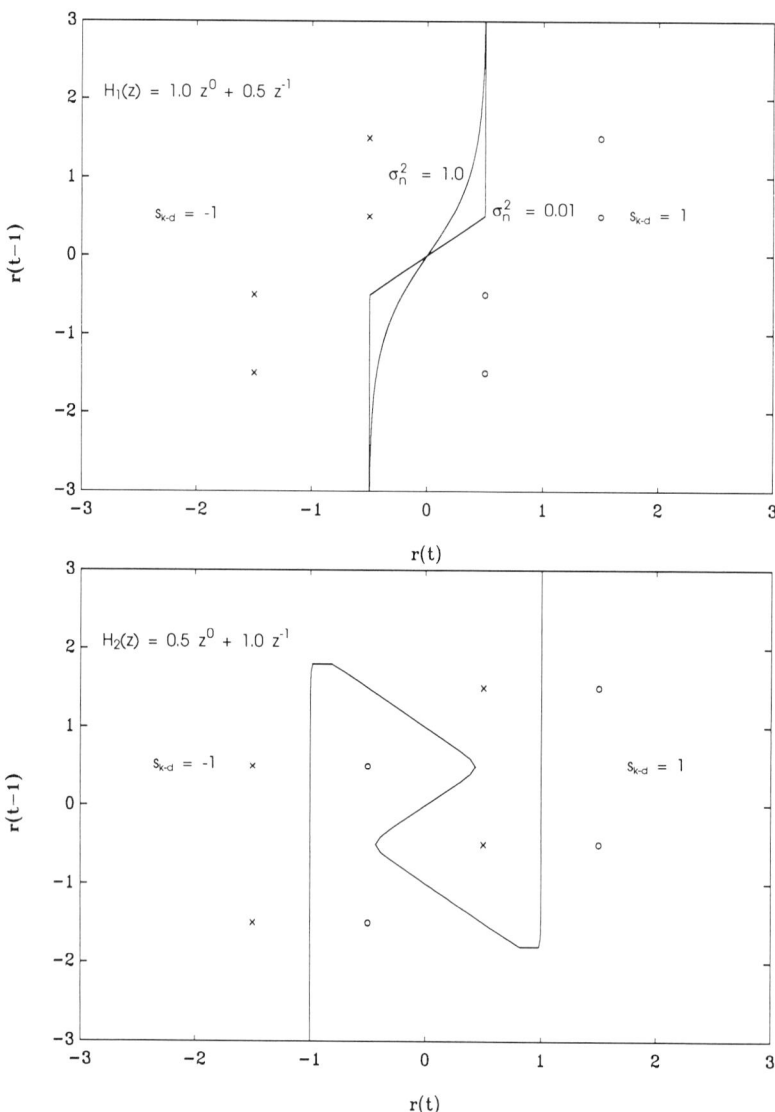

Figure 2 The decision boundaries for $H_1(z)$ and $H_2(z)$ for $m = 2$ and $d = 0$

e.g. +1, and that the other nodes should be at a low level, e.g. −1. There has been some indication [13]-[16] that for this type of network the activation levels of each of the M output neurons correspond to the probabilities that the symbol represented by the neuron was transmitted

$$d_{\theta, k-d} \sim P(s_{k-d} = \theta)$$

If this true then in the case of coded transmission it may be advantageous to feed the outputs of the neural network to a decoder rather than make hard decisions on the signal. This would correspond to soft decision decoding and may allow the system to achieve some of the coding gain that may otherwise be lost by the use of hard decisions on the signal. At present, no one has considered the effect of neural network equalizers on code performance.

An alternative mapping is to label each of the M values for s_{k-d} with a binary number with $log_2(m)$ bits of ±1's. The advantage of this is it reduces the number of output nodes for the network, however, the activation levels will no longer correspond to probabilities. For binary level signalling {±1} this produces a network with a single output where the desired output is ±1 which is equal to s_{k-d}. The network thus acts as an estimator for the transmitted signal.

3.2 Bayesian Estimation

An alternative to detecting the signal s_{k-d} is to estimate it and then to make a decision on the symbol based upon the estimate. The estimator that minimizes the mean squared error between the estimate $\hat{s}_{k-d}$ and s_{k-d} is the Bayes estimator [11] given by

$$\hat{s}_{k-d} = \int_\xi s\, p(s|x_k)\, ds \qquad (7)$$

where the range of integration ξ is over all possible values of the transmitted signal s_{k-d}. As s_{k-d} is an M'ary signal this reduces to the summation

$$\hat{s}_{k-d} = \sum_{j=1}^{M} s_j\, p(s_j|x_k) \qquad (8)$$

where $s_j \in \xi$. For this function a neural network with a single output is used and the network is trained to minimize the expected value of the error power of e_k where $e_k = \hat{s}_{k-d} - s_{k-d}$.

This estimator is subject to the same problems as the Bayesian detector; the fact that we need to know how to calculate $p(s_j|x_k)$ and that its computational complexity increases exponentially with the length of the channel and the length of the received signal vector x_k. However, the Bayes estimator for the special case where the signal s_k is only corrupted by white Gaussian noise, $x_k = s_k + n_k$ yields an interesting result. The estimator function can be calculated as

$$\hat{s}_k = \begin{cases} \tanh(x_k/\sigma_n^2) & s_k \in \xi = \{\pm 1\} \\ \dfrac{\sum_{i=1}^{M} s_i e^{-\frac{(x(t)-s_i)^2}{2\sigma^2}}}{\sum_{j=1}^{M} e^{-\frac{(x(t)-s_j)^2}{2\sigma^2}}} & s_i, s_j \in \xi \end{cases} \quad (9)$$

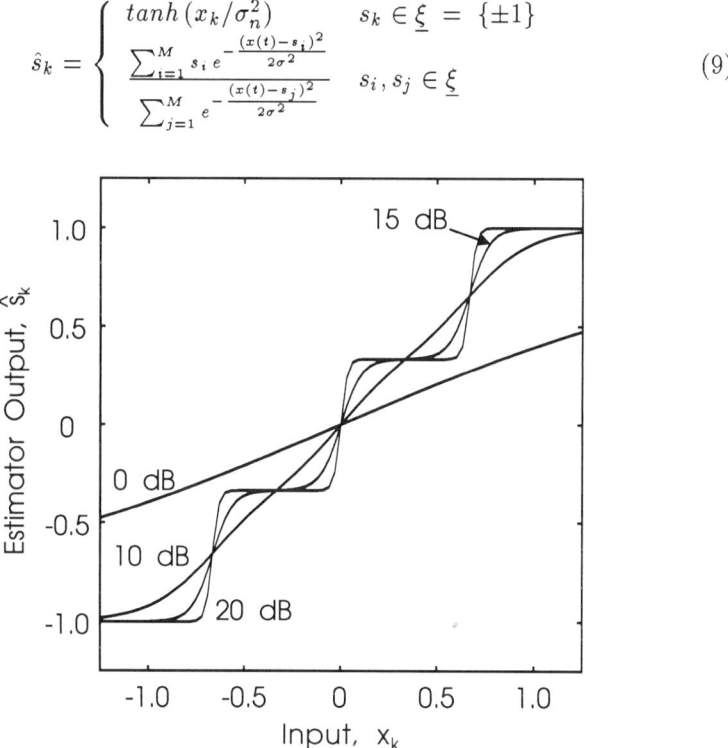

Figure 3 The Bayesian estimator function for 4-PAM with SNR's of 0, 10, 15 and 20 dB.

The $tanh$ function may be recognized as a scaled and shifted version of the sigmoidal nonlinearity commonly used in the back propagation [17][18] al-

Neural Network Channel Equalization

gorithm. The Bayes estimator for $s_k \in \{\pm 1/3, \pm 1\}$ is shown in Figure 3. While the estimator for an M'ary signal is quite complex it may be closely approximated with

$$f(\alpha\, x) = \frac{1}{M-1} \sum_{i=1}^{M-1} tanh\left(\alpha\left(x - \theta_i\right)\right) \qquad (10)$$

where α is a gain term chosen to reflect the noise level present in the system and the θ_i's are thresholds or biases chosen as the mid-points between adjacent signal points in the signal set ξ, e.g. $\{\pm 2/3, 0\}$ for 4-PAM. This function may be seen as the sum of $M-1$ neurons each with a $tanh$ activation function with a gain of α, weights of unity and appropriately chosen bias levels. Evidently the back propagation neural network seems well suited to signal estimation. For low noise levels a minimum network would consist of the linear combination of the outputs of $M-1$ neurons in a single hidden layer. An alternative would be a single neuron using the multi-level nonlinearity given in 10. This would seem to make a case for neurons with multi-level activation functions.

3.3 Linear Equalization Theory

We have already seen how the optimal decision boundary can be nonlinear particularly when the decision delay d is minimal but what has not been established is the effect of the number of taps m and the length of the decision delay on equalizer performance. For this we look at linear filter theory. To illustrate the principles involved let us consider the the minimum and non-minimum phase channels consider previously. These two channels, $H_1(z)$ and $H_2(z)$ are given as:

$$H_1(z) = 1.0z^0 + 0.5z^{-1} \qquad (11)$$
$$H_2(z) = 0.5z^0 + 1.0z^{-1} \qquad (12)$$

If one ignores the constraints of causality and stability and is only concerned with eliminating ISI then the optimum ISI canceling filters for $H_1(z)$ and $H_2(z)$ would be:

$$H_1^{-1}(z) = \frac{1}{1.0z^0 + 0.5z^{-1}} \qquad (13)$$
$$H_2^{-1}(z) = \frac{1}{0.5z^0 + 1.0z^{-1}} \qquad (14)$$

Note that these filters are infinite-impulse response (IIR) filters and that $H_2^{-1}(z)$ is unstable as its pole lies outside of the unit circle.

While the optimum ISI cancellation filter is generally an infinite-impulse response (IIR) filter, adaptive IIR filters are generally not used in practice due to a lack of guaranteed stability, lack of a quadratic performance surface and a minor performance gain over transversal equalizers[1]. Generally, adaptive transversal equalizer structures are used. The output of this filter is the weighted sum of the tap outputs of a tapped delay line, where the output may be written as

$$\hat{s}_{k-d} = w_k^T x_k \tag{15}$$

where w_k is the tap weight vector at time k defined as $w_k^T = \{w_{0,k}, \ldots, w_{m-1,k}\}$. If the criterion for adjusting the tap weights is to eliminate ISI then this is known as a zero-forcing (zf) equalizer[3]. Essentially the transversal zf-equalizer tries to approximate the inverse channel response using a finite number of taps. The problem with a zero-forcing equalizer is that when there are deep nulls in the frequency response of the channel, the zero-forcing equalizer can lead to excessive enhancement of the noise present on the received signals. The least mean-square (LMS) equalizer is more robust as it maximizes the signal to noise plus distortion ratio. That is, it minimizes the mean-squared value of the error signal, e_k, determined from the equalizer's output, i.e. $E[(s_{k-d} - \hat{s}_{k-d})^2]$ is minimized through a linear combination of the components of x_k. This is simply a linearization of the Bayesian estimator.

For real valued data, the adaptation of the ith tap weight at time k, $w_{i,k}$, is governed by the following equations:

$$w_{i,k+1} = w_{i,k} - \eta\, e_k\, x_k$$
$$e_k = \hat{s}_k - s_k$$

where η is an adaptation parameter controlling the step size of the change in weights, e_k is the error signal at time k.

If the values of the channel impulse response at the sampling instants are known then the optimum m element tap weight vector w_o can be obtained by solving a set of m linear simultaneous equations [19] given by

$$R w_o = p \tag{16}$$

where R is the auto-correlation matrix of the tap input vector x, $E[xx^T]$, and p is the cross-correlation vector formed from the desired equalizer response d and the tap input vector x, $E[dx]$. Equation 16 is known as the *normal equation* [19]. The minimum obtainable mean-squared error is given as

$$J_{min} = \sigma_d^2 - p^T w_o. \tag{17}$$

where σ_d^2 is the variance of the desired response.

The performance of this filter is limited by three factors[20]: the presence of noise on the received signal, the necessity for the inclusion of a delay in the equalizer response, and the number of taps used to approximate the ideal response. The reason for the inclusion of a delay in the equalizer response is that the transmitter, receiver and channel filters are causal systems. This means that the transmitted signal will be delayed in time as it passes through these filters. The use of a delay of d samples permits much lower values of minimum mean-squared error and causes the converged adaptive impulse response, when convolved with that of the overall channel response, to approximate an impulse with a delay of d. The optimum value for the delay is a function of the phase of the channel transfer function. For a minimum phase channel the delay may be minimized but as the channel moves towards a maximum phase response the delay must be increased in order to maximize the performance of the equalizer. Generally, provided that a sufficient number of taps (e.g. 9) are used, the choice of the delay time is not critical as long as it corresponds to roughly the middle of the tapped delay line[20].

The work of Gibson and Cowan *et al.* [6, 21, 7] is useful in showing why the optimum transversal structure is nonlinear. For the two channels $H_1(z)$ and $H_2(z)$ the plots of r_{k-1} versus r_k from section 3.1 (Figures 2a and 2b) show that the decision boundary is nonlinear and that the signal points are only linearly separable by a single line for the case of the minimum phase channel $H_1(z)$. However, it can be shown that with the aid of equations (16) and (17) that for $H_2(z)$ the signals become linearly separable as the decision delay is allowed to increase. This is not to say that the optimal decision boundary is a linear one when the optimum decision delay is chosen, rather that even though the signals may be linearly separable a nonlinear equalizer may still offer improved ISI and noise performance. What must be noted is that the optimum decision delay is a function of the phase of the channel.

The use of the Bayes signal detector forms the optimum signal detection scheme and allows for a minimum number of taps and minimum decision delay but at the cost of complexity. With the addition of a few more taps and a slightly increased decision delay the complexity of the system can be greatly decreased with only a slight degradation in system performance provided that the decision delay is appropriately chosen.

The performance of the linear equalizer may be further improved through the use of decision feedback equalization (DFE) [22]. In this case the signals preceding r_{k-d}, $[r_{k-d-1}, \ldots, r_{m-1}]$, are replaced with the detected symbols of $[\hat{s}_{k-d-1}, \ldots, \hat{s}_{m-1}]$. Provided the detected values of $[\hat{s}_{k-d-1}, \ldots, \hat{s}_{m-1}]$ are correct the output noise is only a function of $[r_0, \ldots, r_{k-d}]$ rather than of $[r_0, \ldots, r_{k-m+1}]$. The effect of this on Bayesian detection/estimation is to reduce the number of states in X thereby reducing the complexity of the detector/estimator. A DFE generally performs better than a linear transversal equalizer (LTE), however, the DFE suffers from the problem of error propagation. If the output symbol of the DFE is in error then feeding it back into the equalizer will likely produce another symbol error. Hence the output of a DFE is subject to bursts of errors. The theoretical aspects of the DFE are beyond the scope of this chapter. The interested reader is referred to [22].

3.4 Neural Networks

The goal of using neural networks, be they based upon back propagation or radial basis functions etc., is to achieve the performance of the optimal signal detector or estimator without the complexity associated with these structures. While neural networks are more complex and exhibit slower learning times than linear equalizers they offer the advantage of being able to form nonlinear mappings and hence should be able to achieve improved performance. The networks that seem particularly suitable to this problem are feed-forward networks such as the multi-layered perceptron (MLP) and radial basis function (RBF) networks. Most of the literature on neural network equalization has been dedicated to these two structures with some other work being done on the use of Petri Nets[23], Kohonen networks[24] and Madaline structures[25][26].

The advantage of the RBF network is that it has only a single hidden layer of basis functions which are linearly combined to form the output of the network. As such, RBF networks are generally less complex and exhibit faster learning times than MLP networks that achieve the same functional mapping. In addition, MLP networks are more sensitive to the choice of learning parameters and network topology than RBF networks. Though, both networks are capable of generating arbitrary complex nonlinear decision regions.

The problem that both networks share is in determining the size/configuration of the network. Chen et al.[8] have used an orthogonal least-squares (OLS) algorithm to solve this problem for the RBF network and have extended it to multi-output RBF networks. Chen et al. [8] indicate that the basis functions should be the Gaussian probability density function of width σ_n^2 with the centers for the basis functions chosen as the channel states in X. For MLP networks there have been numerous algorithms proposed that determine the allocation of neurons and the pruning of network weights though none of these schemes have been applied to adaptive equalization.

In section 3.2 it was demonstrated that the minimum MLP network configuration for the estimation of an M'ary signal in white Gaussian noise consists of the linear combination of $M - 1$ neurons arranged in a single hidden layer or alternatively, a single neuron with the multilevel activation function of (10). If a network of this type is used in conjunction with the tapped delay line it may be seen that the linear parts of the neurons in the first hidden layer appear as a system of linear transversal equalizers with the output of each equalizer being fed to the activation function used in each neuron. The system thus appears to be a linear transversal filter followed by a Bayesian estimator for s_{k-d} in white Gaussian noise. The purpose of the weights in the first hidden layer is to approximate the inverse channel filter to remove the effects of the channel from the transmitted signal. The rest of the network then may be viewed as the Bayesian estimator or detector for the received signal, s_{k-d}, depending upon how the output of the network is configured. By looking at the weights of the first hidden layer as weights of a FIR filter it is possible to interpret these weights in the frequency domain in exactly the same ways as it is possible to do so for the weights of a linear transversal equalizer.

The problem with this interpretation is that the weights of neurons in the first hidden layer will colour the noise on the received signal through their filtering action. Thus the noise on the signal at the input to the neural activation functions will be coloured (correlated) noise rather than white. Hence the optimum Bayesian estimator function is no longer the $tanh(s_{k-d}/\sigma_n^2)$ function, however, for moderate to high signal to noise ratios it is a good approximation. Such is usually the case in practice where usable error rates demand moderate to high signal to noise ratios even when there is no channel disturbance. Evidently the effects of linear filter theory should not be underestimated.

If both RBF and MLP networks are capable of achieving the same functional mappings then the use of MLP networks allows us to achieve some insight into the behavior of neural networks through looking at the weights of the first hidden layer of a MLP network as described in the previous paragraphs. This can not be done with an RBF network as the inputs to the RBF network are fed directly to its nonlinear basis functions. However, if the behavior of MLP neural network equalizers is as described, then it is reasonable to expect RBF networks to behave in a similar manner.

4 TWO DIMENSIONAL SIGNALLING AND COMPLEX NEURAL NETWORKS

The previous sections on channel equalization have concentrated on the use of simplified transmission models to illustrate the principles involved in the equalization of a transmitted signal. Gibson and Cowan *et al.* [6]-[8] have used such schemes to demonstrate the superiority of neural network equalizers over conventional linear and decision feedback equalizers. They have demonstrated the ability of neural network equalizers to approach the performance of the optimum Bayesian signal detector. Their work has however been restricted to PAM signalling through a 2-3 tap, T spaced channel filter and is based upon using a minimal decision delay and minimizing the number of taps in the equalizer. In particular they have not consider the effects of pulse shaping on neural network equalizer performance. What is needed is to extend this work to more accurately reflect an actual transmission system. Hence a more sophisticated transmission model is needed, one that allows for the use of pulse shaping, different signal constellations and a more realistic channel model. Such a channel model would allow for multipath delays other than at just multiples of the symbol period.

A general transmission scheme is shown in Figure 4. In describing this system we will concentrate on QAM (quadrature amplitude modulation) transmission [27][28] but the ideas presented are easily extended to other modulation formats. Essentially, QAM transmission makes use of two PAM signalling channels, both occupying the same frequency allocation. This is achieved by using two carriers that are in quadrature (90 degrees apart) to each other through the use of $cos(2\pi f_c t)$, the in-phase channel, and $sin(2\pi f_c t)$, the quadrature channel, where f_c is the carrier frequency. Such a system is most easily simulated in baseband ($f_c = 0$). In this case, the transmitted data values are represented as complex numbers, hence the need for complex valued adaptive equalizers. Such an equalizer is diagrammed in Figure 5. In the equalizer, the real parts of the tap weights combat the intersymbol interference in the in-phase and quadrature channels and the imaginary parts of the tap weights counteract the cross-interference between the two channels [1].

Neural Network Channel Equalization

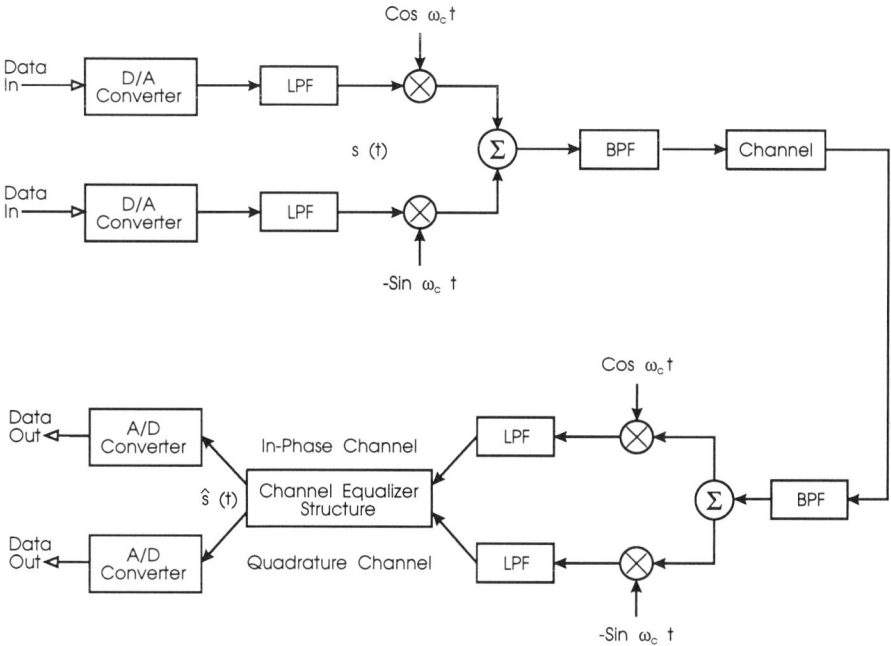

Figure 4 General QAM transmission scheme

The adaptation of this equalizer structure is based on minimizing the average total error power, $E[e\,e^*]$ where e is the complex valued error signal and $*$ denotes complex conjugation. The complex LMS algorithm [29] is given as:

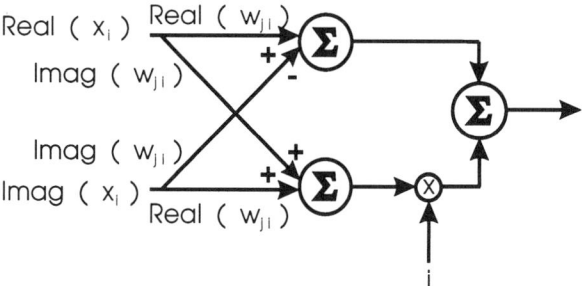

Figure 5 Complex Equalizer Structure

$$w_{j+1} = w_j + 2\eta e x^* \qquad (18)$$

This equation may be extended to complex back propagation provided a suitable activation function is chosen [30][31], i.e. a complex function $f_c(z)$ which is bounded in the complex domain [31]. A suitable function may be formed by combining the activation function used in a real valued network (e.g. $f(x) = tanh(x)$) in the following way

$$f_c(net) = f(net_R) + j\, f(net_I) \qquad (19)$$

where the subscripts R and I denote real and imaginary components. In effect, the real parts of the output neurons correspond to the in-phase data channel while the imaginary outputs correspond to the quadrature data channel.

The adaptation equation for the weight w_{jk} at time $n+1$ is

$$\begin{aligned}
w_{jk}(n+1) &= w_{jk}(n) + \eta\, \delta_k o_j^*(n) & \\
\delta_k(n) &= e_{kR}\, f'_{kR}(net_{kR}) + j\, e_{kI}\, f'_{kI}(net_{kI}) & \text{output layer} \\
\delta_j(n) &= f'_{kR}(net_{kR})\, Real[\sum_k \delta_k(n)\, w_{jk}^*(n)] + & \\
& \quad f'_{kI}(net_{kI})\, Imag[\sum_k \delta_k(n)\, w_{jk}^*(n)] & \text{hidden layers} \\
net_k &= \sum_j w_{jk}(n)\, o_j(n) &
\end{aligned} \qquad (20)$$

where w_{jk} is the complex weight linking neuron j to neuron k, $f_k()$ is the complex nonlinear function for neuron k, o_k is the complex output of neuron k, e_k is the complex error between the desired response, d_k, and the output of the kth neuron, o_k, * denotes complex conjugation and ' denotes differentiation. Similar equations can be derived when the momentum factor is considered. The RBF network may also be extended to the complex domain but as yet there has been no work published on the complex RBF network applied to adaptive equalization [1].

Complex multi-layered perceptron equalizers have been studied using QAM and PSK (phase shift keying) modulation in [33] and [34]. In [33] a digital microwave radio (DMR) channel was used which allowed for comparisons to be made with work presented by Amitay and Greenstein [35] on finite-tap

---

[1] Chen *et al.* [32] have submitted a paper to *Signal Processing* on this subject.

linear equalizers. The channel model was Rummler's [9] multipath fading model which is given as

$$a(1 - be^{-j2\pi(f-f_0)\tau}) \tag{21}$$

This channel displays both a flat fading component (constant level of frequency attenuation over the whole channel) of depth $A = 20\,log(a)$ (dB) and a frequency selective component in the form of a notch located at f_0 MHz with a depth of $B = 20\,log\,(1-b)$ (dB). The constant τ is chosen as approximately 5 times the inverse of the channel bandwidth, e.g. Rummler [9] uses $6.3\,nS$ for a 30 MHz channel. In practice the flat fade or median fade component of the channel is compensated for by the automatic gain control (AGC) of the receiver and only serves to change the signal to noise ratio (SNR). For this reason a may be set to unity and one may concentrate on the notch characteristics of the channel, namely, the parameters B and f_0 at various signal to noise ratios.

For transmission, a baud rate of 22.5 MBaud was used for 4 and 16-QAM with a channel bandwidth of 30 MHz. This was achieved with square root raised cosine spectral pulse shaping with a roll off factor of 0.33. The signal to noise ratio was chosen as 63 dB, a typical value for such a channel. This may seem like an excessively high speed implementation for neural networks but it should be remembered that even though the neural networks were used in this explicit channel it is only the ratios (baud rate, bandwidth and SNR) that matter, e.g. this could just as easily represent 22.5 KBaud at 30 KHz. The time/frequency parameters merely serve to specify the links between various components of the transmission model.

The outage performance of the system was used to evaluate the performance of the equalization schemes [36]. An outage event occurs when the BER (bit error rate) exceeds a specified threshold value for not more than ten consecutive seconds. If the BER exceeds this threshold for more than ten consecutive seconds the data link is said to be unavailable. The outage threshold chosen here is a BER of 10^{-3}. Thus, outage performance is based upon the combinations of B and f_0 where the BER is 10^{-3}.

The performance of various networks is shown in Figures 6a, 6b and 6c (clockwise starting at the top left). These networks are based upon a 5 tap T-spaced delay line with a decision delay of 2 symbols, i.e. the reference tap is at the center of the delay line. Note that this is after the timing offset due to the channel has been compensated for. This was done as in the work of Amitay and Greenstein [35].

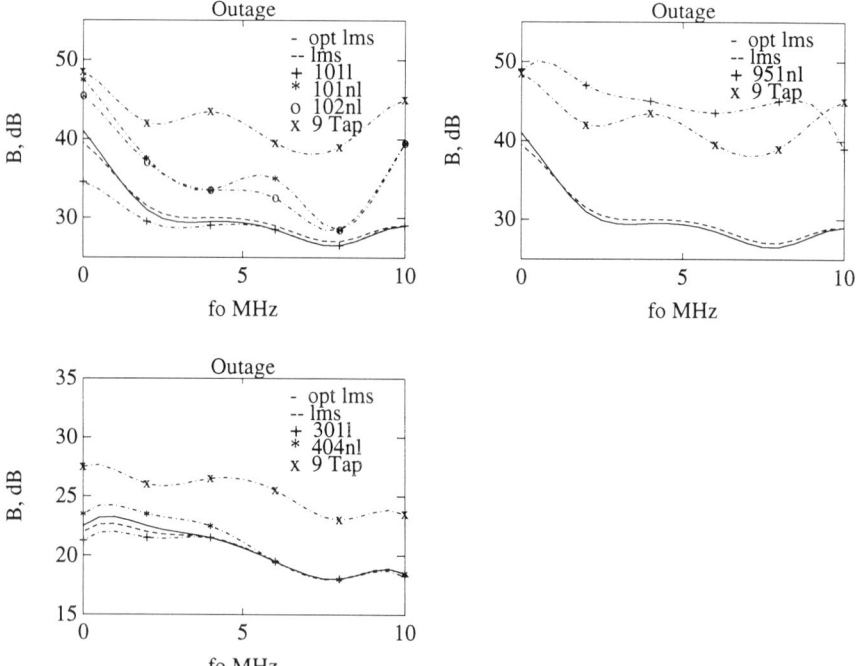

Figure 6 Outage performance for 4-QAM (top two graphs) and 16-QAM (bottom left graph). Each curve shows the notch depth, B (dB), as a function of notch frequency, f_0 (MHz), where the bit error rate for the receiver is 10^{-3}. (Notch frequency is measured relative to the center of the channel, 0 MHz). The outage region for a curve is represented by the area above the curve. Performance of the optimum LMS equalizer (opt lms) is shown by the solid curve while the performance of the simulated LMS equalizer (lms) is shown by the dashed curve. These results are accurate to ±1 dB.

The configuration of the neural networks is denoted by a sequence of three numbers followed by a suffix of nl or l. The three numbers correspond to the number of neurons in the first and second hidden layers and the output layer of the neural network. The suffixes l and nl designate the output of the network as having either a linear or nonlinear activation function. The optimum LMS equalizer as calculated in [35] is denoted as "Opt. LMS".

The results for 4-QAM yielded some surprising results. Although the 1 0 1l network performed similarly to the LTE, the 1 0 1nl network shows substantial outage improvement over the LTE and the 1 0 1l network. The only difference between the 1 0 1l and the 1 0 1nl networks is in the output activation function, one being linear, the other nonlinear ($tanh$). This might suggest that it is better to use a nonlinear output activation function, however, a 9 5 1l network was evaluated and yielded comparable performance to the 9 5 1nl network. Although the 9 5 1nl and 9 5 1l networks offer greatly improved outage performance for 4-QAM, the cost for this is in increased complexity and training times.

The performance of a 9 tap LTE is also shown. With the addition of four more taps the performance of the linear transversal equalizer is greatly improved though it is still less than that achieved with the 9 5 1 neural networks. This improvement is accomplished with far less complexity and training time than that of the neural networks. Amitay and Greenstein [35] indicate that the performance of a 9 or 11 tap equalizer is quite close to that of an equalizer with an infinite number of taps for this channel. Thus one would expect that there would not be much more improvement in outage performance through the use of more than 9 taps for an LTE. Hence, any improvement through the use of more taps is likely to be less than that achieved with the 9 5 1(n)l networks. The results for multi-layered Perceptron (MLP) equalizers with higher level signal constellations tell a different story. Figure 6c shows the results for 16-QAM. Here there is very little improvement in outage performance for the MLP equalizers. Larger networks were tried, i.e. 9 5 1l, 9 5 3nl, 9 5 4nl, and although their performance is not displayed they were found to yield no improvement over the networks depicted here. Nevertheless, note the performance improvement obtained with the 9 tap LTE over the 5 tap LTE for 16-QAM, approximately a 5 dB improvement in notch depth. This seems to indicate that it is better to increase the number of taps in a LTE rather than to consider neural networks for use with higher level signal sets. Note that further improvement for all of the systems may also be obtained with the use of a fractionally spaced tap delay line, e.g. taps spaced at intervals of $T/2$ rather than T.

Ungerboeck [5] offers an interesting comment on the application of nonlinear equalizers to higher level signal sets:

> The investigation in this paper was restricted to binary antipodal signals. One might think of a generalization to multilevel signals. However, it is felt that then nonlinear methods will lose much of their attraction, not only because of the complexity of such schemes but also because the intersymbol interference will more and more approach a Gaussian distribution as the number of levels is increased. In the latter case a linear equalizer will be optimum.

This is particularly true in channels where the data pulse extends over numerous symbol intervals (e.g. the raised cosine pulse) and may help to explain the performances differences between 4 and 16-QAM.

The frequency response of the networks under moderate fading conditions was evaluated. It was found that the response of the linear part of the first hidden layer was similar to that of the LTE. Any magnitude discrepancies between the two responses is likely due to the nonlinearities and the multiple layers of neurons used in the MLP equalizer. In trying to determine why the 1 0 1nl and 9 5 1nl networks exhibited the dramatic improvement in outage performance their frequency response (first hidden layer) was evaluated for a 40 dB notch at six channel locations equally spaced from $f_0 = 0$ MHz to $f_0 = 10$ MHz for comparison to the response of the LTE. The frequency responses of the LMS, optimum LMS, and the 1 0 1nl MLP equalizers is shown in Figure 7 for $f_0 = 0$ MHz. Most of these channel conditions correspond to outage conditions for the LTE but not for the neural network. It was found that there is a dimple (a slight notch) in the LTE frequency response at f_0 for notch locations of 0.0, 2.0 and 4.0 MHz. This contrasts sharply with the MLP equalizer response and the response of the channel. The channel's frequency response, which is not shown here, has no dimple in it so why should the response of the equalizer. Obviously the MLP equalizer provides for a better frequency interpolation for the 5 tap delay line than the LTE for 4-QAM. This behavior was not seen when 16-QAM was used. There the MLP equalizers were unable to provide a better interpolation of the frequency response than the LTE. Note that it was only at these deep notches that the LTE displayed this problem. The LTE did not display this problem around the outage region for 16-QAM as the outage region occurs at much smaller notch depths. At these notch depths the frequency response of the MLP equalizers and the LTE were quite similar, hence there was little or no performance improvement with the use of MLP equalizers.

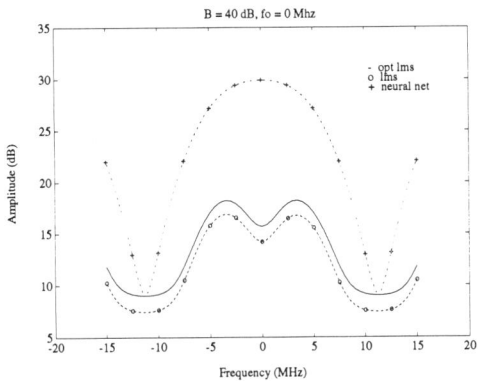

Figure 7 Frequency response of the optimum LMS LTE, LMS LTE and the linear part of the neuron in the first hidden layer of the 1 0 1nl MLP equalizer for B = 40 dB and f_0 = 0 MHz.

Similar work has been presented by Lo and Hafez [34] where they evaluated the performance of multi-layer perceptrons (MLP), conventional linear and decision-feedback equalizers in the presence of ISI, additive noise and co-channel interference (CCI). They used QPSK (quadrature phase shift keying, e.g. 4-PSK) signalling with a data rate of 48 $kbits/s$ over a 30 KHz channel. Pulse shaping was achieved with square root raised cosine spectral filters at the transmitter and receiver with 25% excess bandwidth, a roll off of 0.25. The transmission system was modelled in complex baseband with a $T/2$-spaced model. The channel was a three tap, $T/2$ spaced FIR filter corresponding to a fixed multipath delay profile with paths having relative powers of 0, -5 and -15 dB. The basic result was that the performance of the multi-layer perceptron based DFE matched that of the conventional DFE under all noise and interference conditions. At very low signal to interference ratios (BER > 10^{-2}) the MLP DFE showed a slight improvement of a fraction of a dB over the conventional DFE. These results contrast sharply with the results of Gibson and Cowan *et al.* [6]-[8] which indicate that MLP and RBF based DFE's offer superior performance over conventional linear and decision feedback equalizers. However, Lo and Hafez did not restrict themselves to minimizing the decision delay and size of the tap delay lines as in the case of the work of Gibson and Cowan *et al.*.

Lo concludes that the convergence rate of the MLP DFE is much slower than that of the LMS based DFE but with a sufficiently small adaptation step size, both of these equalizers are robust to decision-directed error propagation during data transmission. Lo indicates that the learning parameter should be sufficiently large during training to achieve convergence, while small enough during transmission so that the MLP DFE is robust to decision-directed error propagation. In order to meet these requirements he makes use of a gear-shifted BP algorithm to achieve much better performance than a network with a fixed learning parameter.

The training times and sensitivity of the MLP DFE to network configuration may be over-come by replacing the MLP structure with that of an RBF network, but as both RBF and MLP networks are capable of achieving the same functional mappings one would expect there to be little improvement in the bit error rate performance of the RBF DFE. It should be noted that in [34] a complex valued activation function of the form

$$f(s) = \left(\frac{1}{\sqrt{2}}\right) \frac{1 - e^{-s}}{1 + e^{-s}}$$

is used where the weight sum input, s, must be confined to a suitable part of the complex plane so that singularities are avoided. This function is not like those discussed in [31] and [30] and used in [33].

5 SUMMARY

At present the problems associated with neural network equalizers (computational complexity and long training times) has made them more of an academic interest than a practicality. Present work has shown their advantages over conventional equalization schemes but has not dealt with the practicalities of implementation. What has been shown is that when one is interested in minimizing the decision delay of the equalizer neural network structures offer superior performance over linear equalization schemes through their ability to achieve the functional mappings of the optimal Bayesian detector or estimator. However, much of the performance of the optimum detector/estimator may be achieved without the complexity of neural networks through the use of a linear transversal equalizer or decision feedback equalization by simply increasing the size of the tapped delay line(s) (number of taps) within reasonable limits and using a suitably chosen decision delay. This is particularly so when data pulse shapes are used that extend over numerous symbol intervals such as the raised cosine pulse. Such pulses give rise to a very large number of channel states, reducing the effectiveness of practical Bayesian detection/estimation systems over conventional equalization systems, especially when the number of signally levels is large. Neural network equalizers seem particularly well suited to binary signalling schemes, 2-PAM, 4-QAM and 4-PSK (binary in the in-phase and quadrature channels) or systems where the data pulse shape exists for only a single symbol period or a small number of periods. Such pulse shapes lead to significantly fewer and more defined channel states. Hence they are more suitable for Bayesian detection and thus the application of neural network equalizers. It is apparent that what really needs to be studied is the performance differences between linear equalization, decision feedback equalization, Bayesian detection/estimation and maximum likelihood sequence estimation in a realistic transmission model in order to accurately define the benefits of using Bayesian detection/estimation and hence neural network based equalizers.

In extending neural networks to the use of complex data, the use of complex neurons in MLP structures has allowed an interpretation of the weights of the first hidden layer in the frequency domain. This has shown that the frequency response of this section tends to have an inverse channel response that is characteristic of the response of a LTE. The exception is in the case of very deep notches when 4-QAM is used. Here the frequency response of the MLP has a better inverse channel response than the LTE. This suggests some possibilities for increasing the performance of the MLP such as preceding the MLP with a LTE or initializing the weights of the neurons in the first hidden layer to be those of a LTE (within an amplitude scaling factor) for a given channel. The latter method poses several problems. Such as: the channel response is generally unknown and thus so is its inverse, how to choose the bias levels for the neurons, what should the relative amplitude scaling factor be and how does one organize the other hidden layers and the output layer?

While neural networks offer performance comparable to Bayesian detection without the complexity they are still more complex and have significantly longer training times than conventional equalization schemes. Yet, they suffer from some other problems as well. The MLP is very sensitive to its adaptation parameters and network configuration. Network configuration for the MLP is still a relatively ad hoc process as is the case for RBF networks. Gibson and Cowan et al.[8] indicate that the centers for the basis functions should be chosen as the channel states for the tapped delay line but in practice these are never known. A further problem is that in many instances one is interested in blind equalization. That is, the equalizer must adapt without a training sequence or any knowledge of a desired signal other than the format of the signalling constellation. For the linear equalizer there are several algorithms [37]-[40] that offer good performance but none of them are applicable to neural networks due to the nonlinear nature of neural networks. At present no one has considered this problem nor the effects of the nonlinear nature of the networks on the performance of systems following the equalizer. This is particularly applicable in the case of coding where significant coding gain may be lost due to the quantization effects of neural networks.

Neural network structures will likely be significant in their application to nonlinear channels. In many cases nonlinearities in the transmission system limit the efficiency of the system. Neural networks with their associated nonlinear properties would seem to be likely candidates to achieve improved system performance. Such applications are in the linearization of power amplifiers and in the areas of magnetic recording media and fiber optics. In looking towards the future, neural networks may be of use in a combined coding-modulation-equalization or equalization-demodulation-decoding system. Ungerboeck has already shown the benefits of combined coding and modulation. It is only a natural step forward to add further systems to this. This will obviously lead to increased computational complexity and require more detailed theory, but neural networks may offer a desirable short cut.

REFERENCES

[1] S. U. H. Qureshi., "Adaptive equalization," Proc. IEEE, September 1985, pp. 1349–1387.

[2] W. D. Rummler, R. P. Coutts, and M. Liniger., "Multipath fading channel models for microwave digital radio," In Microwave Digital Radio, IEEE Press, 1988, pp. 67–77.

[3] R. W. Lucky, J.Salz, and E. J. Weldon Jr., "Principles of Data Communication," New York:McGraw-Hill, 1968.

[4] G. D. Forney Jr., "Maximum-likelihood sequence estimation of digital sequences in the presence of intersymbol interference," IEEE Trans. Inform. Theory, IT-18, May 1972, pp. 363–378.

[5] G. Ungerboeck., "Nonlinear equalization of binary signals in gaussian noise," IEEE Trans. Comm., COM-19(6), December 1971, pp. 1128–1137.

[6] G.J. Gibson and C.F.N. Cowan., "Multilayer perceptron structures applied to adaptive equalisers for data communications," IEEE Proceedings, ICASP, May 1989, pp. 1183–1186.

[7] S. Chen, G.J. Gibson, C.F.N. Cowan, and P.M. Grant., "Reconstruction of binary signals using an adaptive radial-basis-function equalizer," Signal Processing 22, 1991, pp. 77–93.

[8] S. Chen, P. M. Grant, and C. F. Cowan., "Orthogonal least-squares algorithm for training multioutput radial basis function networks," IEE Proceedings-F, 139(6), December 1992, pp. 378–384.

[9] W.D. Rummler, R.P. Coutts, and M. Liniger., "Multipath fading channel models for microwave digital radio," IEEE Communications Mag., 24(11), November 1986, pp. 30–42.

[10] G. Ungerboeck., "Trellis-coded modulation with redundant signal sets part i: Introduction," IEEE Comm. Mag., 25(2), February 1987, pp. 12–21.

[11] H. L. Van Trees., "Detection, Estimation, and Modulation Theory," John Wiley & Sons, 1968. Part I.

[12] S. Chen and B. Mulgrew., "Overcoming co-channel interference using an adaptive radial basis function equaliser," Signal Processing, 28, 1992, pp. 91–107.

[13] Michael D. Richard and Richard P. Lipmann., "Neural network classifiers estimate Bayesian a posteriori probabilities," Neural Computation, 3, 1991, pp. 461–483.

[14] F. Kanaya and S. Miyake., "Bayes statistical behavior and valid generalization of pattern classifying neural networks," IEEE Trans. on Neural Networks, 2, July 1991, pp. 471–475.

[15] E. Barnard., "Comments on "Bayes statistical behavior and valid generalization of pattern classifying neural networks"," IEEE Trans. on Neural Networks, 3(6), November 1992, pp. 1026–1027.

[16] F. Kanaya and S. Miyake., "Author's reply," IEEE Trans. on Neural Networks, 3(6), November 1992, pp. 1027–1028.

[17] R.P. Lippman., "An introduction to computing with neural nets," IEEE Acoust. Speech Signal Process. Mag., 4, 1987.

[18] D. E. Rumelhart, J. L. McClelland, and the PDP Research Group., "Parallel Distributed Processing, volume 1," MIT Press, 1986.

[19] S. Haykin., "Adaptive Filter Theory," Prentice-Hall, 1986.

[20] B. Widrow., "Adaptive Signal Processing," Prentice-Hall, 1985.

[21] S. Siu, G.J. Gibson, and C.F.N. Cowan., "Decision feedback equalisation using neural network structures and performance comparison with standard architecture," IEE Proceedings, 137(4), 1990, pp. 221–225.

[22] A. B. Carlos and John Park Jr., "Decision feedback equalization," Proceedings of the IEEE, 67(8), August 1979, pp. 1143–1156.

[23] S. Arcens, J. Cid-Sueiro, and A. R. Figueiras-Vidal., "Pao networks for data transmission equalization," IJCNN, June 1992, pp. II–963 – II–968.

[24] T. Kohonen, K. Raivio, O. Simula, O. Venta, and J. Henriksson., "Combining linear equalization and self-organizing adaptation in dynamic discrete-signal detection," In IJCNN 90, 1990, pp. 223–228.

[25] B. Widrow and R. Winter., "Neural nets for adaptive filtering and adaptive pattern recognition," Computer, 21(3), March 1988, pp. 25–39.

[26] B. Widrow and M. A. Lehr., "30 years of adaptive neural networks pp. perceptron, madaline and backpropagation," Proceedings of the IEEE, 78(9), September 1990, pp. 1415–1441.

[27] P.R. Hartmann and D.P. Taylor., "Telecommunications by microwave digital radio," IEEE Communications Mag., 24(8), August 1986, pp. 11–16.

[28] T. Noguchi, Y. Daido, and J.A. Nossek., "Modulation techniques for microwave digital radio," IEEE Communications Mag., 24(10), October 1986, pp. 21–30.

[29] B. Widrow, J. McCool, and M. Ball., "The complex lms algorithm," Proc. IEEE, 63(4), April 1975, pp. 719–720.

[30] N. Benvenuto and F. Piazza., "On the complex backpropagation algorithm," IEEE Trans. on Signal Processing, 40(4), April 1992, pp. 967–969.

[31] G. M. Georgiou and C. Koutsougeras., "Complex domain backpropagation," to appear in IEEE Trans. on Circuits and Systems.

[32] S. Chen, S. McLaughlin, and B. Mulgrew., "Complex-valued radial basis function network. part II: Adaptive Bayesian equalizer with decision feedback," Submitted to Signal Proc., 1992.

[33] W. R. Kirkland and D. P. Taylor., "The application of feed forward neural networks to channel equalization," In IJCNN, June 1992, pp. II 919–924.

[34] N. W. K. Lo and H. M. Hafez., "Neural network channel equalization," In IJCNN, June 1992, pp. II 981–986.

[35] N. Amitay and L.J. Greenstein., "Multipath outage performance of digital radio receivers using finite-tap adaptive equalizers," IEEE Trans. Comm., COM-32(5), August 1984, pp. 597–608.

[36] L. J. Greenstein and M. Shafi., "Outage calculation methods for microwave digital radio," IEEE Communications Mag., 25(2), February 1987, pp. 30–38.

[37] Y. Sato., "A method of self-recovering equalization for multilevel amplitude modulation systems," IEEE Trans. on Communications, COM-23, June 1975, pp. 679–682.

[38] A. Benveniste and M. Goursat., "Blind equalizers," IEEE Trans. on Communications, COM-32(8), August 1984, pp. 871–883.

[39] S. Bellini and F. Rocca., "Blind deconvolution: Polyspectra or bussgang techniques ?," In E. Biglieri and G. Prati, editors, Digital Communications, pp. 251–263. Elsevier Science, B. V. (North-Holland), 1986.

[40] O. Shalvi and E. Weinstein., "New criteria for blind deconvolution of nonminimum phase systems, (channels)," IEEE Trans. on Information Theory, 36(2), March 1990, pp. 312–320.

9

NEURAL NETWORKS AS EXCISERS FOR SPREAD SPECTRUM COMMUNICATION SYSTEMS

Richard Bijjani and Pankaj K. Das*

Center for Computer and Information Security, The Bear Group, Inc.

*Electrical, Computer and Systems Engineering Department
Rensselaer Polytechnic Institute*

1 INTRODUCTION

Spread-Spectrum Communications Systems have many applications; the most important being that of jamming suppression. The interference rejection capability of a system depends upon its processing gain, or the ratio of the transmission bandwidth to that of the data. The interference immunity would suffer if the power of the interfering signal is increased. Signal processing techniques that have the potential of improving the anti-jamming resistance without increasing the transmission bandwidth are normally employed. If the interference signal is narrowband, then notch filters are ordinarily used to improve performance. Three such methods for suppressing narrowband jammers in a direct-sequence spread spectrum receiver have been studied and successfully implemented.

1.1 Transform Domain Processing

The first notch filter is implemented in the transform domain. The basic concept of transform domain processing is to make a transform of the received signal, process this transform and then go back to the time domain [1, 2]. This is shown in Figure 1, where F and F^{-1} denote the transform and its inverse respectively, and where H(w) represents the excision processing transfer function. Sometimes it is not necessary to return to the time domain, since one could extract the desired information (e.g. make a bit decision) right in the transform domain. The advantage of filtering in the transform plane for some cases is obvious. For an example, consider a narrowband jammer or narrowband noise which will show up as a spike in the transform plane, if Fourier Transform is used. As the desired signal is wideband, it will be easy to discriminate between the two signals and, subsequently, remove the interference without distorting the signal too much. For many cases, it is quite possible to improve the signal to noise ratio as the clutter and the signal might have different characteristics in a particular transform domain. However, it is important to find the optimum transformation which can best characterize the difference between the desired signal and the unwanted noise, clutter, etc. For this, one might need to perform multiple transform domain processing. Note that, when using a digital receiver, this type of transform domain processing can be easily implemented, as the digital receiver can be built with an enormous amount of computing power inside it.

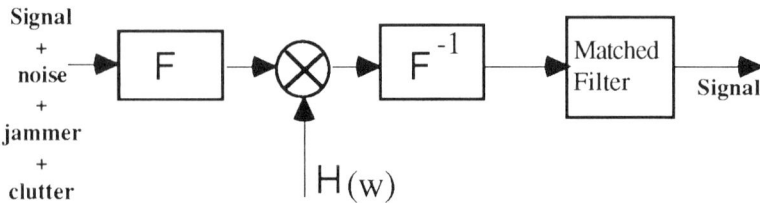

Figure 1 Simplified Transform Domain Processor

The most widely-used transform is the Fourier Transform, but there are many others of importance such as the Hadamard, Fresnel, Hartley, Mellin, and Hilbert Transforms, to name but a few. In communications and radar applications, particularly ones using spread spectrum techniques, transform domain processing can be utilized to suppress undesired interference and, consequently, improve performance. Here, the basic idea is to choose a transform such that the jammer or the undesired signal is nearly a delta function in the transform domain, while the desired signal is transformed to a waveform which is very "flat" or "orthogonal," with respect to the transformed interference. A simple exciser can then remove the interferer without removing a significant amount of desired signal energy. An inverse transform then produces the nearly interference-free desired signal.

Because of the importance of the Fourier Transformation and its ability to represent interference spectra and its discrete representations, most of the research and development related to the application of transform domain signal processing has been restricted to Fourier Transform. Most of the practical implementations use "short-time" Fourier Transforms which observe only a segment of the input signal to produce an estimate of the transform. While, the Fourier Transform uses infinite-duration sinusoids as the basis functions for the transformation, the "short-time" transforms, such as the Fast Fourier Transform (FFT) uses truncated sinusoids.

The transform domain processing can be extended to include Wavelets as the basis functions. Because the signals to be processed in radar and communication systems are both time and band limited, the main advantage of using wavelets as opposed to the standard Fourier basis functions is the reduction of sidelobes. These sidelobes are the result of the windowing of the input signal that is required to use the "short-time" transform. The reduction of these sidelobes can improve our ability to remove the interference without distorting the desired signal, with a resultant improvement in performance. For the application of encoding, the efficiency increases as the signals to be encoded match the wavelet basis functions more closely, resulting in data compression. Wavelet transforms are linear and square integrable transforms having a kernel (called the primitive or mother waveform) which is not fixed. Using this mother wavelet, daughter wavelets are formed by dilation and shift. These mother and daughter wavelets form a complete orthonormal set. The fundamental difference between the wavelet transform and other transforms is that any square integrable function can become the primitive or mother waveform. This ability to choose the function is vital to the success of the wavelet transform in separating desired and undesired components of a signal in a radar or communication system [3].

Frequency Transform Domain

To perform frequency transform domain processing, we require real time Fourier transformers. This could be accomplished by using a tapped delay line surface acoustic wave (SAW) device, that have a chirp impulse response built into the taps. An anti-jamming neural network capable of recognizing the presence of multiple narrowband interferers could be implemented to act as a real time notch filter, capable of excising multiple narrowband and swept-tone jammers. The input to the net will be the frequency domain signal $F(\omega)$, with the output being the jammer excised signal.

Wavelet Transform Processing

A typical wavelet transform exciser in block diagram form is shown in Figure 2. The choice of the wavelet basis functions is very important and for obvious reasons should be orthonormal.

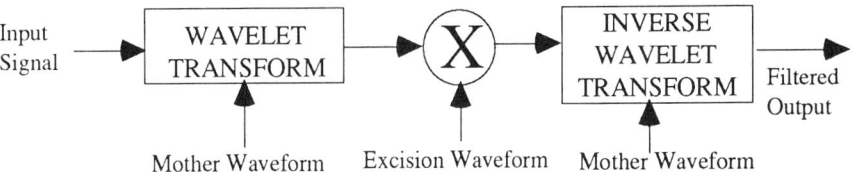

Figure 2 A wavelet Excision System.

Figure 3(a) shows the unit discrete wavelet transformer in detail. h(n) and g(n) are constants, coefficients of a low pass and a high pass digital filters respectively. This transformer is equivalent to the 2-point short term Fourier Transform analysis filter where the input signal is split into two frequency bins, where one corresponds to the lowpass elements and the other corresponds to the highpass elements. The output values are obtained using undersampling by a factor of 2 following the filters. Using the unit blocks repeatedly, one can obtain higher resolution frequency bins.

The inverse discrete wavelet transformer accomplished by reversing the above transformation is denoted as UIWT. For excision, some coefficients are removed where the jammers might be present. Actually, the analysis/synthesis filters can be modified as shown in Figure 4 to reduce the number of computations. If we estimate that the jammer is in the high frequency bin in a particular stage, we need not continue to decimate that bin as shown in Figure 4.

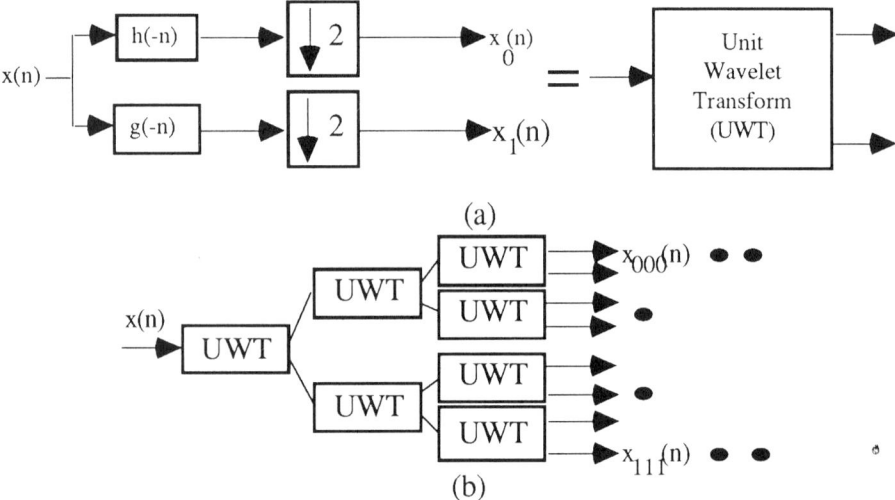

Figure 3 Analysis Filter Bank

A complete treatment of the wavelet transforms and of perfect reconstruction filters could be the subject for a whole other chapter. However, we mention in the following some salient points:

- Perfect reconstruction multirate digital filters consist of an analysis filter and synthesis filter. Any signal can be exactly reconstructed after passing through these filters. There are many possible solutions to the design of these filters. The one discussed in this section is a simple and elegant one. Physically it splits the frequency components of the signal into two halves. However it does this in such a way that there is no aliasing error as well as no phase or amplitude error.

- The realization of wavelet transformer discussed so far is computationally insufficient as redundant processing is performed. This is obvious from figure 4 as half the calculated values are not used for further processing. However, one can use polyphase implementation which is computationally as efficient as Fast Fourier Transform.

- Some numerical results on the performance of wavelet exciser are discussed in the literature [4].

1.2 Time-Space Domain Processing

The second method for implementing the notch filter is to utilize deconvolution to reduce signal distortion. It is well known that an adaptive transversal filter with adjustable weights can be used to implement the inverse of the channel response, making it possible to remove the channel distortion [5, 6]. This technique is also known as both a multipath equalizer and, also, as a deconvolver. The usual way of performing deconvolution is to pass a known training signal through the channel and adjust the weights of the tapped delay line to minimize a cost function such that the received signal approaches the undistorted signal [7]. There are a number of different algorithms available for this purpose, the most common of which is the LMS algorithm. However, it is also possible to perform what is known as blind deconvolution. In this case the training signal is not used and the cost function is minimized by using some known property of the transmitted signal. This technique makes it quite possible under certain circumstances to correct the variations of the signal for each pulse. The main problem, of course, is to choose the proper cost function and to achieve a global minimum without being hung up at a local minimum. In many instances this is possible. In place of using a one-dimensional delay element, a multilayer perceptron model including non-linear functions in the weight updating algorithm can be used. This represents an extension of adaptive filtering to so-called neural-network processing [8, 9]. Although more computationally intensive, neural networks have a number of advantages, which will be discussed in details in Section 2.

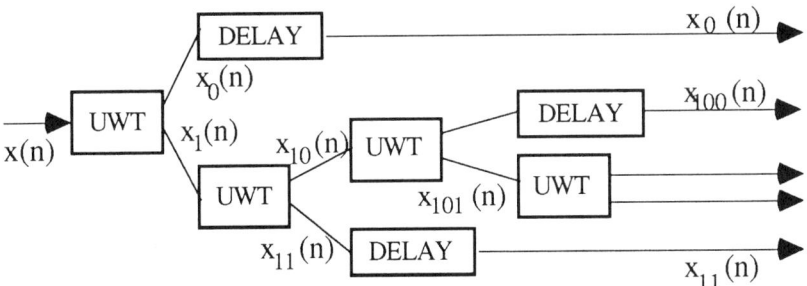

Figure 4 Reduced Computation wavelet analysis filter for excision of narrowband jammers.

Another such application combines time and space processing by using an adaptive phased array antenna for the detection of a broadband signal in the presence of multiple interference sources. In order to null an antenna pattern in the directions of narrowband jammers, the antenna outputs are multiplied by complex-valued weights which are computed adaptively. A neural network architecture has been introduced to steer the phased array antennas systems [10]. In Section 2.3, a quadrature phase 2 antenna element system will be presented. The case where only a single antenna is used, the system reduces to an LMS Widrow-Hoff adaptive filter.

Non-Linear Processing

Non-linear processing can be used for excision, both in the time and the frequency domains. Until recently, the subject of signal processing for communication and radar has been dominated by linear processing. Nonlinear processing, however, has been known for quite some time but did not obtain widespread use due to the fact that it requires an enormous amount of computing power, making it impossible to implement in real time. However, the breakthrough in signal processing chips such as *TMS320C40* has completely changed that picture. One particular aspect of nonlinear processing is what is known as "locally optimum" detection. The fundamental idea behind locally optimum detection is that if the undesired part of the received signal dominates the probability distribution function (pdf) of the received signal, then this pdf can be taken as an estimate of the pdf of the interference. Expanding it in a Taylor series, the second term will contain the derivative of pdf multiplied by the desired signal itself. The nonlinear processing of the signal consists of passing the noisy signal through a filter which uses this first derivative of the pdf to get rid of the dominant noise or clutter. We believe that this nonlinear processing is well suited to remove undesired components of the received signals. It is to be mentioned that neural network processing also includes non-linear processing and will be discussed later.

2 TIME DOMAIN PROCESSING

The adaptive transversal filter is a common tool used for broadband signal enhancement, and is the basis upon which the neural network application is built on. Figure 5 shows a block diagram of a spread spectrum receiver which uses this type of adaptive suppresser.

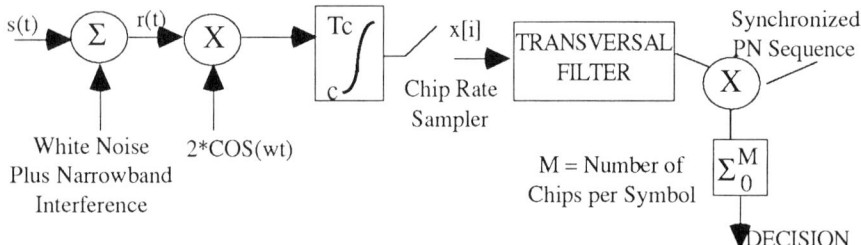

Figure 5 Spread spectrum receiver block diagram.

The received signal $r(t)$ consists of three additive components: the PN code, the wide-band noise and the narrow-band jammer: $r(t) = s(t) + I(t) + n(t)$ (1)

The transmitted signal $s(t)$ is a modulated wide band signal given by

$$s(t) = Ac(t)d(t)\cos(\omega_o t) \quad (2)$$

where A is a constant amplitude, ω_O is the carrier frequency, $d(t)$ is a binary data sequence taking on the equiprobable values of ±1 each of which lasts for T seconds and $c(t)$ is the spreading sequence, usually a Pseudorandom Noise (PN) code, which also takes the values of ±1 but which lasts for T_C seconds, where $T_c << T$.

The narrowband interfering signal $I(t)$ is given by:

$$I(t) = \alpha\cos\left[(\omega_o + \Omega)t + \theta\right] \quad (3)$$

where the phase θ is a uniformly distributed random variable in [0, 2Π], α is a constant amplitude, and Ω is the frequency offset of the interfering signal.

The noise $n(t)$ is an Additive White Gaussian Noise (AWGN) with zero mean and a two-sided power spectral density $N_0/2$.

The received signal $r(t)$ is first demodulated as shown in figure 5. The discrete time signal x_i ($x[i]$) entering the transversal filter is

$$x_i = \int_{(i-1)T_c}^{iT_c} 2r(t)\cos(\omega_o t)dt = d_i + V\cos(\Omega i T_c + \Phi) + n_i \quad (4)$$

where $\quad d_i = Ac[(i-1)T_c]d[(i-1)T_c]T_c = \pm AT_c \quad (5)$

with $c(t)$ and $d(t)$ taking the values ± 1.

$V = \alpha T_c Sinc\left(\Omega T_c / 2\right)$ where $Sinc(a) = Sin(a)/a$ and $\Phi = \theta - (\Omega T_c)/2$.

d_i and n_i in equation 4 are wide band signals, and since the noise is AWGN, they are poorly correlated. Therefore, when the transversal filter tries to estimate the next sample of the signal, it succeeds only in estimating the highly correlated jamming signal and consequently manages to suppress it. The output of the single-sided transversal filter y_i is given by

$$y_i = x_i - \sum_{j=1}^{L} a_j x_{i-j} \quad (6)$$

where L represents the number of the delay elements or the *dimension* of the filter. The output y_i is then correlated with the local PN reference code $c(t)$, and an estimate of the transmitted data $d(t)$ is obtained.

The weights a_j are adjusted according to the LMS algorithm to obtain minimum $E[y_i^2]$, or the minimum expected value of y_i^2. The optimum set of weights can be obtained by solving the Wiener-Hopf equation, which involves the calculation of a $L \times L$ correlation matrix and of its inverse. This requires a knowledge of the system's statistics (which might not be available) and a lot of calculations. In a practical system this is usually not feasible and an adaptive solution such as the Widrow-Hoff is applied. The LMS Widrow-Hoff algorithm is given by

$$a(k+1) = a(k) + 2\mu y_k x_k \quad (7)$$

where μ is a positive valued scalar and k is the discrete time.

In this chapter a new method based on the multi-layer back-propagation perceptron neural network model is presented. Its performance is reported and is compared to that of the Widrow-Hoff LMS single-side transversal filter in respect to the bit error probability improvement, notching capability, speed of convergence and robustness of the weights.

2.1 Multi-Layer Neural Network

The Neural Network approach [11] can be viewed as an extension of the estimation type filter. The first modification is to introduce a non-linear element in the transversal filter after the adder, as shown in Figure 6, where $f(\ )$ is a non-linear operation and θ is an appropriately defined threshold. A system with 10 taps $L = 10$ is considered, where the frequency offset comes at $\Omega T_c = 60^o$, $\sigma_n = 0$ and $J/S = 100$. The linear adaptive filter is allowed to run for a long time till the weights virtually attain their optimal values. The transfer function is shown in Figure 7 to have a deep notch at 60°. Figure 8 shows the effect of introducing the non-linearity. Here, the two systems are allowed to run for the same amount of time, and their transfer functions are computed. The transfer function of the partially trained non-linear system clearly shows a notch at 60° while the linear Widrow-Hoff does not. It appears therefore that the introduction of the non-linearity speeds up the convergence rate and improves performance.

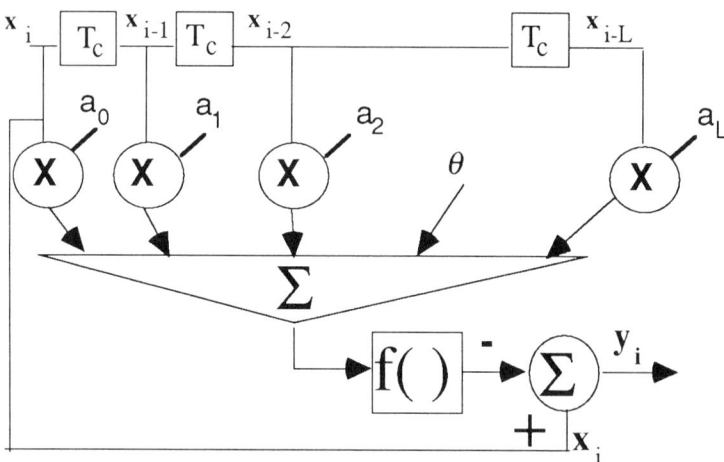

Figure 6 Transversal filter with non-linearity.

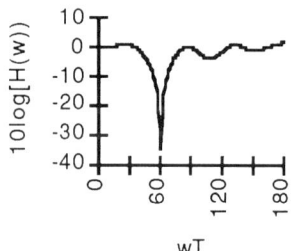

Figure 7. LMS Widrow-Hoff solution.

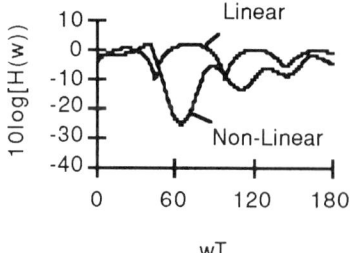

Figure 8 Perceptron versus LMS, after partial convergence.

The second modification is to introduce an additional layer of M adders as shown in Figure 9. The input x_{i-j} is multiplied by a weight factor $w_{j,k}$ before it feeds into the kth adder of the first layer. The output of the kth adder is then operated upon by a non-linear element with a threshold value θ_k, and the resulting output is denoted as q_k. This is then multiplied by a weight factor W'_k and fed into an accumulator whose output is operated on by the non-linear function with threshold θ'.

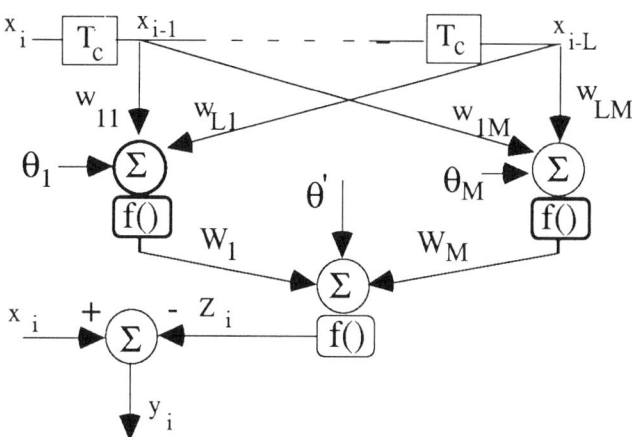

Figure 9 Two-layer non-linear adaptive filter.

Mathematically, we summarize the above by the following equations:

$$q_k = f\left(\sum_{j=1}^{L} w_{j,k} x_{i-j} + \theta_k\right), \text{where } k = 1, 2, ...M, \tag{8}$$

and
$$z_i = f\left(\sum_{k=1}^{M} W'_k q_k + \theta'\right) \tag{9}$$

$$y_i = x_i - z_i \tag{10}$$

If the non-linear element is replaced by a linear one (that is $f(z) = z$), then equations 8, 9 and 10 could be reduced to:

$$y_i = x_i - \sum_{j=1}^{L} a_j x_{i-j} \tag{11}$$

where
$$a_j = \sum_{k=1}^{M} W'_k w_{j,k} \tag{12}$$

Hence for linear systems, equation 11 becomes identical to equation 6. Note that for linear systems and since $E\{x_i\} = 0$, for all i, then the optimum thresholds θ_i are equal to zero.

However, if $E\{x_i\} \neq 0$ (such is the case if the input is scaled to be in the region [0,1] as is customary in Neural Networks), then the Linear non-homogeneous estimation provides an improvement upon the homogeneous case of equation 6, and the optimum values for the thresholds are usually non-zero.

In general we choose the non-linear function $f()$ to be a sigmoid function, where

$$f(z) = \frac{1}{1+e^{-\gamma z}}, \text{ where the gain, } \gamma \in \Re^+ \tag{13}$$

In a Neural Network, the system is usually presented with a set of training data, and the appropriate desired outputs. During the training period, the weights are updated using the back propagation algorithm until convergence is achieved. A SS system provides no training data, and therefore the actual data x_i is used for training and the weights are continuously updated. The system in Figure 9 depicts a two-layer network where the thresholds and the weights are periodically updated according to the following back-propagating algorithm.

$$W_k'(t+1) = W_k'(t) + \zeta \delta' q_k \tag{14}$$

where ζ is a constant learning parameter whose value determines the speed of convergence, and δ' is

$$\delta' = z_i(1-z_i)(x_i - z_i) = z_i(1-z_i)y_i \tag{15}$$

where x_i denotes the desired output while z_i is the actual output.

$$w_{j,k}(t+1) = w_{j,k}(t) + \zeta \delta_k x_{i-j} \tag{16}$$

where $\quad \delta_k = q_k(1-q_k)\delta' W_k'(t+1) \tag{17}$

For the adaptation of the threshold values, we use the same set of equations with the corresponding inputs set to one.

Now that the Neural Network algorithm has been presented, an additional measure of performance is introduced; that of *Robustness*. This is defined as the immunity of the system to the failure or corruption of some of its weights. The LMS error in the presence of corrupted weights could be used as a measure for the robustness of the system, where the error ρ is defined as: $\rho \equiv E\{\hat{y}_i^2\}$ \hfill (18)

and where $\hat{y}_i$ corresponds to the output of the system when presented with a corrupted set of weights $\hat{a}_j$. In the event where $\hat{a}_j = a_j$, then $\rho = 0$, and no statement can be made regarding the robustness. In general $\rho > 0$. A large value of ρ indicates a higher susceptibility to weight variations. We shall first consider the case where x_i is not orthogonal to all x_{i-j}. x_i can then be expressed as: $x_i = \sum_j a_j x_{i-j}$. Let the corrupted weights be expressed as

$$\hat{a}_j = a_j(1+e_j) \tag{19}$$

where e_j is a Gaussian noise with zero mean and variance equal to σ_e^2. For the ensuing analysis to hold true, the noise has to have a zero mean and be independent and identically distributed, but can otherwise have any distribution.

First we find ρ_{LMS}, the LMS error for the transversal filter of equation 6.

$$\hat{y}_i = x_i - \sum_j a_j(1+e_j)x_{i-j} = -\sum_j a_j e_j x_{i-j} \tag{20}$$

and after some algebra we determine $\rho_{LMS} = \sigma_e^2 \sum_j a_j^2 x_{i-j}^2$. $\tag{21}$

For a two-layer linear network, it can be shown that

$$(a_j - \hat{a}_j) = -\sum_k W'_k w_{j,k}(e_k + e_{j,k} + e_k e_{j,k}) \tag{22}$$

It follows that the factor ρ_2 for this system is

$$\rho_2 = \sigma_e^2(2+\sigma_e^2)\sum_k\sum_j(W_k^2 w_{j,k}^2 x_{i-j}^2) + \sigma_e^2\sum_k\sum_l\sum_{j \neq l}(W_k^2 w_{j,k} w_{l,k} x_{i-j} x_{i-l}) \tag{23}$$

One possible solution is to set all W_k to $+1$, all the cross-terms $w_{j,k}$, $j \neq k$ to zero, all $w_{i,j}$ to a_j and $M = L$. Then $\rho_2 = (2+\sigma_e^2)\rho_{LMS}$. The preceding argument can be applied to all linear systems, hence multi-layer linear networks are less robust. These systems also do not offer any advantages in terms of performance (equation 11). Therefore the use of a multi-layer linear network is *not recommended*.

The quantitative study of a non-linear system is substantially more involved. We shall defer any such study to a later date, concentrating for the time being on a qualitative interpretation of some computer simulation results. In the following section, we shall present some computer simulation outcomes and attempt to describe the results graphically.

2.2 Results and Analysis

The perceptron model was presented as a possible substitute for the LMS Wiener filter. The introduction of the non-linearity is expected to have a positive effect on the performance of the system. To demonstrate proof of concept, we selected a non-linear one-layer delay line as in Figure 6, and compared its performance to that of the traditional linear estimation filter. Randomly, we selected the following parameters for our test. The number of delay lines is 10,

$L = 10$. The interfering frequency is at $\Pi/3$ or $\Omega T_C = 60^\circ$. The jammer to signal power ratio (J/S) is equal to 100. No noise was introduced initially. The parameters remain unchanged for subsequent simulations.

The simulations were carried out on an IBM-PC-486, using a proprietary software package. The back-propagation algorithm as given by equations 14 to 17, was implemented. The differentiable sigmoid function of equation 13, was used for the non-linear function. The weights were originally set to small random variables. The reference signal used was x_i scaled down to [0,1], while its 10 delayed versions were used as the inputs to the network. The PN code selected was a 31-bit code and $d(t)$ was a random binary sequence lasting $T = 31T_C$ seconds. Both the linear and the sigmoid systems were allowed to run on only 500 such samples. After the training was discontinued, the frequency response $H(\omega)$ of both systems was calculated. $H(\omega)$ was computed by setting the input $x_i = e^{ji\omega T}$, where $j = \sqrt{-1}$ and ωT spans the frequency range 0°-180°. The output y_i was then computed according to equations 6, 8, 9 and 10, to be $y_i = H(\omega T)e^{ji\omega T}$. Figure 8 displays the plot of $10logH(\omega T)$ versus ωT for the one layer case. The plot for the sigmoid function reveals a deep notch at the jamming frequency of 60°. The linear system shows some notching capability but apparently did not have sufficient time (or samples) to generate a deeper notch. Note that if the linear system is operated indefinitely, then the transfer function looks like Figure 7, which represents the optimum performance of the filter under the constraint that the weights are expressed to only three decimal places. Therefore we conclude that the sigmoid function improves the performance of a partially trained system ($\approx 20dB$ in notching power) and offers faster convergence.

Next, we investigate the effect of adding a hidden layer of neurons. It has already been shown in section 2.1, that for the linear case, the performance will not improve with added layers, but the robustness of the system will be decreased. It was also argued that the robustness of a non-linear system will also be decreased (because of the propagation of error). Therefore we eliminated the multi-layer linear network from consideration. As for the sigmoid system, we shall weigh the decrease in robustness versus any potential increase in performance. The system's performance improves substantially upon the addition of an extra layer of hidden elements. Using the same parameters as before, an additional layer of 30 neurons is introduced ($M = 30$ in Figure 9). Again the weights are initialized to small random variables, and the system was trained on the same 500 data vectors and in the same order as before. The result is shown in Figure 11. The resulting notch is substantially narrower and $\approx 10dB$ deeper than that of Figure 8, and is comparable to the *ideal* Widrow-Hoff notch of Figure 7. Therefore an arrangement such as that in Figure 9 has a definite advantage over that in Figure 6.

We also evaluated the robustness of the one and two-layer non-linear systems. The weights were calculated to the third decimal place. The accuracy was reduced to 2 and then to 1 decimal place. Then the weights were substituted for, by a binary value of ±1. Having binary weights is extremely well suited for electronic and optical implementations, and would offer distinct advantages over the analog systems currently in use. For the 1-layer network we reduced accuracy of the weights translated into virtually no change on the frequency response, see Figure 10. The clipping of the weights into binary values, changes the frequency response of the system as shown in Figure 10. Varying the initial set of random weights, the response to the binary weights, remained within ±6dB, always showing a deep notch at or around the interfering frequency.

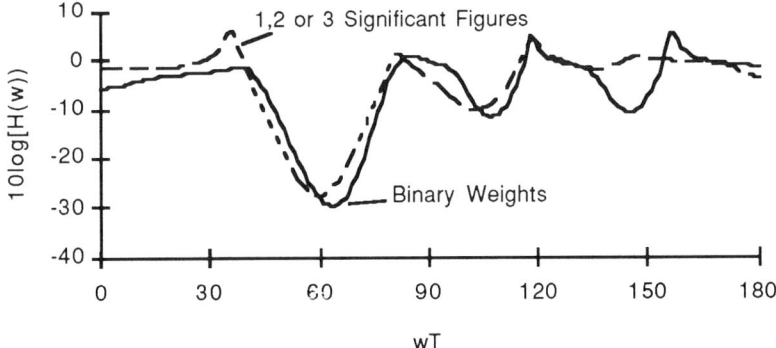

Figure 10 One-layer perceptron versus weight accuracy.

The 2-layer network however, suffered tangible degradation due to the weights inaccuracies. As shown in Figure 12, the *10 dB* advantage was eliminated after the weights dropped to 1 decimal place. The notch remained narrow, still showing an advantage for the 2-layer networks. The narrower notch might translate into lower error probability. The binary weights, on the other hand, failed miserably in showing any knowledge of the presence of a jammer.

One area of research could be to experiment with different values for the binary weights, other than ±1. The increasing number of operations might warrant a change in the values. The performance of the 2-layer sigmoid with binary weights might competitively improve if the number of inputs L increases. Another area of research is to study the robustness quantitatively by applying equation 18.

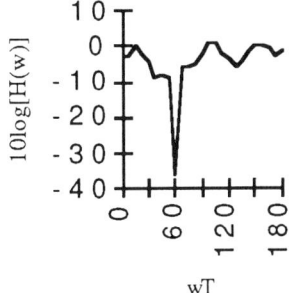
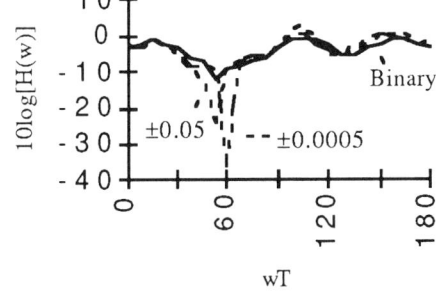

Figure 11 Two-layer perceptron 30 hidden neurons.

Figure 12 Two-layer perceptron versus weight with accuracy.

The error probability was experimentally computed. First the network was trained by running a few hundred data vectors until the weights converged. The signal corrupted with white noise and narrowband interference, was then continuously fed into the system while the weights continued being updated. The output of the filter was then multiplied by a synchronous spreading PN sequence and a decision was made as to the data sequence. Keeping count of the number of errors, we approximated the error probability as, P(e) = N(e)/N(R), where N(e) represented the number of errors detected and N(r) the total number of runs.

A 7-chip PN code and an 8-tap delay line were used for calculating the bit error probability. The results showed that after the system converged, the neural network provided no improvement upon the LMS Widrow-Hoff system. However, as it was noted earlier, the rate of convergence of the system improved with the introduction of the non-linearity. The plot of the mean-squared error versus time in Figure 13, shows that the neural network requires substantially fewer iterations than the LMS system in order to attain the same minimum energy state. Figure 13 also shows that the convergence rate increases as the gain factor γ is increased from 4 to 40. The higher convergence rate appears to be an important factor in the case where the system is presented with a fast moving swept-tone jammer.

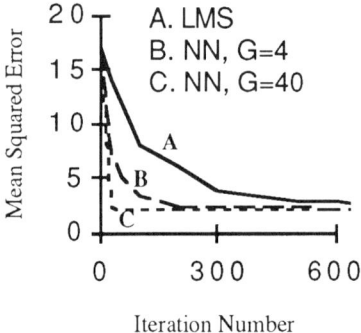

Figure 13 Convergence rate: Mean squared error versus time.

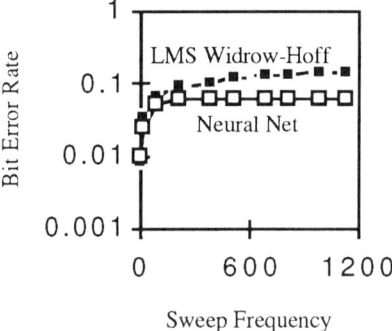

Figure 14 Bit error rate for a swept-tone jammer.

Preliminary results are shown in Figure 14, where the bit error rate was plotted versus the sweep frequency for both the LMS and the neural net. Here, a 7-bit PN code, a carrier frequency of 300 MHz and a tap delay equal to 150 nsec. were considered. The neural network clearly outperforms the LMS Widrow-Hoff.

2.3 Adaptive Phased Array Antennas

Jammer excision could be also achieved by using an array of M antennas, instead of the single antenna as was the case in the preceding section. This will add the extra dimension of space to the time domain processing. The individual antenna elements are spatially separated by a distance d, and are co-linear. The received signal is incident to the plane normal at an angle θ. The phase difference Φ between any two adjacent elements is:

$$\Phi = 2\Pi\left(d/\lambda\right)\sin(\theta) \qquad (24)$$

where λ is the wavelength of the received signal $r(t)$. In fact the received signal at each antenna element is identical except for the phase difference component and the corresponding time delay τ.

$$\tau = \left(\frac{d}{v}\right)\sin(\theta) \qquad (25)$$

where v is the velocity of the received wave.

A two element system, shown in Figure 15, was chosen in order to demonstrate the neural-network processing. Please note that in order to null or excise N narrowband jammers, we will require N+1 antenna array elements. In this case a 2 element system can null only one jammer.

In Figure 15 the weights W_1 and W_2 are complex valued and can be calculated to null the broadband or narrowband jammer effect. In the case of two elements, the received signals $r_1(t)$ and $r_2(t)$ at antennas 1 and 2, respectively, can be individually processed using the LMS algorithm. The outputs of the adaptive filter are added to produce the phase array system output, as shown in Figure 16. The two sets of weights a_i^1 and $a_j^2, \forall i,j = 1,2,..L$, are calculated in a similar fashion as in equation 7.

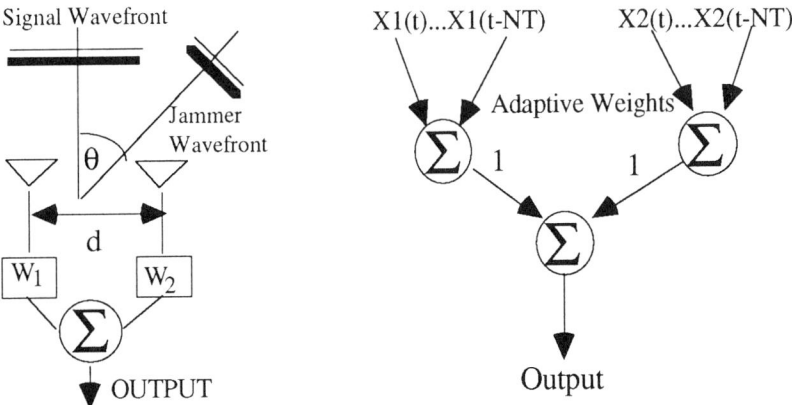

Figure 15 Two-element adaptive antenna system.

Figure 16 Adaptive Two Antenna System using the LMS Algorithm.

Just as we expanded the LMS into a neural-net for the one antenna case, we do the same here. In addition to having non-linear processing for each of the antennas, we combine the individual outputs into a non-linear weighted sum as shown in Figure 17.

In conventional antenna systems, both the signals and the weights have complex values. Whereas it is feasible to modify the learning algorithm (equations 14 to 17) to handle complex numbers, it is simpler to envision a quadrature phase shifter at each antenna output. This will cause all signals and weights to be real, and the two-antenna system will be similar to having four different signals, each being separately processed as in section 2.1. The output of all four systems will be combined into a weighted non-linear summation process, as in figure 17.

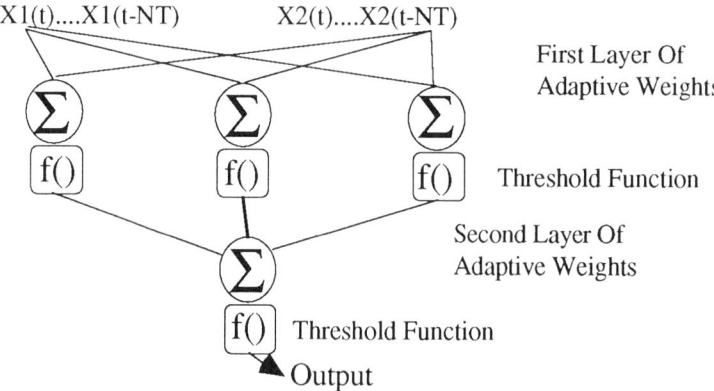

Figure 17 Adaptive Two Antenna System using the Multi-Layer Neural-Networks.

An adaptive Neural-Network processor offers additional degrees of freedom in the adaptation process and should provide improved performance over the conventional LMS method. We have presented in a previous section how a neural-network processor could improve the bit error rate for a DS-SS communication system which is subjected to narrowband jamming. We feel that the phased array antenna receiver employing neural-network processing techniques could excise both the narrowband and the wideband interferences. The trade-off for this increased immunity, is of course the added complexity.

3 CONCLUSIONS

The general subject of excision of narrowband jammer for a spread spectrum communication receiver has been discussed. For the time domain processing, we considered transform domain processing, adaptive processing and neural network processing. The space domain processing included use of neural networks in adaptive phased array antennas. The topic of wavelet transform excision, a subset of transform domain processing has been introduced.

Some BER results using numerical simulation have been presented. However it does not include a thorough analysis on different noise due to the error terms such as misadjustment (of weights) error, coefficient quantization error and Stalling error which arises when the coefficient updates becomes less then 1/2 least significant bit.

For the case of gradient perceptron model, there is the important question of achieving global minimum rather than local minima in some cases. However, at least for this case one can use the simulated annealing methods to avoid it, although at a heavy cost of computational burden and complexity. Even a few years back this would have been unthinkable. But as we all know, computing is cheap and fast and it is getting cheaper and faster. There is no reason not to put an INTEL pentium chip (or one of its successors) in the communication receiver of the 21st century.

REFERENCES

[1] Gevargiz, J., P.K. Das and L.B. Milstein, "Performance Of A Transform Domain Processing Ds Intercept Receiver In The Presence Of Finite Bandwidth Interference," IEEE Global Telecommunications Conference, December 1986.

[2] Das, P.K. and C. DeCusatis, Acouto-Optic Signal Processing, Artech House, 1991.

[3] Rioul, O. and G. Gitterly, "Wavelets and Signal Processing," IEEE Signal Processing Magazine, 1991.

[4] Das, P.K., M.J. Medley and G.J. Saulnier, "Jamming Excision of Spread Spectrum Signals Using Wavelet Transforms," to be published.

[5] Saulnier, G and P.K. Das, "Suppression Of Narrowband Interference In A PN Spread Spectrum Receiver Using A CTD-Based Adaptive Filter," IEEE Trans. Comm., vol. COM-32, November 1984.

[6] Milstein, L.B., "Interference Rejection Techniques in Spread Spectrum Communications," Proc. IEEE, vol.76, no.6, June 1988.

[7] Higbie, J.H., "Adaptive Suppression of Interference," Proc. MilCom, 1988.

[8] Lippman, R.P., "An Introduction to Computing with Neural Nets," IEEE ASSP Magazine, April 1987.

[9] Grant, P.M. and J.P Sage, "A Comparison of Neural Network and Matched Filter Processing for Detecting Lines in Images," AIP Conference Proceedings, Snowbird Utah, 1986

[10] DeCusatis, C. and P.K. Das, "Optical Controller for Adaptive Phased Array Antennas Using Neural Network Architecture," Proc. SPIE, vol 1217, January 1990.

[11] Bijjani, R. and P.K. Das, "Rejection of Narrow-Band Interference in a Spread Spectrum System, using Neural Networks," Proc. IEEE Global Telecommunications Conference, December 1990.

10

STATIC AND DYNAMIC CHANNEL ASSIGNMENT USING SIMULATED ANNEALING

Manuel Duque-Antón, Dietmar Kunz
and Bernd Rüber

Philips GmbH Research Laboratories

1 PROBLEM DESCRIPTION

When planning a radio network, the operator has to assign frequencies, or more general channels, to base stations in, such a way that both, the call quality and the channel availability are good enough. This means that for an incoming call request the probability that an idle channel can be found is sufficiently high, and the probability that the signal-to-interference-ratio S/I falls short of a predefined value is sufficiently low.

Calculating the RF propagation from topography and morphostructure and using it together with the spatial density of the expected traffic, leads to compatibility constraints stating which base stations may use the same channel or adjacent channels. The traffic can also be used to calculate for each base station the number of required channels. Furthermore, technical or legal restrictions must also be taken into account. For example, parts of the allocated frequency band have to be blocked at national boundaries.

1.1 Definition

Next, we give a formal definition of the basic channel assignment problem. For the carrier frequencies to be assigned we assume that they are evenly spaced and therefore can be written in the form $f = f_0 + k \cdot \Delta f$ with $k > 0$ integer. Since this implies a one-to-one correspondence between f and k, we will identify f with k in the sequel. We assume to have a mobile radio network of n radio cells, each of them capable to carry any of the m channels that are available for the whole system. Usually, m is given by the available radio spectrum which is reserved for building up the network. Furthermore, we assume that the channel demand vector and the interference compatibility matrix are known. Usually, these data are calculated using standard planning tools, like e.g. GRAND [1]:

- For each radio cell j a non-negative channel demand t_j is calculated which gives the number of required radio channels in j,
- For each pair j, j' of radio cells a minimum allowable channel separation $c_{jj'}$ is calculated which is required for compatible operation in j and j'.

Essentially, the *channel assignment problem* deals with assigning channels to fulfill the traffic demand without violating any interference condition, thereby using only the available channels. Formally, this could be defined as follows:

Definition 10.1 *Given a channel assignment problem $P = (J, I, T, C)$ where $J = \{1, ..., n\}$ defines the number of cells, $I = \{1, ..., m\}$ defines the number of available channels, $T \in \{1, ..., m\}^n$ defines the traffic demand vector, and $C \in \{0, ..., m\}^{n \times n}$ defines the compatibility matrix. Then, a channel assignment $F = (F_j)_{j=1}^n$ is called admissible, if the following conditions are satisfied:*

- *Each $F_j \subset I$ contains exactly t_j channels.*
- *If $i \in F_j$ and $i' \in F_{j'}$, then $|i - i'| \geq c_{jj'}$.*

It follows that there is no guarantee to find admissible channel assignments.

1.2 Radio Input Data

So far we have assumed that the traffic demand vector and compatibility matrix are given. This is true for the macro-cellular networks which have been used up to now. In this case, these data are estimated during the planning phase. It is obvious that the quality of the found channel assignments depend mainly on the estimated values. As these values are highly uncertain, it would be better to consider gradual differences. If, in particular, an admissible channel assignment fulfilling all conditions could not be found, it should be possible to ignore those interference constraints that are less severe.

Besides the basic requirements, interference and traffic, in practice there often evolve a whole series of additional engineering wishes. They may stem from the task of extending an existing network, where the operator wants to preserve as many of the already installed channels as possible. On the other hand, building up a new network, one may want to leave some liberty of action for future traffic changes. In order to be prepared for future traffic growth one may wish to estimate the number of not needed channels within the allocated frequency band. This, in particular, necessitates the ability to compute the minimal number of consecutively assigned channels. A quite common task in any case is the need to trade-off all these design objectives against each other while keeping their individual importance in mind.

Finally, it is expected that the number of subscribers will increase dramatically, whereas the radio frequency resources will remain nearly constant. The only way to handle this is by means of micro cells. But, in this case, the required degree of detail for a proper description of traffic and interference may not be predictable with sufficient accuracy. Consequently, the static channel assignment has to be replaced by a dynamic one which observes the radio data during system operation.

2 EXISTING APPROACHES

The classical solution of the frequency assignment problem in cellular radio systems has been given by the theory of regular hexagonal networks, where each frequency is re-used in a doubly periodic fashion [2, 3]. Several variants of this theory exist, among which those using sectorisation [4, 5] and re-use partitioning [6] deserve special mention. But usually, propagation conditions and traffic density are very irregular, so that suitable solutions can not be expected from the hexagonal theory.

On the other hand, [3, 7] showed the close analogy between channel assignment and graph coloring. Every channel assignment problem P could be mainly represented by an interference graph. For that, each radio cell, respectively each channel requirement determine a node within the interference graph and each interference relation determine an edge between two nodes. Then the task is to find a coloring of the nodes such that adjacent nodes are colored differently. Unfortunately, graph theoretic approaches [1, 8, 9, 10, 11] do not allow the consideration of gradual differences, and cannot account for a trade-off between different requirements. For example, graph theoretic approaches need as input hard interference decision indicating whether the use of the same channel by two radio cells is permitted or not. But if a channel assignment fulfilling all interference constraints cannot be found, it would be appropriate to violate those interferences first that are less severe.

The reformulation of the channel assignment problem as a cost minimization problem allows the design of flexible objective functions. Having defined the specific cost function for the CAP, a variety of discrete optimization methods could be applied. The neural network approach of Hopfield and Tank [12, 13, 14] was shown to be inappropriate for the channel assignment problem [15], as it yields bad solutions, even in simple cases: it favors sub-optimum channel assignments. However, under some restricted conditions does the Hopfield and Tank approach yield good solutions [16, 17]. Simulated annealing is a generally applicable method for solving cost optimization problems. Furthermore, it gives enough liberty of action to consider problem-specific heuristics. In particular, it can incorporate existing graph-coloring heuristics.

3 APPLYING SIMULATED ANNEALING TO THE STATIC CAP

3.1 Simulated Annealing

Simulated annealing, a general method for the approximate solution of difficult (i. e. NP-complete) combinatorial optimization problems, was originally proposed by Kirkpatrick et al. [18] and Černy [19]. It can be viewed as a simulation of the physical annealing processes found in nature, the settling of a solid to its ground state (the state with minimum energy). Theoretical studies of the algorithm have shown that a global optimum of the optimization problem can be reached with probability one, provided a set of conditions on the annealing schedule is satisfied, see for example [20].

Given a combinatorial optimization problem specified by a finite set S of configurations or solutions and by a cost function C defined on all solutions $s \in S$, the simulated annealing algorithm is characterized by a rule to generate randomly a new solution with a certain probability, and by a random acceptance rule according to which the new solution will be accepted or rejected. More precisely, the generation mechanism defines a neighborhood structure N on S. $N(s)$ determines for each solution s a set of possible transitions which can be proposed by s. Depending on a control parameter (the actual temperature), the acceptance rule occasionally allows "uphill moves" to solutions of higher cost.

Hence, simulated annealing offers the possibility to escape from local minima: a solution s' such that $C(s') \leq C(s) \; \forall s \in N(s')$.

In order to speed up simulated annealing, the annealing process has to be controlled carefully. The annealing process consists of an initial temperature, the equilibrium condition, the temperature decrement rule and the stopping criterion. The following pseudo-code program describes a general approach for a simulated annealing procedure:

>**Simulated Annealing** (s_{start}, T_0):
> $T := T_0$;
> $s := s_{start}$;
>   *repeat*
>   *repeat*
> $s' := $ **generate** (s)
>   *if* $C(s) \leq C(s')$ *then* $s := s'$
>   *else*
>   *if* $e^{\frac{C(s)-C(s')}{T}} > random[0,1)$ *then* $s := s'$
>   *until* **equilibrium condition valid**
> $T := $ **update** (T)
>   *until* **stop criterion valid**.

Initially, one starts with an arbitrary configuration s_{start}. The corresponding initial temperature has to be chosen high enough in order to guarantee that most of the proposed transitions are accepted. Hence, at the start of the algorithm, a free search in configuration space is intended. As T decreases, ever fewer proposed transitions are accepted, and finally, at very small values of T, now being in a local minimum, no proposed transition is accepted at all. Having arrived at this point, there is no use in further continuing the annealing process. The algorithm may stop if no further substantial improvements in cost can be expected. Thus, the annealing process converges to a final configuration s_{final}, which could be interpreted as a solution of the discrete optimization problem.

3.2 Application

In order to apply simulated annealing to CAP, we have to formulate CAP as a discrete optimization problem. For that reason we have to define the corresponding discrete configuration space S, the cost function C and the neighborhood structure N. This could be done in a more or less straightforward manner.

Configuration Space

According to section 1.1 we assume to have a mobile radio network of n radio cells, each of them capable to carry any of the m channels that are available for the whole system. In a configuration one wishes to represent the current state of the channel assignment. So, a natural choice is given by a binary matrix (s_{ij}) of dimension $m \times n$ with the following interpretation of the solution entries:

$$s_{ij} = \begin{cases} 0 \\ 1 \end{cases} \text{if channel } i \text{ is } \begin{cases} \text{not used} \\ \text{used} \end{cases} \text{at radio cell } j. \qquad (1)$$

Cost Function

Basic requirements for a mobile radio system are the avoidance of interference and the ability to serve the expected traffic, see section 1.1. Thus, a generic choice for a cost function is

$$C(s) = \frac{1}{2} A \cdot \sum_{\substack{(i,j),(i',j') \\ (i,j) \neq (i',j') \\ |i-i'| \leq c_{jj'}}} s_{ij} \cdot s_{i'j'} + \frac{1}{2} B \cdot \sum_j \left(\sum_i s_{ij} - t_j \right)^2, \qquad (2)$$

which penalizes the violation of each such constraint in the following way: The first term becomes positive if two interfering cells j and j' are assigned two channels i and i' within their interference bandwidth of $c_{jj'}$, whereas the second term penalizes traffic violations. It becomes positive, if the number of channels momentarily employed at cell j (which is $\sum_i s_{ij}$) differs from its demand t_j. Thus, $C(s)$ reaches its minimum of zero if all constraints are satisfied.

If a channel assignment satisfying all interference relations cannot be found, their violation should be ranked according to their individual importance. An appropriate modification of the interference term in the cost function is given

by
$$C_{int} = \frac{1}{2}A \cdot \sum_{\substack{(i,j),(i',j') \\ (i,j) \neq (i',j')}} \tilde{c}_{jj'ii'} \cdot s_{ij} \cdot s_{i'j'}, \tag{3}$$

where $\tilde{c}_{jj'ii'}$ models the interference strength between channels i and i' at cells j and j'.

In extending existing networks it may be useful to preserve the already installed radio channels in order to avoid hardware replacements. For that, the following preset term could be added to the cost function which punishes preset violation:

$$C_{pres} = \frac{1}{2}C \cdot \sum_j \sum_i p_{ij}(1 - s_{ij}), \tag{4}$$

where the strength p_{ij} determines the weight with which preset channel i is demanded at base station j.

Optionally, the part of the frequency spectrum that is needed should be minimized. This means minimizing the total number of assigned channels. In this case we need an upper bound m for the number of required channels. Thereby, m depends mainly on the maximal degree of the corresponding interference graph: i.e. $m \approx max_{j \in J} |\{j' \in J | c_{jj'} > 0\}|$, see [13]. The following term minimizes the total number of used channels:

$$C_{min} = \frac{1}{2}D \cdot \sum_i \sum_{i'} \frac{|i - i'|}{m} s_i s_{i'}$$

where $\quad s_i = \begin{cases} 1 & \text{if channel } i \text{ is used at some cell } j \\ 0 & \text{otherwise} \end{cases}. \tag{5}$

Neighborhood Structure

Simple choices for the neighborhood of a configuration s are produced by performing the following transitions:

- a *single flip*, i. e. just switching on or off one channel i in one cell j or
- a *flip-flop*, i. e. replacing at cell j one used channel i_1 with one unused i_2.

We use a generation probability that proposes new configurations equally in $N(s)$.

Obviously, the flip-flop is designed to preserve the number of channels used at each base station. Consequently, the configuration space has to be restricted to channel assignments with the required channel numbers, and the traffic term in the cost function (second term in equation (2)) can be omitted.

3.3 Cooling Schedule

Now, the hard work of applying simulated annealing to a definite problem consists mostly in the fine tuning of the algorithm's components in order to optimize run time efficiency. As for the cooling schedule, this work can be done in a consistent way for a substantial class of problems. The results in this paper have been obtained by implementation of a mixture of different cooling schedules which show a polynomial-time approximation behavior [20, 21, 22].

3.4 Heuristic Neighborhood Structures

This section describes the solution generation mechanism of the simulated annealing. The generation mechanism allows the consideration of constructive elements. This could be done by optimizing the neighborhood structure with respect to the definite problem, in conjunction with the configuration space and cost function. In the case of the channel assignment task this gives us enough liberty of action to combine the simulated annealing procedure with graph-coloring heuristics.

In the following, we exemplify two of our approaches used for regular networks on the one hand, and for non-regular ones on the other (cf. section 3.5).

Dense Packing

In regular hexagonal networks (cells of equal size, homogeneous propagation conditions and constant channel demand per cell), the optimal channel assignments are produced by repeating a basic pattern (cluster) characterized by the fact that in this way the same channel is re-used as closely as possible. Let us consider a coordinate system with axes inclined at 60 degrees. If we assume that cells whose centers have mutual distance not less than $\sqrt{d}$ may use the same channel, MacDonald [5] shows that an optimal channel assignment will need $d \cdot t$ channels, with t the channel demand per cell. It is clear that the number of interferers depends mainly on the cluster size d: The higher the cluster size d, the higher the number of interferer per cell. In the 3-cell cluster case, each cell interferes only with its 6 direct neighbors. Figure 1 shows the six interferers of cell j which are not allowed to use the same channel with j (the cross-hatched ones) respectively the six nearest cells j' to j that are allowed to use the same channel with j. In order to imitate this behavior for the simulated annealing algorithm, we have modified the basic flip-flop transitions in the following way:

1. Choose at random a cell j.
2. Find all the nearest cells j' to j that may share the same channel with j.
3. From all the channels i currently not used at j switch on the one which is used most of all within cells j'. (This introduces the preference of densely packed channel structures as found in the hexagons.)
4. Switch off at random one of the channels previously used at j.

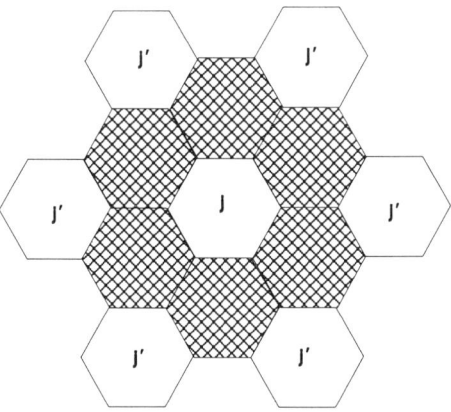

Figure 1 "Nearest re-usable" cells j' to cell j in the 3-cell cluster

Because of theoretical reasons [20], these dense-packing transitions cannot be used exclusively, but have to be mixed with the basic flip-flops. More precisely:

$$g_{ss'}(t) := a \cdot g_{ss'}^{flipflop}(t) + b \cdot g_{ss'}^{dense}(t), \text{ with } a, b \in [0, 1] \text{ such that } a + b = 1.$$

The parameter setting $a = b = 0.5$ yielded excellent results for the regular hexagons (cf. section 3.5).

Heuristic-Flip-Flops and Heuristic-Chains

A serious problem with the non-regular networks was to be trapped in an overflowing number of local minima. Also, it proved impossible to concentrate computing time on "relevant regions of the cooling curve." We almost always encountered the following behavior: Once being trapped in a poor minimum (by a misplaced transition), the chance to randomly get out of it, being just one of the numerous other possible transitions, is much too low.

Now, in contrast to the regular case, there is no additional knowledge available about the non-regular networks. A simple heuristic neighborhood structure would propose preferably flip-flops that resolve existing interferences while being disadvantageous for those that introduce new violations. In the sequel, we will call this kind of flip-flop transitions heuristic flip-flops. This way, some improvements in the algorithm's performance could be reached.

It is known from literature [23, 24] that one can preserve a fast cooling schedule if the algorithm is allowed to occasionally propose arbitrary long jumps. These long jumps offer the possibility to detrap from any minimum in a single

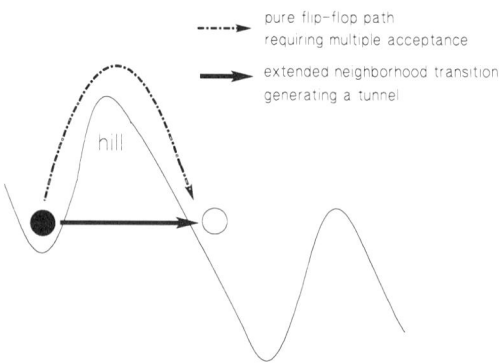

Figure 2 "Tunneling" within the cost function landscape

transition, without being questioned by an occasional long-chain of acceptance decisions. Bearing this in mind, we have extended the heuristic-flip-flop transition to a chain of consecutive ones, thus producing a long jump. Moreover, each flip-flop member of the chain is selected to systematically resolve the existing interferences while introducing the smallest possible number of new ones. Figuratively speaking, this provides something like the possibility of tunneling through a hill of the cost function landscape instead of painfully working, one's way to its top, to just be precipitated into the next valley (see figure 2). In the sequel, this kind of transitions will be called heuristic-chains.

3.5 Results

To assess the capabilities of simulated annealing in network planning, several cases of channel assignments have been extensively studied. To highlight our results, we present two representative examples:

1. An artificial regular hexagonal network, and
2. A typical inhomogeneous (non-regular) network as found in hilly terrains.

The regular network consists of $n = 14 \times 14$ cells and $m = 14$ channels. Each cell was assumed to interfere with the first ring (6 cells) and the second ring (12 cells) of its neighbors which corresponds to a cluster size 7. The network was cyclically continued in both the two spatial directions and the channel domain to avoid boundary effects. There the channel assignment task is to equip each cell with 2 out of 14 channels.

The inhomogeneous network consists of $n = 239$ cells and $m = 38$ channels (or more exactly: frequency groups) originated from the planning process in an European country. Interference is inhomogeneously distributed as caused by a typical hilly terrain with the number of interference partners of a cell ranging from about 20 to 70. Additional complexity is added by the wish to preserve as much as possible of an already existing channel assignment, which means that each cell typically has one preset channel. The channel demand ranges from 1 to 3.

In order to appreciate the following results an additional word has to be said concerning the statistics. As mentioned in the previous sections there are several components of simulated annealing depending on random decisions, thus rendering it a stochastic algorithm. Accordingly, the details of a single run and especially its final configuration are to some extent arbitrary. It is our experience that for the channel assignment application approximately 25 random number generator settings should be collected to come to statistically significant results as required for a sound analysis of the method. But, as one will see from the following data, high-quality results can already be achieved with much fewer repetitions. Before the results are presented, the implementation environment is described.

Implementation

The algorithm was implemented in C both on a VAX environment running under VMS and on an attached transputer network using the 3L software. For the transputers, the routines were not parallelized but the network is only used as a processor farm, i.e. each transputer runs its own simulated annealing program evaluating one start point out of an ensemble. This way, the statistics for the evaluation of the stochastic algorithm are supplied in parallel.

Simple Versus Sophisticated Neighborhood Structures

Table 1 presents a comparison of the main performance figures for the simulated

Problem	Hit rate		Final cost		Computing time	
	FF	HC	FF	HC	FF	HC
7-cell cluster	16%	52%	48 ± 22	28 ± 29	(3.5±1)h	(3±2)h
inhom. network	0%	32%	3.07 ± .61	1.28 ± .44	(202±21)m	(51±7)m

Table 1 Main performance figures for the different neighborhood schemes: pure flip-flop (FF) vs. heuristic chains (HC).

annealing algorithms using simple and sophisticated neighborhood schemes (dense-packing and heuristic-chains), respectively. Results were obtained on a T800. For the final costs (C respectively $C_{int}+C_{pres}$) and the computing times, their averages and spreads are computed over the set of 25 runs. The hit rate refers to the percentage of runs that reach a final configuration with zero violated interferences. This corresponds for the 7-cell cluster to the global minimum, and for the inhomogeneous network it is at least a near optimum, as the presets are weighted ten times less than the interferences. To obtain the results of table 1 we used the following selection for the sophisticated neighborhood schemes, which was for the 7-cell cluster a mixture of 50% dense-packing to the flip-flops and for the inhomogeneous network 2% heuristic-chain to 98% ordinary flip-flops (cf. section 3.4). Of course, the elaboration of these mixtures required numerous searches in the parameter space. But according to our experience, the values found are stable over a broad range of channel assignments, implying that they are also applicable to new problems.

Computing time for the run set of table 1 was selected such as to result in a reasonable trade-off between effort (computing time) and expected solution quality (final cost). In the case of the inhomogeneous network, using the pure flip-flop variant, it was not possible to find an optimal solution without going up to the power limit of our computing environment.

More detailed information on the statistical distribution of the different simulated annealing runs with respect to the finally reached cost can be found in Figure 3. Figures 4 and 5 show the typical convergence behavior of the average cost, i. e. the average of all cost values encountered on one temperature level is plotted against temperature. It is obvious that the sophisticated schemes provide a much higher solution quality on average and strive much more directly for the final low-cost states.

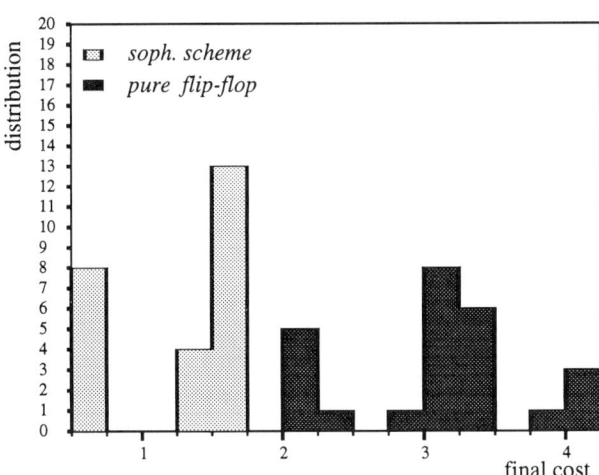

Figure 3 Distribution of the single runs with respect to final cost for the inhomogeneous network.

204   CHAPTER 10

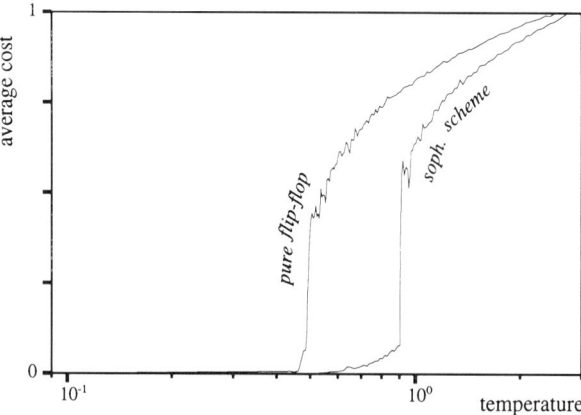

Figure 4 Convergence histories of (normalized) average cost for both neighborhood schemes on the 7-cell cluster.

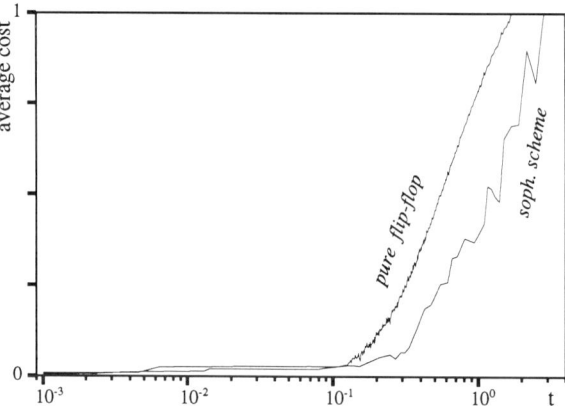

Figure 5 Convergence histories of (normalized) average cost for both neighborhood schemes on the inhomogeneous network.

4 SIMULATED ANNEALING IN A RUNNING SYSTEM

It is expected that the number of subscribers will increase dramatically, whereas the radio frequency resources will remain nearly constant. The only way to handle this is by means of micro cells. But, in this case, the required degree of detail for a proper description of traffic and interference may not be predictable with sufficient accuracy. Consequently, the static channel assignment has to be replaced by a dynamic one which, in particular, observes the interference matrix from the operating system. As a first step, again, one can use simulated annealing to compute fixed channel assignment lists, using an estimation of the interference matrix. During operation, the network itself can use such an approach for optimizing its performance. This optimization will be done in a parallel and distributed way so that no central controller needs to know the status of the whole network.

Now, we can interpret the radio network as a Boltzmann machine [20] where each state is determined by the corresponding element $s \in \{0,1\}^{m \times n}$ of the configuration space. This means, the elements s_{ij} of this machine are indicating whether a channel i at base station j is currently enabled or not. But in contrast to the situation when using a known interference matrix for the interconnection strengths, the amount of interference induced by using channel i both at stations j and j' is not known. So the result of any transition must be estimated via trial-and-error.

4.1 Radio Channel Usage

We assume that each base station is equipped with a number of transceivers that is sufficient to accommodate the expected traffic. These transceivers can be tuned to any of the available radio channels, although this may take some time so that it will not be done during a call. Please recall that $J = \{1, ..., n\}$ defines the set of all base stations, see section 1.1. The channels that are open for use at base station j are given by $F_j := \{i | s_{ij} = 1\}$ and we have

$$|F_j| = t_j = \sum_i s_{ij}. \tag{6}$$

We assume that each base station is able to exchange information with each of its potential interferers. Formally, we define $J_j \subset J$ the set of all base stations that may interfere with j.

4.2 Optimization Criterion

We now have to define the optimization criterion. For that we introduce a quality measure q_{ij}, $0 \leq q_{ij} \leq 1$, for each used channel $i \in F_j$. This value should estimate the probability that a call cannot be serviced by channel i over the whole period of time, the corresponding mobile is located in cell j. In other words, q_{ij} estimates that a call starting at or being handed over to cell j and using initially channel i will have to perform an intra-cell hand-over to continue. The importance of avoiding these intra-cell hand-overs lies in the fact that at peak hour all other channels may be occupied so that the intra-cell hand-over will fail. In order to measure q_{ij}, the received radio field strength and the bit error rate at the base station, respectively at the mobile station have to be taken. Then, the capacity of radio cell j will reach its maximum if $\sum_{i \in F_j} q_{ij}$ is minimized. Hence we define the energy E of the system as follows:

$$E := \sum_{j=1}^{n} \sum_{i \in F_j} q_{ij}. \tag{7}$$

The system should work such that E is minimized.

Generally, when performing a transition, the new channel will have a different quality compared to the old one. But also the interference level at the neighboring base stations will change as long as they use either the old or the new channel. If at base station j channel i is replaced by i', a transition will result in a change of E:

$$\Delta E = q_{i'j}^{new} - q_{ij}^{old} + \sum_{j' \in J_j\ i \in F_{j'}} q_{ij'}^{new} - q_{ij'}^{old} + \sum_{j' \in J_j\ i' \in F_{j'}} q_{i'j'}^{new} - q_{i'j'}^{old}. \tag{8}$$

4.3 Transitions and Their Generation

Again, as in the static case, we consider the flip-flop transitions, see section 3.2. This is in contrast to the classical Boltzmann machine where each element changes its state independently [20]. Here we assume that two elements at the same base station change their state simultaneously. This may be interpreted as a very high coupling between those elements at the same station which result in an immediate correction, if the number of channels is changed.

Since the algorithm is to run on independent base stations, each base station may generate its own transitions in parallel. This can be done by generating a pseudo-random exponentially distributed waiting time. Thus the Boltzmann machine works asynchronously and in parallel. As usual, parallel actions may lead to erroneous transitions if two transitions affect the same quantities. Since we have asynchronous actions we have to choose the density of transitions low enough that it becomes sufficiently unprovable that a base station $j' \in J_j$ begins a transition while a transition on base station j is still in progress. Alternatively, one may foresee a locking mechanism that prevents all $j' \in J_j$ from starting another transition while the transition at j has not finished.

One problem with the optimization in a real-world environment is that the relation between change of interference level and use of a certain channel is not known. It therefore has to be measured. Let us assume that a base station j intends to switch over from channel i to i'. Then it will inform its potential interferer $j' \in J_j$. These base stations will store the old value for q_{ij} and $q_{i'j}$ and reset their measurement of these quantities. Then j will temporarily switch over to i'. After a specific period of time the old and new values that are required in order to evaluate the change of ΔE (equation 8) are reported to j. As usual the acceptance decision depends on the temperature parameter T and is based on the Metropolis criterion, see section 3.1.

4.4 Temperature Control and Diffusion

The simplest case will be that the temperature is centrally controlled. Thus the temperature is initialized to a value high enough to allow most of the proposed transitions. If the central temperature control detects that the system is in thermal equilibrium, the temperature can be gradually decreased. This would be in analogy to the situation in the static case, see section 3.1. Unfortunately, this approach needs a central control of the temperature, respectively of the energy. Furthermore, the global decrease of the temperature would make it impossible to react on any change in the environment once the optimum has been found. Any need for local correction will result in a global 'melting' to rerun the search for the new optimum.

In order to avoid these drawbacks, the temperature T_j can be determined independently in each base station. This allows the network to converge rapidly where this is possible and to spend additional time only in 'difficult' areas. Problems may arise at the borders. Assume that there is a base station which is newly installed, or which suddenly faces additional traffic demand. In this case not only the channel assignment at the corresponding base station but also at the neighboring stations must be revised. To this end, a temperature exchange mechanism must be employed.

5 CONCLUSIONS

Simulated annealing has been systematically applied to some forms of the channel assignment task of radio network planning. After a relative straightforward reformulation of channel assignment, problems were encountered concerning run time efficiency and solution quality. But these could be cured by the careful design of appropriate neighborhood structures. This design of specialized components for simulated annealing can be further exploited for the treatment of additional engineering requirements within the basic channel assignment task.

Furthermore, simulated annealing in a running system was proposed on a conceptual level. Before it can be implemented in a real system, many practical questions have to be answered. It is too early to speculate on a possible outcome, but it seems reasonable that an algorithm which behaves well for a known interference matrix will prove to be reasonable for cases in which the interference matrix is observed while the network is running.

After all, simulated annealing appears to be quite a valuable approach for practical radio network design and operation, worthwhile to be studied further (cf. also [25]).

REFERENCES

[1] Gamst, A., Beck, R., Simon, R., Zinn, E.-G., "An Integrated Approach to Cellular Radio Network Planning," in Proceedings of the 35th IEEE Veh. Techn. Conf., May 21–23, 1985, Boulder, Co., pp. 21–25.

[2] Araki, K., "Fundamental Problems of Nation-Wide Mobile Radio Telephone Systems," Rev. El. Commu. Lab., Vol. 16, 1968, pp. 162-166.

[3] Box, F., "A Heuristic Technique for Assigning Frequencies to Mobile Radio Nets," IEEE Trans. Veh. Techn., Vol. VT–27, May 1977, pp. 57–64.

[4] Lorenz, R.W., "Kleinzonennetze für den Mobilfunk," NTZ, Bd. 31, 1978, pp. 192-196.

[5] MacDonald, V. H., "Advanced Mobile Phone Service: The Cellular Concept," Bell Systems Techn. J., Vol. 58, 1979, pp. 15-41.

[6] Halpern, S.W., "Reuse Partioning in Cellular Systems," 33rd IEEE Veh. Techn. Conf., 1983, Toronto, Canada, pp. 322-327.

[7] Metzger, B. H., "Spectrum Management Technique," 38th National ORSA Meeting, 1970, Detroit.

[8] Gamst, A., "A Resource Allocation Technique for FDMA Systems," Alta Frequenza, Vol. LVII, Feb.–Mar. 1988, pp. 89–96.

[9] Hale, W. K., "Frequency Assignment: Theory and Applications," in Proceedings of the IEEE, Vol. 68, 1980, pp. 1497–1514.

[10] Pennotti, R. J., "Channel Assignment in Cellular Mobile Telecommunication Systems," M. S. thesis, Polytechnic Inst. of New York, Brooklyn, 1976.

[11] Zoellner, J. A., Beall, C. L., "A Breakthrough in Spectrum Conserving Frequency Assignment Technology," IEEE Trans. El. Magn. Comp., Vol. 19, 1977, pp. 313-319.

[12] Hopfield, J. J., Tank, D. W., "'Neural' Computation of Decisions in Optimization Problems," Biol. Cybern., Vol. 52, 1985, pp. 141–152.

[13] Duque-Antón, M., Kunz, D., "Channel Assignment Based on Neural Network Algorithms," in Proceedings of the DMR IV, June 26–28, 1990, Oslo, Norway, pp. 5.4.1–5.4.9.

[14] Kunz, D., "Channel Assignment for Cellular Radio Using Neural Networks," IEEE Trans. on Veh. Tech., Vol. 40, No. 1, Feb. 1991, pp. 188–193.

[15] Kunz, D., "Suboptimum Solutions Obtained by the Hopfield-Tank Neural Network Algorithm," Biol. Cybern. 65, 1991, pp. 129–133.

[16] Aiyer, S. V. B., Niranjan, M., Fallside, F., "A Theoretical Investigation into the Performance of the Hopfield Model," IEEE Trans. on Neural Networks, Vol. 1, No. 2, June 1990.

[17] Funabiki, N., Takefuji, Y., "A Neural Network Parallel Algorithm for Channel Assignment Problems in Cellular Radio Networks," IEEE Trans. on Veh. Techn., Vol. 41, No. 4, November 1992.

[18] Kirkpatrick, S., Gelatt, C. D., Vecchi, M. P., "Optimization by Simulated Annealing," Science, Vol. 220, No. 4598, May 13, 1983, pp. 671–680.

[19] Černy, V., "Thermodynamical Approach to the Travelling Salesman Problem: An Efficient Simulation Algorithm," J. Opt. Theory Appl. 45, 1985, pp. 41–51.

[20] Aarts, E., Korst, J. , "Simulated Annealing and Boltzmann Machines," John Wiley & Sons, 1989.

[21] Huang, M. D., Romeo, F., Sangiovanni-Vincentelli, A., "An Efficient General Cooling Schedule for Simulated Annealing," in Proceedings of the IEEE ICCAD-86, Nov. 11–13, 1986, Santa Clara, Ca., pp. 381–384.

[22] Johnson, D. S., Aragon, C. R., McGeoch, L. A., Schevon, C., "Optimization by Simulated Annealing: An Experimental Evaluation, Part II (Graph Coloring and Number Partitioning)," to be published.

[23] Sue, H., Hartley, R., "Fast Simulated Annealing," Phys. Lett. A, Vol. 122, No. 3–4, June 1987, pp. 157–162.

[24] Romeo, F. I., "Simulated Annealing: Theory and Applications to Layout Problems," Memo No. UCB/ERL M89/29, Univ. of California, Berkeley, March 1989.

[25] Hegedüs, K., Zsuffa, Z., Bozsóki, I., "A Computer-Aided Method for Solving of Frequency Assignment," EMC, June 1990, Wroclaw, Hungary.

[26] Duque-Antón, M., Kunz, D., Rüber, B., "Channel Assignment Using Simulated Annealing,", in Proceedings of the MRC '91, November 13–15, 1991, Nice, France, pp. 121–128.

11

CELLULAR MOBILE COMMUNICATION DESIGN USING SELF-ORGANIZING FEATURE MAPS

Thomas Fritsch

Institute of Computer Science
University of Wuerzburg

1 INTRODUCTION

1.1 Preliminary

[1] An increasing number of subscribers of mobile communication radio networks like the analog C-Net in Germany and the recently introduced digital pan-European GSM-Net will have a considerable effect on the design and planning of radio networks. The introduction of the GSM-Net gives rise to hope, that this market will be one of the most expanding branches of telecommunication technologies. The design of a mobile radio communication system not only concerns the geographical planning but also the optimal usage of the frequency spectrum and moreover juridical questions.

The first important task in the design of a radio network is the choice of suitable locations for the base stations which are necessary to provide communication links in the radio network. These locations must be determined taking the topographical and morphological structure of the network's coverage area into consideration. But in reality the radio wave propagation is never homogeneous and isotropic, resulting from reflections, deflections and refractions caused by obstacles. The real coverage areas of the base stations are irregularly bounded, depending on the topography and the transmitting power.

The second task is the determination of the estimated traffic demand, i.e. the traffic density distribution in the affected region. To a certain degree the traffic distribution also depends on the topography. This problem is strongly related

---

[1]This chapter is an extended version of [1]

to the choice of the radio base station locations, but usually the time-varying characteristics of the traffic flow are not considered. Furthermore a realistic estimation of the traffic demand is essential for the planning of frequency reuse patterns. The concept of frequency reuse is crucial for an optimal usage of the existing radio channel spectrum. Hereby frequency reuse means that a frequency which is used in a certain cell can be reused in a spatially distinct region, where distortions, caused by co-channel or adjacent-channel interferences are weak. Since the cells have no regular form it is very complicated to reconstruct a new frequency reuse pattern due to a new frequency assignment to the cells. In this case dynamic channel allocation (DCA) has been applied. DCA can prove to be necessary, if traffic characteristics are time-dependent. Last but not least it should be noted that regulatory problems can also influence the planning decisions.

1.2 Mobile communication systems

The principal function of a mobile communication system is to provide mobile subscribers with communication facilities, such as free and distortionless communication channels for the call holding time. Consequently, a hierarchical system structure is assumed. Mobile units are moving in a coverage area of a radio base station, hereby having contact with this station and a corresponding telephone partner. A mobile switching center (MSC) has the purpose to provide each mobile unit with a free channel over which it can communicate.

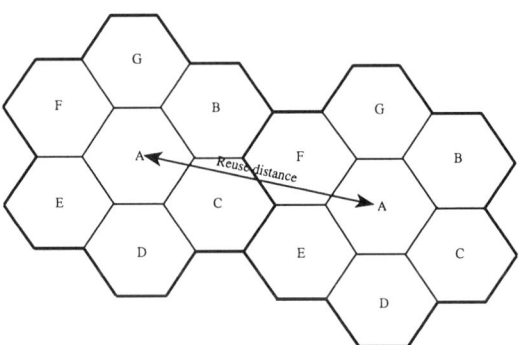

Figure 1 Frequency reuse pattern and reuse distance

To each base station there is assigned a supplying area where the mobile unit can receive a call with a certain probability, depending on the signal to noise ratio at the current location of the mobil unit in th is area. For reasons of simplicity, these areas are assumed to be of hexagonal shape up to the moment.

The MSC is connected to a wired-switch network and by this way to other MSCs, which also serve a certain number of base stations. If a mobile unit leaves the supplying area of a base station, a handover to the MSC of a neighboring cell has to take place. The disposable spectrum for mobile radio communication is restricted by other communication services and constrained by regulatory and military problems. Moreover since this spectrum is also bounded by the number of mobile subscribers, the mobile system has to take these restrictions into account by its structure. The idea is to reuse frequencies at a certain distance, depending on the power of the base stations and the topographical situation. Thereby groups of frequencies are assigned to supplying areas in a static scheme, the frequency-reuse-pattern (see Fig. 1, where frequency groups are denoted by capital letters A to F). Thus it is possible to serve a larger number of mobile units which are spatially distributed by means of a bounded number of frequency channels.

2 MOBILE NETWORK PLANNING

In the planning phase of radio networks we have to distinguish four major design steps:

1. *Radio network definition*, where the locations of the radio base stations are determined and furthermore antenna height, antenna type, beam directions and transmitting power are defined;
2. *Cellular analysis*, where the supplying level, the traffic assignment and the interference situation are analyzed;
3. *Frequency channel assignment*, where available radio channels are assigned to different radio cells;
4. *Radio network analysis*, where quality features of the whole communication system are investigated, e.g. net supplying probabilities, net interference probabilities and net blocking probabilities.

In the following we restrict ourselves to the problems of the network definition in the planning phase. The main question is: where can a base station be placed, so that a sufficient signal to noise ratio at every reception point in the coverage area of this base station will be guaranteed?

The answer depends on the radio wave propagation in the coverage area. But the propagation itself depends on the topography and morphostructure of this coverage area in relation to the power of the transmitting station and the location of this station.

Topography and morphostructure are responsible for dispersions, reflections, refractions and other signal-distorting influences. As a result the cell boundary appears irregular shaped (see Fig. 2). There are holes, for instance, where no radio wave can be received or the reception is attenuated.

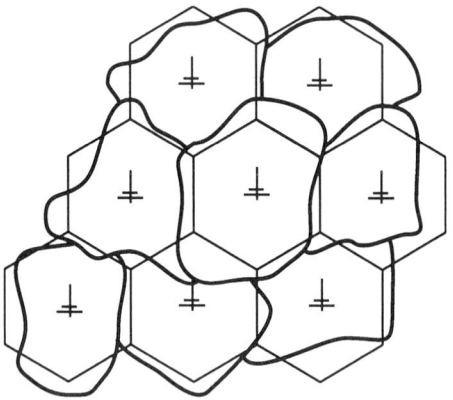

Figure 2 Irregular shape of base stations supplying areas

Moreover, if a mobile unit gets closer to the border of the supplying area a handover has to take place at a certain time. Determining this time can be difficult if interferences to frequencies of neighboring cells occur.

Available design tools include some simplifying assumptions on the calculation or prediction of the electrical field strength in the coverage area. In most cases the Okumura statistical model is applied (see [2]). As the name indicates the model was developed in Japan, where flat urban area was assumed to dominate and so hexagonal cell shape was not far from reality. But the radio wave propagation remarkably changes, if some obstacles appear, like mountains, lakes or trees. Furthermore, traffic is time-dependent and in rush hours the need for more channels or transmitting power will be bigger than at night. As the base stations themselves are not movable, the increased demand can be satisfied by increasing the power of the base stations or by increasing the number of cells (splitting to microcells). Both methods change the field strength distribution. It is obvious, that this bundle of problems is not tractable in a reasonable computing time, even when simplifying assumptions are made.

Facing the problems described above, we looked for an appropriate neural net solution, which would be able to represent the radio field strength according to topographical information. At first the idea was to learn the topography of the chosen area by a self-organizing feature map [3] and then to propagate a simulated radio wave from a transmitting station, resulting in a coverage area, where at all points a field strength beyond a critical value would be detectable. The value of the field strength F is calculated following the formula in [4]

$$F/(dB(\mu V/m) = 74.8 + 10\ log_b\ (P_s/W) + 10\ log_b\ (G_s) - 20\ log_b\ (d/km) \quad (1)$$

with F as the field strength in decibel (dB), P_s as the transmitting power in Watt (W), G_s the antenna gain and d as the distance to the transmitting station in km. The transmitting power P_s and the antenna gain values can vary to powers of ten, thus P_s and G_s are calculated in a logarithmical scale, where the base b for the log-function was set to $b = 10$. For more details see [4].

As mentioned above first experiments were carried out dealing with the learning of the topography of a terrain. The Kohonen net learned artificial terrain data well, but calculating the field strength distribution caused severe computing-time problems, since each rearrangement of base stations has to be followed by a complete new learning process. Thus this approach was rejected and we were looking for a dynamical algorithm which would need less computing time. The following integrated approach to radio network planning includes the approximate consideration of the traffic density in the concerning coverage area and provides in conjunction with some restrictions a quasi-optimal arrangement of the base station locations. The restrictions refer to the estimation of the traffic density and to the radio wave propagation. These two problems can be treated independently.

Each new arrangement of base stations makes necessary a recalculation of the field strength distribution. This implies a very high computing time. Thus we made a tenable simplification: The signal distortion is set proportional to the interruption of the direct sight connection line between the base station location and the position where the field strength is calculated according to eqn. 1. The sight connection line is determined using the well-known Bresenham algorithm ([5]) from computer graphics methods. The intersection of this line with the vertical section of the terrain surface is calculated (see Fig. 3) and the percentage of this intersection on the total sight connection line is interpreted as the field strength attenuation at the position where the effective field strength shall be determined.

This method can be improved further by using more sophisticated computer graphic methods like ray-tracing (see [6] and [7]) or radiosity (see [8]).

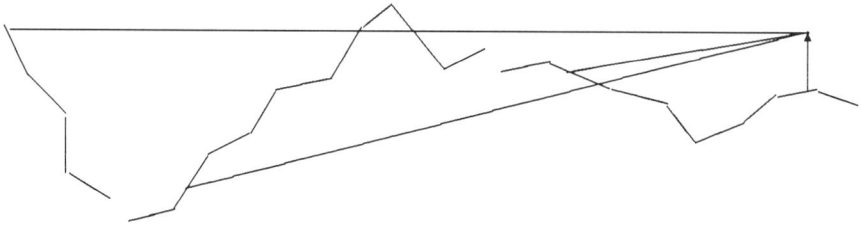

Figure 3 Simplified description of radio wave propagation

3 TRAFFIC DENSITY DETERMINATION USING A SELF-ORGANIZING FEATURE MAP

Traffic density in a coverage area is one of the most important parameters in determining locations of transmitting stations because supplying an area without any possible user of the radio network would be uneconomical. Normally, the traffic density is obtained by measurements.

The method presented in this chapter uses a fixed set of sensory neurons, which are distributed over the coverage area according to an approximated traffic density. This density essentially influences the positioning of the transmitting stations. Consequently, we need a method which associates the mean traffic density $E[V] = V^*$ to the spatial distribution of the sensory neurons. This implies that a high traffic density in a certain part of the coverage area must be represented by a high density of sensory neurons at this location.

Self-organizing feature maps are well suited for topological mappings. These neural nets reflect the topological neighborhood relationships of the input space in their neighborhood structure. In the following a brief description of the algorithm is given to outline the process of non-linear adaptation. Details of the self-organizing feature maps can be found in [3] and [9].

The neural net consists of 2 layers of neurons, which are fully interconnected (see Fig. 4). Each neuron of the input layer is connected with each neuron of the mapping (output) array. The number of input neurons is determined by the dimension of the input vectors. The number of neurons in the mapping array has to be chosen suitably. Since it is not known in advance which number of neurons will be adequate for the special problem, some experiments have to be carried out until the proper number is found. A dynamically growing net structure was proposed in [10], but not yet considered in our application.

3.1 Self-organizing feature map algorithm

The algorithm can be characterized by the following steps:

1. Present a new input vector **v(t)**

2. Compute the distance d_j between all input neurons and all mapping array neurons j according to

$$d_j = \sum_{i=1}^{N}(v_i(t) - w_{i,j}(t))^2 \qquad (2)$$

where $v_i(t)$ is the i-th component of the N-dimensional input vector and $w_{i,j}(t)$ is the connection strength between input neuron i and mapping array neuron j at time t.

3. Select the mapping array neuron j^* which has the minimal distance d_{j^*}.

4. Update all weights, within the actual topological neighborhood $N_{j^*}(t)$

$$w_{i,j}(t+1) = w_{i,j}(t) + \eta(t)(v_i(t) - w_{i,j}(t)) \qquad (3)$$

for $j \in N_{j^*}(t)$ and $1 \leq i \leq N$. Here $\eta(t)$ represents a monotonically decreasing function.

5. Iterate the above steps until a predetermined error criterion is met.

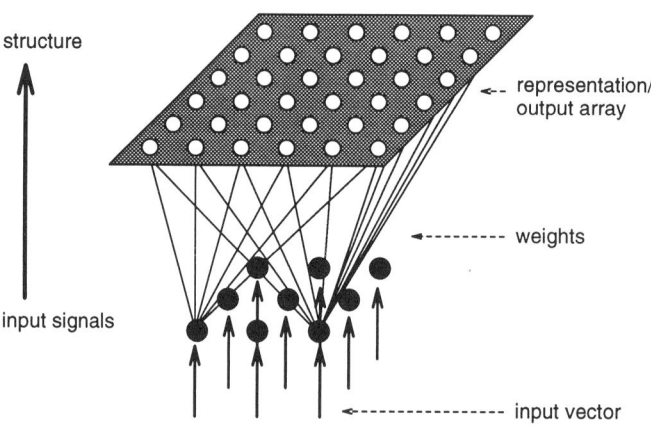

Figure 4 Basic structure of self-organizing neural net

3.2 Modified algorithm

In our case the input vectors presented to the net are three-dimensional, but the weights of the neurons are two-dimensional. An input vector v looks like this:

$$v = (x, y, V^*) \tag{4}$$

where V^* is the mean traffic density at the point (x, y) in the coverage area. The learning rate η is now chosen proportional to V^*:

$$\eta(V^*) = c \cdot V^*, \; c \text{ fixed.} \tag{5}$$

The selection probability of input vectors must be uniformly distributed. The choice of η proportional to V^* results in very good mappings. Because exact

Neural Nets in Cellular Mobile Communication Planning

measurements of the real traffic are not available we need an alternative method to achieve useful sensory neuron distributions. Examining a geographical map of the coverage area (see Fig. 13) it is obvious that most of the car traffic takes place on roads or highways or in towns. The locations of highways or towns depend on the topography and the morphostructure. Most of them are located in lower terrain. So a modified input vector v has the following form:

$$v = (x, y, h) \qquad (6)$$

where h now represents the topographical height at coordinates (x, y). Now we choose η inversely proportional to h with the following constraint. At the lowest points of the coverage area there is a river (see Fig. 13 on the left side), where of course only low traffic can be detected. Therefore the traffic distribution should contain a gap at this location. If such a point is chosen in the course of the algorithm, η will be set to a low value, so that the corresponding distribution of the sensory neurons will be low. During the adaptation process of the neural net, positions of the neurons will be attracted by low terrain, like valleys or river borders, where most of the car traffic takes place. Thus the mean traffic density problem is replaced by a spatial learning. The weight vectors of the neurons are taken as the positions of the sensory neurons for the method outlined in the next section. A resulting sensory neuron density map of this modified learning phase is shown in Fig. 5.

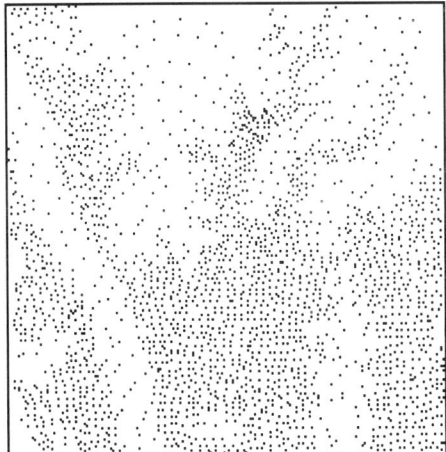

Figure 5 Traffic density learned by a self-organizing feature map

4 THE MAIN ALGORITHM

The algorithm takes advantage of "simulated annealing" which is well known from statistical physics [11]. In this method, a thermodynamic system is assigned an energy value for every system state and a corresponding system temperature which is reduced slowly at each discrete time step. The goal is to reach a state with minimal energy without getting stuck at a local minimum of the energy function. Transitions between system states take place with a certain probability, in our case

$$prob\{Z^{new} = Z^{act}\} = e^{-\frac{\Delta E}{\tau}} \quad (7)$$

with Z denoting the system state and ΔE the energy difference between the old state energy and the current state energy at temperature τ.

The mobile communication system model consists of a set of transmitting stations (base stations) which are able to change their positions and transmitting power, and a set of sensory neurons, which are fixed according to a predefined distribution density. The main idea of the algorithm is that all transmitting stations are competing for a large number of sensory neurons which shall be covered by the transmitting station's supplying areas. In the progress of the algorithm the transmitting stations permanently change their positions, attracting and repelling each other until an optimal placement is found. To reach a situation where more sensory neurons are covered by a supplying area than before, the transmitting stations have two means at their disposal. First, they can change their position and second they can increase their transmitting power.

Clusters of sensory neurons cause an attractive displacement of the transmitting station (see Fig. 6). Sensory neurons which lie in the intersection of two or more supplying areas cause a repellation of transmitting stations (see Fig. 7).

The second possibility is to change the power of the transmitting station, so that a larger number of neurons are covered by the transmitting station's supplying area (see Fig. 8.) Hereby it is assumed that these neurons are assigned only to this transmitting station. If decreasing power becomes necessary, then the number of neurons which lie in the intersection of different supplying areas will be lessened (see Fig. 9.)

Algorithm:

1. (a) Transmitting Stations

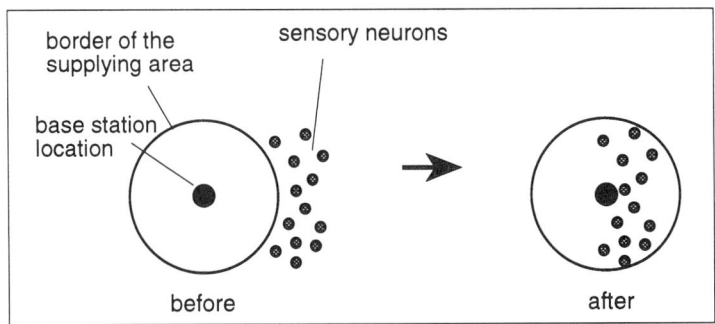

Figure 6 Displacement - case of attraction

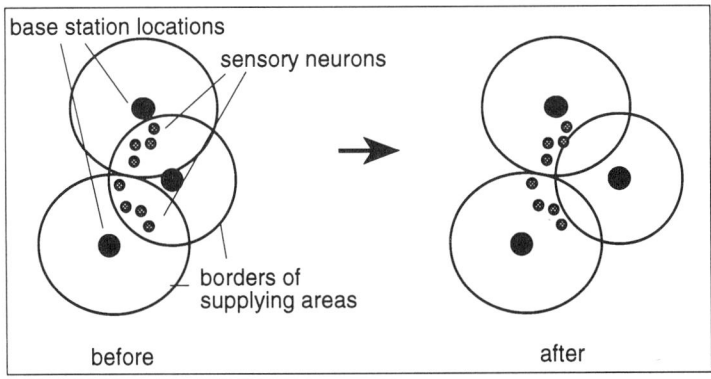

Figure 7 Displacement - case of repellation

A transmitting station T is defined by

$$T := (Pos(T), P(T)) = (x, y, P) \tag{8}$$

with (x, y) representing the position and P the power of T.
Additionally a vector $F(T) := (\Delta x, \Delta y)$ is introduced, called the position error of T. This vector can be interpreted as a resultant of several direction vectors. $F(T)$ is used to determine the direction of the next possible displacement of the transmitting station in the adaptation step. Hereby the difference of the position vectors of a sensory neuron and the transmitting station T is calculated and added

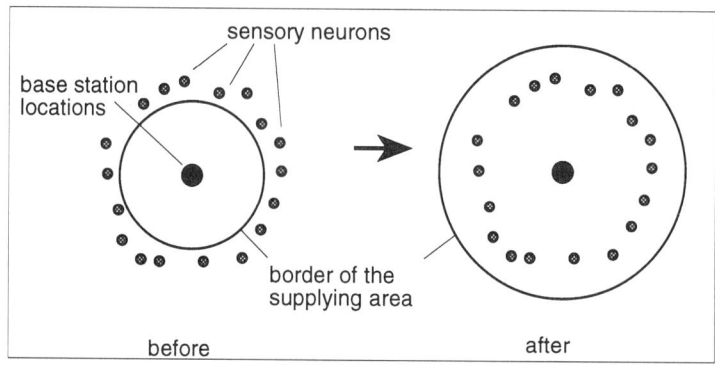

Figure 8 Power - case of increasing

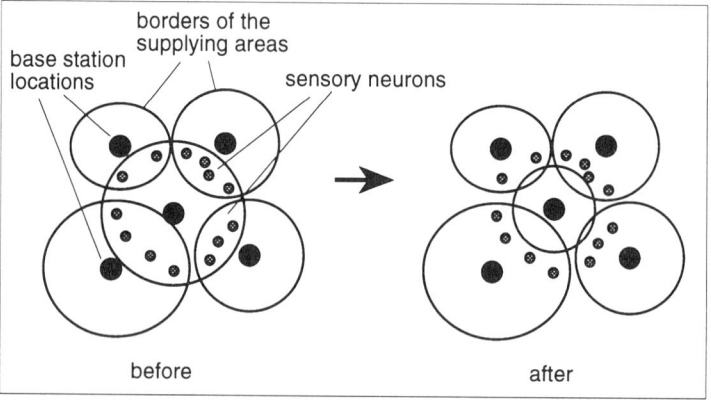

Figure 9 Power - case of decreasing

to or subtracted from the former position error depending on the direction of the displacement. A variable $n(T)$ counts the number of free sensory neurons which are added in the selection step. On the other hand the number of sensory neurons which were assigned to multiple transmitting stations is subtracted from $n(T)$ where $n(T)$ denotes the number of sensory neurons covered by the transmitting station T. Furthermore this criterion and the values of $F(T)$ are used to decide whether the position or the power of the transmitting station shall be changed. The set of all transmitting stations is called T.

(b) Sensory Neurons

A sensory neuron S is defined by

$$S := (Pos(S), H) = (x, y, H) \tag{9}$$

with $H := (H_1, ..., H_k)$ as the field strength values detected from the transmitting stations $T_1, ..., T_k$ by sensory neuron S.

The set of all sensory neurons is called S.

2. Initial Conditions

   (a) Distribute n sensory neurons $S \in S$ according to the traffic density determined by the self-organizing feature map.

   (b) Assign all base stations $T \in T$ with initially low transmitting power P_0. This initialization allows more flexible changing possibilities for the transmitting stations at the start of the algorithm.

   (c) Initialize the temperature parameters τ_0 (initial value) and the decrement parameter $\Delta \tau$.

   (d) Initialize $\Delta x, \Delta y$ and n with zero for all transmitting stations.

3. Determination of the supplying areas $V(T_j)$

   Let $H_{receive}$ be the minimal reception field strength level (signal to noise ratio) in the communication system. Below this field strength level there's no satisfying reception quality available. The supplying area $V(T_j)$ of a transmitting station T_j is then defined by

   $$V(T_j) := \{S_i \in S : H_j(S_i) \geq H_{receive}\},$$
   $$\forall i, j = 1 \ldots k \quad \text{and} \quad V := \bigcup_{T_j \in T} V(T_j) \tag{10}$$

   where $H_j(S_i)$ are the field strength levels at the positions of S_i. Then the supplying area $V(T_j)$ is the set of all neurons S_i which fulfill this condition. In the undistorted case all supplying areas are circular shaped. Actual irregular forms as well as disconnected regions result from attenuations caused by obstacles.

4. Selection Step (see Fig. 10)

   Select a sensory neuron S_i randomly under the assumption of uniform distribution. Choose one of the following cases.

(a) S_i belongs to no supplying area

$$S_i \notin V$$

i. Find the transmitting station T_{next} with minimal spatial distance to S_i.
$$T_{next} := T_j \quad \text{with}$$

$$\|Pos(S_i) - Pos(T_j)\| \le \|Pos(S_i) - Pos(T_k)\|$$
$$\forall T_k \in T,\ T_k \ne T_j \tag{11}$$

ii. Add to the position error $F(T_{next})$ the difference vector of the positions of S_i and T_{next}. This causes an attraction of T_{next} by S_i (see Fig. 6).

$$F^{(new)}(T_{next}) := F^{(old)}(T_{next}) + (Pos(S_i) - Pos(T_{next})) \tag{12}$$

iii. Increase $n(T_j)$ by 1

(b) S_i lies in exactly one supplying area

$$S_i \in V(T_j) \wedge S_i \notin V(T_l) \quad \forall T_l \ne T_j \tag{13}$$

In this case no changes take place.

(c) S_i is assigned to more than one supplying area of the transmitting stations

$$S_i \in V(T_{j_1}) \cap ... \cap V(T_{j_l}),\ 2 \le l \le k \tag{14}$$

i. Subtract the difference between the positions of S_i and T_{j_s} from the position error $\forall s = 1, .., l$. This causes a repellation of the transmitting stations from S_i (see Fig. 7).

$$F^{(new)}(T_{j_s}) := F^{(old)}(T_{j_s}) - (Pos(S) - Pos(T_{j_s}))$$
$$\forall T_{j_s} \quad \text{with} \quad S_i \in V(T_{j_s}) \tag{15}$$

ii. Decrease $n(T_{j_s})$ by 1

5. Adaptation Step (see Fig. 11)

A system state Z at time step (t) is characterized by the assignment of all sensory neurons S_i according to the current distribution of the transmitting stations T_j and their corresponding supplying areas $V(T_j)$ to one of the following sets:

Neural Nets in Cellular Mobile Communication Planning

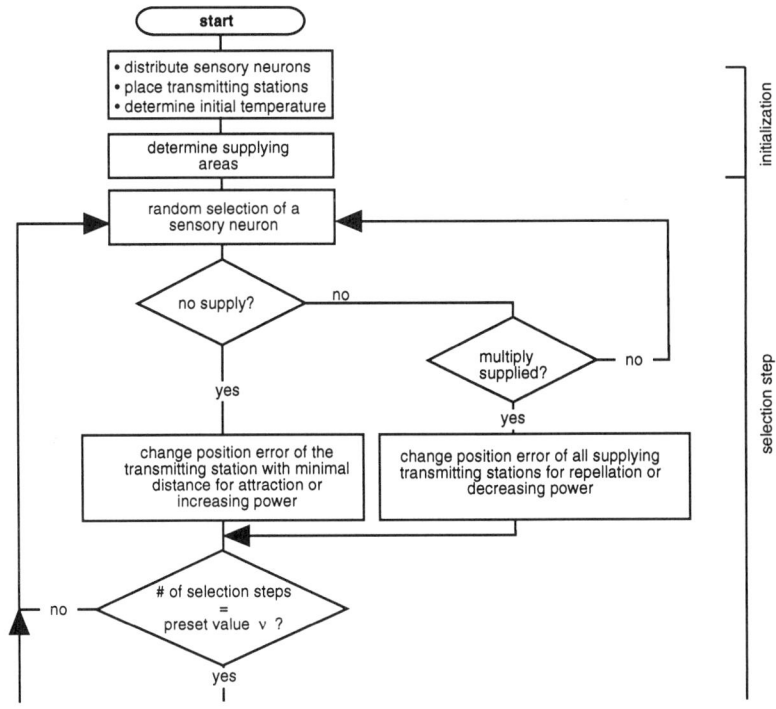

Figure 10 Part 1 of the algorithm - Initialization and selection

i. free sensory neurons
$$S_f := \{S_i | S_i \notin V\}$$

ii. multiply assigned sensory neurons
$$S_m := \{S_i | S_i \in \bigcup_{l,j}\{V(T_j) \cap V(T_l)\}\},$$
$$\forall l, \forall j, l \neq j$$

iii. definitely assigned sensory neurons
$$\overline{(S_f \cup S_m)}$$

(a) Determine T_{worst}, the transmitting station with maximum $||F||$,

$$T_{worst} := T_j \text{ with}$$

$$\|F(T_j)\| > \|F(T_k)\|, \ \forall j, k, \ T_j \in T \tag{16}$$

We now introduce the concept of the energy of a system state. We identify the order of the system by the magnitude of $\overline{(S_f \cup S_m)}$. The category of the system's energy which is decreasing with increasing order is introduced similarly. The system's energy is hereby associated to the magnitude of $(S_f \cup S_m)$. Therefore a maximization of the definitely assigned neurons and thus the minimization of $(S_f \cup S_m)$ corresponds to the minimization of the system's energy.

(b) Save the system's state before changing the position or the power of the transmitting station.

Let $E^{(old)}$ be the energy and $Z^{(old)}$ the system's state at time step t

(c) If $\|F(T_{worst})\|$ takes on values that make a displacement necessary

$$\|F(T_{worst})\| \geq t_{move},$$

adapt the position:

$$Pos^{(new)}(T_{worst}) := Pos^{(old)}(T_{worst}) + \epsilon * F(T_{worst}) \tag{17}$$

The threshold t_{move} is found by experience and prevents small displacements.

The transmitting station T_{worst} is moved into that direction, where many attracting sensory neurons are located, and in the opposite direction in the case of repellation. The parameter ϵ plays the role of a scaling factor also preventing extreme displacements.

(d) If $F(T_{worst}) < t_{move}$
adapt the transmission power

$$P^{(new)}(T_{worst}) := P^{(old)}(T_{worst}) + n(T_{worst}) * p \tag{18}$$

where p is a fixed value for the magnitude of change of power.

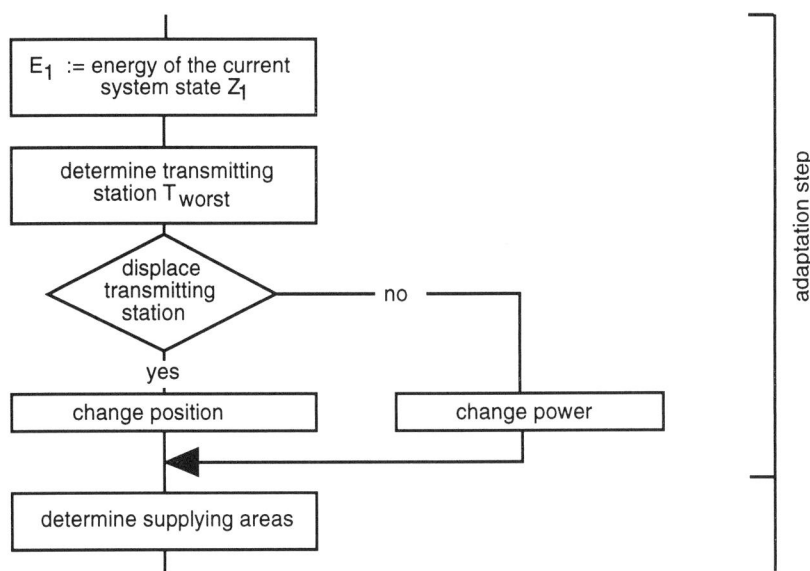

Figure 11 Part 2 of the algorithm - Adaptation

6. Optimization Step (see Fig. 12)

    (a) Determine the energy $E^{(cur)}$ of the current system state $Z^{(cur)}$ at time step $(t+1)$
    (b) Check, if a state transition can take place
        i. Determine $\Delta E := E^{(cur)} - E^{(old)}$
        ii. If $\Delta E < 0$ set $\Delta E := 0$
            In this case the state transition shall happen with probability 1
        iii. A state transition into state $Z^{(new)}$ happens with probability

        $$prob\{Z^{(new)} = Z^{(cur)}\} = e^{-\frac{\Delta E}{\tau}} \tag{19}$$

        Remaining in state $Z^{(old)}$ takes place with probability

        $$prob\{Z^{(new)} = Z^{(old)}\} = 1 - e^{-\frac{\Delta E}{\tau}} \tag{20}$$

    (c) Decrease the temperature τ

    $$\tau^{(new)} := \tau^{(old)} - \Delta\tau \tag{21}$$

228   CHAPTER 11

(d) Reset Δx, Δy and n for all transmitting stations

7. If $\tau^{(new)}$ reaches zero and ΔE is about zero this situation can be interpreted as a possible state of minimal disorder. Then the algorithm terminates, otherwise repeat from step 4.

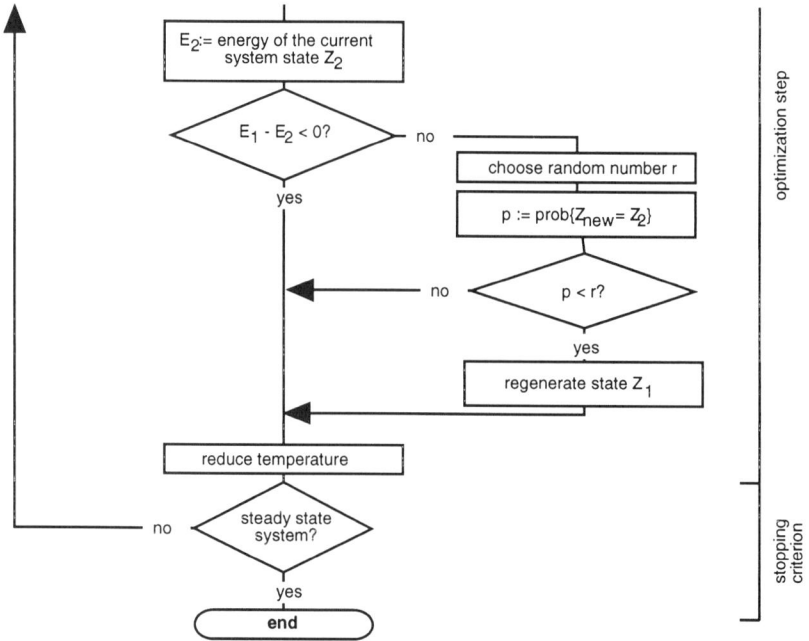

Figure 12 Part 3 of the algorithm - Optimization

5 EXPERIMENTAL RESULTS

Fig. 13 shows the topographical terrain of a 10km x 10km area in the north of Wuerzburg. Dominating the region is the river Main on the left side. The resolution of the data of the digital terrain model, delivered by the Bavarian Department of Surveying [12] is 100 meters in each direction. The figure is printed out with eight gray scales, where dark regions correspond to low terrain and bright regions to high terrain.

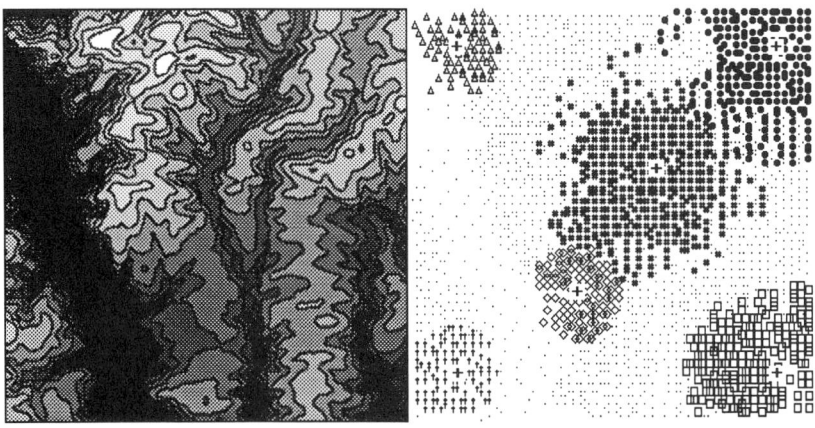

Figure 13 Terrain data   **Figure 14** System state after 100 steps

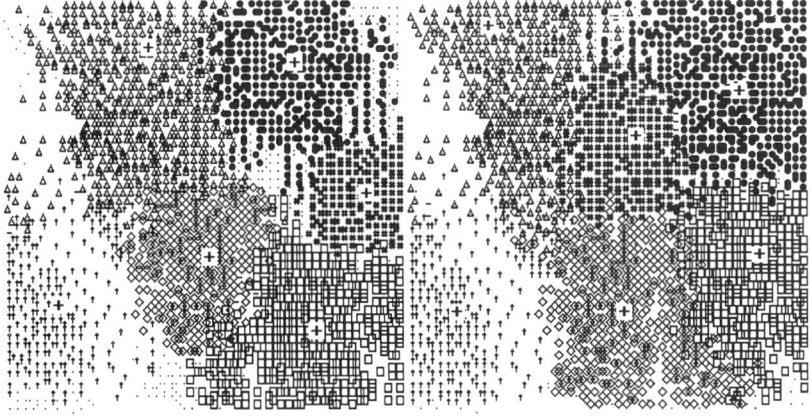

Figure 15 System state after 1000 steps **Figure 16** System state after 10000 steps

Fig. 16 shows the final arrangement of the six transmitting stations. The supplying areas of the 6 transmitting stations are denoted by different symbols. The base stations are denoted by a '+'-symbol, whereas free sensory neurons are denoted by a point symbol and multiply assigned sensory neurons by a '-' Symbol.

Steps	Free sensors	Fraction in %	Multiply assigned sensors	Fraction in %
0	1892	75.68	0	0
100	1330	53.20	0	0
300	623	24.92	0	0
500	359	14.36	0	0
1000	272	10.88	7	0.28
3000	154	6.16	16	0.64
10000	103	4.12	19	0.76

Table 1 Free and multiply assigned neurons at different system states

Fig. 14-16 show a visualization of different system states during the progress of the algorithm. An early system state is shown in Fig. 14.

The system appeared stable after 500 steps, but there were still free neurons, which forced the algorithm to continue. Fig. 15 shows that after 1000 steps one base station had been pushed to the right side of the coverage area, hereby releasing free space for the base station with the diamond symbol, which could conquer back its old position 2000 steps later. Furthermore it can be noted that the first multiply assigned neurons had appeared.

Fig. 16 shows the final arrangement of base station locations after 10000 steps. Since no state transitions took place anymore, the algorithm was terminated.

The corresponding percentage of free and multiply assigned sensory neurons at different system states is shown in Table 1.

The final placement of transmitting stations was plausible to a human expert. Nevertheless it is necessary to compare the algorithm to other existing methods, e.g. the fuzzy-logic-based program GRAND [13] which already yielded excellent results for highly complicated areas in Switzerland. Furthermore, the problems concerning field strength prediction and traffic density measurement can influence the performance of the algorithm significantly. Future research is needed to check the quality of our approach.

6 CONCLUSIONS

In this chapter we presented an algorithm which is able to arrange the locations of the transmitting stations of a cellular mobile communication network according to an approximated traffic distribution, learned by a self-organizing neural net. The algorithm is well suited for the determination of radio base station locations even when the prediction of field strength in the coverage area was calculated under simplifying assumptions for reasons of computation time. In contrast to planning concepts based on hexagonal cells, it is possible with this algorithm to support the planning of radio networks more realistically, if efficient methods for field strength prediction and traffic measurement are provided. Most important in this context is the improvement of the field strength prediction. A new idea in this direction is applying computer graphic methods like ray tracing or radiosity algorithms to approximate the radio wave propagation. In addition, the learning of a road graph of the coverage area by a neural net would modify the algorithm such that the fine-tuning of the radio base station locations could significantly be improved. Furthermore the algorithm can take advantage of exact measurements of traffic data. Alternative approaches like fuzzy logic could also be used to improve the quality of the achieved solution, since coverage areas can be regarded as fuzzy cells with uncertain borders. Future research shall focus on the combination of neural nets and fuzzy sets to provide a realistic mapping of the coverage area and on the consideration of time-varying channel demand for dynamic channel allocation. Taking these improvements into account, a future application of the algorithm could be the planning of mobile inhouse LANs.

REFERENCES

[1] Fritsch, T. and Hanshans, S., "An integrated approach to cellular mobile communication planning using traffic data prestructured by a self-organizing feature map," Proc. 1993 IEEE Int Conf on Neural Networks at San Francisco, CA, Vol II, March 28 - April 1, 1993, pp. 822-D–822-I

[2] Okumura, Y., Ohmori, E., Kawano, T. and Fukuda K., "Field strength and its variability in VHF and UHF land-mobile radio service," Rev. ECL, Vol.16, 1968, pp. 825–873

[3] Kohonen, T.,"Self-organizing feature maps," tutorial at IJCNN-89, Washington, D.C. Jun 18-22, 1989

[4] Lee, W.C.Y., Mobile communication design fundamentals, Howard W. Sams, Indianapolis, 1986

[5] Foley, J.D. and Van Dam, A., Fundamentals of interactive computer graphics, Addison Wesley, 1983, pp. 433–436

[6] McKown, J.W. and Hamilton, R.L., "Ray tracing as a design tool for radio networks," IEEE Network, Nov. 1991, Vol.5, No. 6, pp. 27–30

[7] Seidel, S.Y. and Rappaport, T.S., "A ray tracing technique to predict path loss and delay spread inside buildings," Proc of the IEEE Globecom 92, Dec. 6-9, 1992, Orlando, FL, Vol II, pp. 649–653

[8] Cohen, M.F. and Greenberg, D.P., "The hemi-cube: a radiosi ty solution for complex environments," SIGGRAPH 85, 1985, pp. 31–40

[9] Ritter, H.; Martinetz, T. and Schulten, K., An introduction to the neural information process of self-organized networks, Addison-Wesley, 1991

[10] Fritzke, B. and Wilke P., "Flexmap - a neural network for the traveling salesman problem with linear time and space complexity," Proc IJCNN Singapore 1991, Vol. II, pp. 929–935

[11] Aarts, E. and Korst J., "Simulated annealing and Boltzmann machines," Wiley, New York, 1990

[12] "Digital terrain model of the bavarian department of surveying, Use allowed 9.4.92 with Az.: Vm 2280 B - 3115" (in German)

[13] Krueger, M. and Beck, R., "GRAND - a program system for mobile radio communication planning," (in German), PKI Tech. Mitt. 2/1990 pp. 7–12

12

AUTOMATIC LANGUAGE IDENTIFICATION USING TELEPHONE SPEECH

Yeshwant K. Muthusamy and Ronald A. Cole

*Center for Spoken Language Understanding,
Oregon Graduate Institute of Science & Technology*

1 INTRODUCTION

1.1 The Problem

Automatic language identification is the problem of identifying the language being spoken from a sample of speech by an unknown speaker. Within seconds of hearing speech, people are able to determine whether it is a language they know. If it is a language with which they are not familiar, they often can make subjective judgments as to its similarity to a language they know, e.g., "sounds like German".

What are the distinctive sound patterns that characterize languages? Languages have been described subjectively as "singsong", "rhythmic", "guttural", "nasal" etc. Languages differ in the inventory of phonological units used to produce words, in their frequency of occurrence, and the order in which they occur. The presence of individual sounds, such as the "clicks" found in some sub-Saharan African languages, or the velar fricatives found in Arabic, are readily apparent to speakers of languages that do not contain these phonemes. Less obvious acoustic patterns are also observed. Mandarin Chinese has a higher frequency of occurrence of nasals than English. Hawaiian is known for its very limited consonant inventory. Prosodic differences also abound between languages. For example, it has been shown that fundamental frequency (F_0) patterns of continuous speech display different characteristics in Mandarin Chinese (a tone language) and American English (a stress language) [4]. The key to solving the problem of automatic language identification then, is the detection and exploitation of such differences between languages.

1.2 Motivation

What makes this problem so challenging and interesting? First, automatic language identification requires research to discover the fundamental acoustic, perceptual and linguistic differences among languages. Second, solutions to the problem will move beyond current approaches to computer speech recognition. In mono-lingual spoken language systems, the objective is to determine the content of the speech. This requires that researchers cue in on small portions of the speech—frames, phonemes, syllables, and words, to determine what the speaker said. In contrast, in text-independent language identification, phonemes and other sub-word units alone are not sufficient cues, since phonemes, syllables and even words are common across different languages. One also needs to examine the utterance as a whole to determine the "acoustic signature" of the language, the unique characteristics that make one language sound distinct from another.

Aside from the fact that it is a challenging area of research, there are several important applications for automatic language identification. Much of the past funding for research in this area has been provided by government agencies interested in communications monitoring for national security purposes. However, there are also important commercial demands. As the global economic community expands, there is increasing need for automatic spoken language translation services. For example, checking into a hotel, arranging a meeting or making travel arrangements can be difficult for non-native speakers. Telephone companies will be better equipped to handle such foreign language calls if an automatic language identification system can be used to route the call to an operator fluent in that language.

Rapid language identification and translation can even save lives. There are many reported cases of 911 operators being unable to understand the language of a distressed caller. In response to these needs, AT&T recently introduced its *Language Line* Interpreter Service, to serve business, the general public and police departments handling 911 emergencies. The service uses trained human interpreters, can handle 140 languages and satisfies an important need in our increasingly cosmopolitan communities. However, tremendous responsibility is placed on the human operator who must route the call to the appropriate interpreter.

Finally, an automatic language identification system could also serve as a front-end for a multi-language translation system in which the input speech can be in one of several languages. The input language needs to be quickly identified before translation to the target language(s) can begin.

2 BACKGROUND

While the past two decades have witnessed tremendous advances in automatic speech recognition, our literature review revealed only thirteen published papers in automatic language identification [3, 7, 9, 10, 11, 12, 13, 14, 15, 16, 22, 23, 24], apart from our recent work on a 4-language identification system using high-quality speech [18]. The studies are summarized in Tables 1 and 2 and reviewed in some detail in [19].

The literature does not produce a coherent picture. The data used in these studies have spanned the range from phonetic transcriptions of text to telephone and radio speech. The number of languages has varied from three to twenty. The approaches to language identification have used "reference sounds" in each language, segment- and syllable-based Markov models, formant vectors, and acoustic, phonetic and prosodic features derived from broad phonetic categories. A variety of classification methods have been tried, including HMMs, expert systems, VQ, quadratic classifiers and artificial neural networks.

In addition to the many differences just noted, meaningful comparisons across studies are virtually impossible, for two main reasons:

- Many of the studies represented classified or sensitive research, so experimental details (e.g., languages used) are often not described.
- There is no common, public-domain database (cf. TIMIT [6, 17]) with which to evaluate different approaches to automatic language identification.

It is clear that basic research using a public-domain, multi-language speech corpus is an essential prerequisite to further advances in automatic language identification.

Table 1 Studies in Automatic Language Identification

STUDY	LANGUAGES	TYPE OF DATA	SPEAKERS	APPROACH	RESULTS
Texas Instruments (1973-80)	7 (not specified)	Read Speech	100 adult males (50 train 50 test)	Detection of "Reference Sounds" and estimation of log likelihoods of the languages	62% (no rejection) 100% (68% rejection)
House and Neuberg (1977)	American English, Chinese, Greek, Korean, Urdu, Japanese, Russian and Swahili (8)	Phonetic trans. of text (no real speech)	-	HMMs trained on sequences of broad category labels	Near-perfect discrimination (no % specified)
Li and Edwards (1980)	2 Asian & 3 Indo-European (not specified)	Read Speech	150 (50 train 50 evaluate 50 test)	Segment-based and syllable-based Markov models	80%
Cimarusti and Ives (1982)	American English, Czech, Farsi, Korean, German, Mandarin, Russian & Vietnamese (8)	Read Speech	40 (train and test sets unspecified)	Acoustic features and a polynomial decision function	84%
Ives (1986)	American English, Czech, Farsi, Korean, German, Mandarin, Russian & Vietnamese (8)	Spoken Speech (100 to 5000 Hz)	122 adult males (train and test sets unspecified)	Expert System Production Rules	92%
Foil (1986)	3 (not specified)	Speech from radio (SNR 5 dB)	Not specified	Processing Pitch & Energy Contours Formant-clustering algorithm	39% 64 % (11% rejection)
Goodman et. al (1989)	Four different sets of languages (not specified)	Speech from radio (SNR 9 dB)	Not specified	Improved Formant Clustering Algorithm	Reduced Foil's error rate in half (no % specified)

Table 2 Studies in Automatic Language Identification (continued)

STUDY	LANGUAGES	TYPE OF DATA	SPEAKERS	APPROACH	RESULTS
Sugiyama (1991)	20 languages (CCITT SG-XII CD-ROM)	Spoken speech (Avg. SNR 49.2 dB)	76 Males 77 Females	Standard VQ VQ histogram Algorithm	65% 80%
Savic et. al (1991)	English, Hindi, Mandarin & Spanish	Read Speech (0 - 4.5 kHz)	Not specified	HMMs & Pitch Contour Analysis	Not specified
Nakagawa et. al (1992)	English, Japanese, Mandarin & Indonesian	Conversational Speech (0 - 6 kHz)	60 Males	HMM-based methods	86.3%
Kwasny et. al (1992)	English & French	Read Speech (0 - 12 kHz)	1 Male 1 Female	Acoustic Features and ANNs	100%
Muthusamy et. al (1992)	English, Japanese, Mandarin & Tamil	Conversational Speech (0 - 8 kHz)	40 Males 40 Females	Broad phonetic category segment-based features & neural networks	79.5% (5.7s of speech) 89.5% (17.1s of speech)

3 A TELEPHONE SPEECH CORPUS FOR AUTOMATIC LANGUAGE IDENTIFICATION

3.1 Motivation

Research in speaker-independent automatic language identification requires a large corpus of multi-lingual speech data to capture the many sources of variability within and across languages. These include variability due to speaker differences (e.g., age, gender, dialect), microphones, telephone handsets, communication lines, background noise and the language being spoken. The corpus should also contain a variety of speech samples from each speaker, ranging from fixed-vocabulary utterances to natural, continuous speech. This makes it useful for both content-dependent and content-independent language identification.

In the real world, an automatic language identification system is likely to be used over some form of communication channel e.g., a telephone line, which may be characterized by low-bandwidth, channel distortion and non-linearities, microphone variability and low SNR. If the system is to perform accurately under these conditions, it needs to be trained on speech recorded under these conditions. It was therefore decided to collect speech in many languages over commercial telephone lines.

3.2 Language Selection

The languages currently in the corpus, English, Farsi (Persian), French, German, Korean, Japanese, Mandarin Chinese, Spanish, Tamil and Vietnamese, were selected based on a combination of linguistic considerations and the availability of native speakers in the United States.

These languages represent a range of unrelated languages (e.g., Vietnamese and Tamil and German) as well as languages from the same sub-family (e.g., Germanic languages such as English and German, Romance languages such as French and Spanish). The languages also include various prosodic features, e.g., Mandarin Chinese and Vietnamese are tonal languages, Japanese uses pitch-accents and syllabic mora. In addition to their linguistic characteristics, the languages represent important geographic and political regions, and many speakers of these languages can be found relatively easily in the U.S.

Since most approaches to language identification rely on discriminators based on patterns of sounds and sound classes, it is important that the corpus include pairs of languages that are phonologically similar and others that are quite distinct. For example, syllable patterns of Vietnamese and Chinese are similar, basically consonant-vowel (CV) or consonant-vowel-consonant (CVC) patterns, with a relatively limited consonant repertoire, and with a characteristic tonal contour associated with each syllable. In contrast, German and English have relatively elaborated syllable structures, potential clusters of half a dozen consonants between vowel nuclei, and no distinctive tonal contrasts at the syllable level. From the point of view of automatic language identification, Chinese and Vietnamese should be more confusable based on phonological sequences, and Chinese and German should be less confusable.

3.3 Data Acquisition

Collection Campaign. Speaker participation was promoted under a "donate your voice to science" theme. Requests for callers were posted on several university bulletin boards and national computer network newsgroups. In addition, a press release describing the research project and the need for volunteers resulted in newspaper and radio coverage. A toll-free telephone number in the U.S. and Canada, open round-the-clock, was provided.

Call Format. A touch-tone phone was needed for the call. Callers received a brief greeting in English followed by a prompt, in each language, to select a language by pressing a digit from 0 through 9. All subsequent instructions and prompts were given in the target language. This procedure helped reduce the number of crank calls by non-native speakers. The instructions and prompts in each language were pre-recorded by a native speaker of that language.

Recording Equipment. Speech was collected using a Gradient Technology Desklab connected via a SCSI port to a Sun 4/110 workstation. The device was programmed to answer the telephone, play digitized files in each of the ten languages requesting the speech samples, and digitize the callers' response for a designated period of time. Speech was sampled at 8000 samples per second at 14 bit resolution. As the Desklab did not have any automatic gain control mechanism, the recording gain was set to 10—the maximum level recommended by the manufacturer.

Recording Protocol. The recording protocol was designed to obtain (a) speech samples that create well-defined, useful vocabularies, (b) domain-specific descriptions, and (c) samples of elicited free speech.

Small useful vocabularies included (a) language names: the speakers' native language and the language they spoke most of the time, (b) the days of the week, and (c) the numbers 0 through 10. Domain-specific descriptions were obtained by asking callers to describe (a) some aspect of their home-town that they liked, (b) the climate in their home-town, (c) the room that they were calling from, and (c) their most recent meal. Elicited free speech was obtained by asking callers to speak for 1 minute on any topic of their choice. Several hints and suggestions about the possible topics were provided: recite a poem, make up a story, describe your favorite sport or hobby, etc. They were then given 10 seconds to organize their thoughts before the actual 1 minute recording. This was done to minimize the number of long pauses and false starts in the free speech. Each speaker contributed 9 utterances, a total of approximately 126 seconds of speech to the corpus. The duration of each call was approximately 5 minutes. A complete transcript of the protocol can be found in [21].

3.4 Corpus Development

Development of this corpus was divided into two phases. Phase I consisted of (a) **preliminary verification**: listening to each utterance and deleting prank or invalid calls (hangups); (b) **chopping**: removing excess noise at the beginning and end of each utterance; (c) **evaluation**: making several judgments about the quality and type of speech; and (d) **broad phonetic transcriptions**: providing time-aligned broad phonetic labels to a subset of the utterances.

Phase II involved (a) **verification and evaluation** of the utterances by native speakers of the individual languages, (b) **orthographic transcriptions** of each utterance, and (c) **time-aligned fine phonetic transcriptions** of a subset of the utterances for each language. Stages (b) and (c) were done by trained human transcribers. The details of these stages are provided in [20, 21].

We received a total of 2490 calls over a period of eight months. Of these, 1044 calls were in English, with an average of 144 calls in the remaining 9 languages. On the average, 22.0% of the calls were rejected in each language, mainly because of hangups. A total of 1987 calls (about 43 hours of speech), 868 in English, and an average of 122 calls in the remaining 9 languages, were judged as useful after verification by native speakers.

3.5 OGI_TS

Ninety calls in each language were then selected to form the OGI Multi-language Telephone Speech Corpus, or OGI_TS, for short. The criterion for including a call in this corpus was that it contain at least 2 spontaneous speech utterances (in addition to the four fixed-vocabulary utterances listed above). Using this constraint, it was found that Korean had the least number of such calls, 90. To ensure that the data from each language was of comparable size, the first 90 calls from each language that satisfied the above constraint were selected for the corpus.

OGI_TS has been placed in the public domain. It has been donated to NIST[1] and LDC[2], and can be obtained from the Center for Spoken Language Understanding at the Oregon Graduate Institute.

---

[1] National Institute of Standards and Technology.
[2] Linguistic Data Consortium.

4 HUMAN LISTENING EXPERIMENTS

How well can human listeners discriminate among these languages from excerpts of speech? To determine human listening performance on excerpts of speech from the 10 languages, 7 male and 4 female monolingual native English speakers were presented with 1-, 2-, 4- and 6-second excerpts of spontaneous speech excised from the 10 languages.

Experimental Procedure

The experiment was conducted using an interactive graphics program that played excerpts of speech chosen at random from each of the 10 languages, and maintained a log of subject responses. Following a brief training session, subjects were presented with 760 different excerpts, 19 at each duration from each language. The subjects could listen to each excerpt as many times as they desired. After responding, they were given feedback on every trial. The subjects could also listen to an excerpt *after* making the choice—a feature that was included to aid in the learning process. Each block of 100 trials was considered a session, and the program automatically quit after every 100 trials, to ensure that the subjects did not get fatigued.

Results

The average listener performance for each language is shown for the four durations in Figure 1. As duration increased from 1 to 2 to 4 to 6 seconds, the average performance over all languages rose from 37.0% to 43.0% to 51.2% to 54.6% respectively. It can be seen that English, spoken by all the subjects, was identified at 91.9%, 94.3%, 100% and 99%, for excerpts of 1, 2, 4 and 6 seconds, respectively. Excluding English, the average performance at each duration was 20.7%, 37.4%, 45.8% and 49.7%. Note the relatively high performance on French, German and Spanish—languages that the listeners were most often exposed to, either through courses or by contact with foreign friends. Performance on Farsi, Korean, Tamil and Vietnamese—languages that the listeners had never been exposed to, was very poor.

Analysis of performance by each block of 190 trials revealed little evidence of learning during the experiment. For example, for 6-second excerpts, the average performance on Korean for the first and last 190 trials was 13.5% and 16.7% respectively.

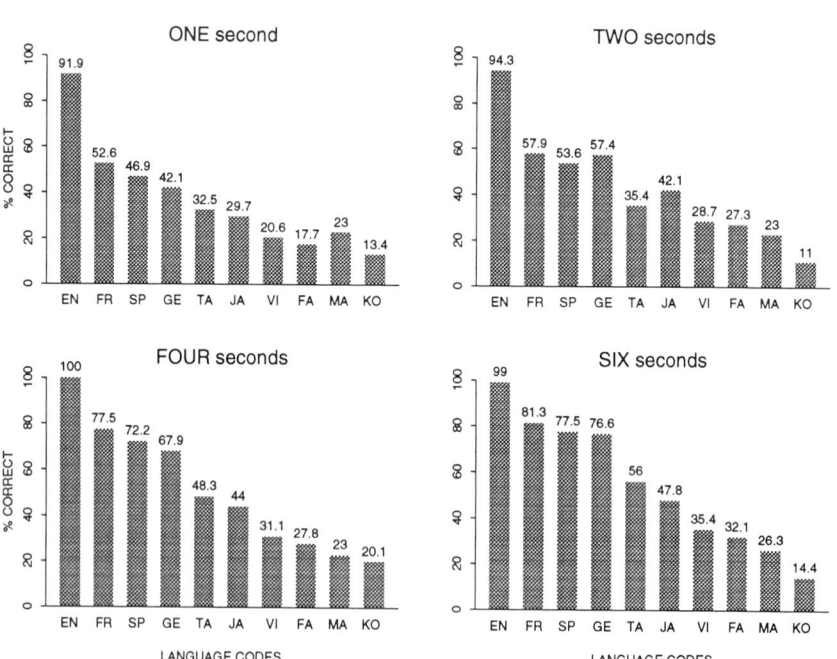

Figure 1 Average Listener Performance for the Four Durations

5 A SEGMENTAL APPROACH TO AUTOMATIC LANGUAGE IDENTIFICATION

A segmental approach to automatic language identification assumes that each language has a unique acoustic structure, and that this structure can be defined in terms of segmental and prosodic features of speech. Segmental features include the inventory of broad phonetic segments such as vowels, fricatives, stops, nasals and closures, and their frequency of occurrence and co-occurrence in speech. Prosodic information consists of the relative durations and amplitudes of sonorant (vowel-like) segments, their spacing in time, and patterns of pitch change within and across these segments.

To the extent that these assumptions are valid, languages can be identified automatically by segmenting speech into broad phonetic categories, computing segment-based features that capture the relevant phonetic and prosodic structure, and training a classifier to associate the feature measurements with the spoken language. The segment-based features are linguistically motivated, i.e., based on an analysis of the language-specific and language-universal properties of the broad phonetic sequences. In addition, the use of segments as anchors for prosodic analysis of speech augments the power of this approach.

5.1 Broad Phonetic Segmentation

Segmentation is performed by a fully-connected, feed-forward, three-layer neural network that assigns 7 broad phonetic category scores to each 3 ms time frame, similar to the one described in [18]. The 7 broad phonetic categories are: vowel (VOC), fricative FRIC), stop (STOP), pre-vocalic sonorant (PRVS), inter-vocalic sonorant (INVS), post-vocalic sonorant (POVS), and silence or background noise (CLOS).

The input to the segmenter consists of 120 spectral features derived from a PLP analysis [8] of the waveform. The features were empirically derived to capture the contextual information in the vicinity of each frame [5]. The number of hidden neurons in the network is determined experimentally.

Training and Test Sets

The segmentation algorithm was trained and tested on utterances from the first 25 valid calls in each language. The training set consisted of 300 utterances;

2 per call for 15 calls from each language. The development test set consisted of 100 utterances; 2 per call from a different set of 5 calls in each language. The final test set also consisted of 100 utterances; 2 per call from yet another set of 5 calls in each language. The average duration of the utterances was 4.0 seconds.

The development test set was used to fine-tune the feature set and to determine the optimal network configuration. The feature set and network configuration that produced the best classification performance on the development set were then used to test on the final test set. This procedure ensured that the reported results were obtained on uncorrupted test data.

Segmenter Training

The training and test utterances were hand-labeled by experts with the seven broad phonetic categories. Since it was not feasible to train the network on each time frame of each utterance, frames were chosen at random from the hand-labeled utterances in the training set. The network was trained using backpropagation with conjugate gradient optimization [1]. The frame-by-frame outputs of the segmenter were converted into a time-aligned sequence of the 7 broad phonetic category labels using a Viterbi search with duration and bigram constraints.

Segmenter Evaluation

The segmenter was scored on the final test set using two different scoring procedures. In the first one, the labels output by the segmenter were compared frame-by-frame with the hand-labels, and the percentage of total frames in agreement was computed. In the second method, a string alignment and scoring program developed by NIST was used. This algorithm treats each segment as a word and measures the number of insertions, deletions and substitutions in the segmenter output with respect to the hand-labels. Word accuracy is given by the equation:

$$Word\ Accuracy = \%Correct - \%Insertions \qquad (1)$$

With the first method, the performance accuracy was 79.8%. This compares favorably with 85.1% for 4 languages using high-quality speech [18]. With the NIST algorithm, the word recognition performance was 81.8% correct, with 6.9% substitutions, 11.2% deletions, and 9.6% insertions. By equation (1), the word accuracy was 72.2%.

5.2 Language Classification

In this stage, a number of features were computed on the waveform and on the time-aligned sequence of broad phonetic labels output by the segmenter. These features were then input to an artificial neural network that classified the languages.

Training and Test Sets

The language classifiers were trained and evaluated on only the spontaneous speech utterances from the first 70 valid calls in each language. The training set consisted of 2714 utterances (from 342 males and 158 females); 2-6 utterances per call for 50 calls in each language. The development test set consisted of 1120 utterances (from 151 males and 49 females); 2-6 utterances per call for 20 calls in each language[3]. The utterances ranged in duration from 1 second to 50 seconds with an average of 13.4 seconds.

To determine the effect of longer duration utterances on the identification accuracy, the networks were also evaluated on just the 50-second "story" files from each call in the development test set. There were 186 "story" files, one per speaker, with an average duration of 45.3 seconds.

Feature Development

The design of the features was aided by statistical analyses of the broad phonetic sequences and phonological knowledge of the languages. For example, the presence of tonal languages like Mandarin Chinese and Vietnamese in the data set led us to design pitch features that captured the large variation in pitch within and across segments for utterances from these two languages. Similarly, the fact that syllables are about equally spaced in Japanese led us to include an "inter-segment duration difference" feature. The segment-pair and segment-triple features were motivated by the informal knowledge that languages differed in the characteristics of pairs and triples of broad phonetic categories.

The final set contained 333 features. These features were global in the sense that they were computed over the entire length of the utterance rather than in

---

[3] Results on the final test set consisting of 1077 utterances from 20 calls in each language are not reported here in deference to the NIST language identification evaluation in progress (see Section 6).

a moving analysis window. Since all the input vectors to the network had to be of the same length, the features were designed to yield the same number of values regardless of the duration of the utterance. A partial list of these features, grouped by type, is given below. The numbers in parentheses refer to the number of values generated. The complete list of 333 features is provided in [21].

- **Segment-pair Features (80 values).** There are 20 legal segment-pairs of the seven broad phonetic categories. Four feature sets based on these 20 segment-pairs were examined:
    - frequency (number of occurrences per second) (20 values)
    - ratio of the number of occurrences to the total number of segments in the utterance (20 values)
    - median duration of each segment-pair in an utterance (20 values)
    - ratio of the total duration of all occurrences of a segment-pair in an utterance to the total utterance duration (20 values)
- **Segment-triple Features (116 values).** There are 63 legal segment-triples of the seven broad phonetic categories. An analysis of variance using the one-way layout model indicated that five of these triples had p-values greater than 0.01. These were discarded. For the remaining 58 triples, the following features were computed:
    - frequency (number of occurrences per second) (58 values)
    - ratio of the number of occurrences to the total number of segments in the utterance (58 values)
- **Pitch-based Features (8 values).** A neural network-based pitch tracker, developed by Burnett [2] was used in the computation of the following features:
    - Intra-segment pitch variation: Average of the standard deviations of the pitch within all sonorant segments—VOC, PRVS, INVS and POVS (4 values)
    - Inter-segment pitch variation: Standard deviation of the average pitch in all sonorant segments (4 values)
- **Frequency of Occurrence (11 values).** These include frequency of occurrence of each of the seven broad phonetic labels (7), all segments (1), sonorants (1), obstruents (1), and voiced obstruents (1). The term "obstruent" refers to STOPs and FRICs combined. An obstruent is considered

voiced if more than half the segment was labeled as voiced by the voicing detector [2].

- **Segment Occurrence Ratios (47 values).** Mere frequency of occurrence of individual broad phonetic categories does not give the complete picture. Ratio of occurrence, i.e., number of occurrences of a particular category divided by the total number of segments, in conjunction with the frequency of occurrence, normalizes speech rate differences among speakers of the same language. Examples of segment occurrence ratios are:
    - Ratio of number of occurrences of each of the seven broad phonetic categories to the total number of segments (7 values)
    - Ratio of number of occurrences of each of the seven broad phonetic categories to the total number of sonorants (7 values)
    - Ratio of number of occurrences of each of the seven broad phonetic categories to the total number of obstruents (7 values)
    - Ratio of number of occurrences of sonorants to that of obstruents (1 values)

- **Segment Duration Ratios (47 values).** Apart from speech rate differences among speakers, one needs to examine intrinsic speech rate differences across languages. To do this, it is necessary to examine segment *duration* ratios in conjunction with segment *occurrence* ratios. The 47 duration ratio features can be obtained from the list of segment occurrence ratios by replacing "number of occurrences" and "total number" with "duration" and "total duration" respectively.

- **Duration (24 values).**
    - Inter-segment duration difference: absolute difference in durations between successive segments. The features computed were the minimum, median, average, standard deviation and maximum of inter-segment duration difference. To avoid using outliers, the minimum and maximum values were obtained by generating a histogram of the duration differences and choosing the 5th and 95th percentile values (5 values)
    - Vowel center distance: distance between the centers of successive vowels. The features computed were the minimum, median, average, standard deviation and maximum vowel center distance. (5 values)
    - Average duration of the seven broad phonetic labels (7 values)
    - Standard deviation of the duration of the seven broad phonetic labels (7 values)

Classification Experiments

The following classification experiments were conducted, all using the above set of 333 features:

- A single network to classify all 10 languages
- A single network to classify English, Japanese, Mandarin Chinese and Tamil (for the sake of comparison with the high-quality system described in [18]).
- Nine English–L' networks, where L' is one of the remaining 9 languages
- Ten $L - Other$ networks, where L is one of the ten languages
- Nine English–L' – $Other$ networks

All the networks were trained with backpropagation with conjugate gradient optimization. Approximately equal numbers of utterances were chosen at random for the *Other* languages.

Networks were trained on features extracted from the utterances in the training set, and after a number of iterations through the training vectors, the network was evaluated on the feature vectors from the development test set. The network which provided the best result on the development test set is reported here.

Results

For the short utterances (13.4 seconds average duration), the 10-language network performed at an accuracy of 48.5% and the four-language network at 69.7%. The corresponding scores for the "story" utterances were 65.6% and 82.2%. In comparison, the corresponding 4-language classifier trained on high-quality speech performed at an accuracy of 89.5% on test utterances that were 17.1 seconds long on the average [18]. The results of the remaining three experiments are shown in Tables 3, 4, and 5. $Accuracy_S$ and $Accuracy_L$ refer to the identification performances on the short and long utterances:

- It can be seen that the English–L' classification (Table 3) is the least difficult, with performances ranging from 69.0% (English-Farsi) to 87.7% (English-Tamil) for the short utterances, and from 81.1% (English-German)

Table 3 Results of the English–L' Experiment

Network	$Accuracy_S$ (%)	$Accuracy_L$ (%)
English-Farsi	69.0	83.8
English-French	79.1	86.8
English-German	73.4	81.1
English-Japanese	80.6	91.4
English-Korean	74.6	86.1
English-Mandarin	75.0	83.8
English-Spanish	80.1	91.9
English-Tamil	87.7	97.3
English-Vietnamese	80.2	91.7

to 97.3% (English-Tamil) for the "stories". The median accuracies were 79.1% and 86.8% respectively.

- Classification of individual languages against all others (L-$Other$) produces about the same level of performance (Table 4), from 63.7% (English-Other) to 86.2% (Mandarin-Other) for the short utterances, and 80.6% (English-Other) to 97.3% (Tamil-Other) for the "stories", with median accuracies of 77.0% and 87.5% respectively.

- English–L'-$Other$ classification is more difficult (Table 5), with performances ranging from 53.3% (English-Farsi-Other) to 64.9% (English-Mandarin-Other) for the short utterances, and from 69.2% (English-Vietnamese-Other) to 81.1% (English-Tamil-Other) for the "stories," with median accuracies of 59.1% and 70.6% respectively.

6 SUMMARY AND FUTURE WORK

We have described the results of automatic language identification experiments using segment-based features and artificial neural networks. The features were computed on sequences of broad phonetic categories as determined by a neural network-based segmentation algorithm. The results of the classification experiments indicate that for utterances of average duration 13.4 seconds, performance on constrained tasks (two languages) is reasonably high, 87.7%, falls to 69.7% for four selected languages, and to 48.5% for all ten. The results

Table 4 Results of the $L - Other$ Experiment

Network	$Accuracy_S$ (%)	$Accuracy_L$ (%)
English-Other	63.7	80.6
Farsi-Other	71.2	86.5
French-Other	76.2	84.2
German-Other	75.7	83.8
Japanese-Other	77.7	88.6
Korean-Other	72.8	83.3
Mandarin-Other	86.2	94.6
Spanish-Other	78.5	94.6
Tamil-Other	85.1	97.3
Vietnamese-Other	81.9	91.7

Table 5 Results of the English$-L'-Other$ Experiment

Network	$Accuracy_S$ (%)	$Accuracy_L$ (%)
English-Farsi-Other	53.3	69.8
English-French-Other	59.9	74.1
English-German-Other	59.4	73.6
English-Japanese-Other	58.4	70.6
English-Korean-Other	55.1	73.1
English-Mandarin-Other	64.9	69.8
English-Spanish-Other	58.6	69.8
English-Tamil-Other	64.4	81.1
English-Vietnamese-Other	60.8	69.2

of the human perceptual experiments, 49.7% for 6-second excerpts of speech (excluding English), provide an interesting benchmark on this difficult task (although the average duration of speech was over twice as long for machine classification).

There are obvious limitations to a broad phonetic approach to automatic language identification. Given that humans often recognize a language by cuing into specific sounds (or phonemes) of the language, it is clear that exploiting differences between languages at the phonemic and phonetic levels is the key to more accurate language identification. This requires the availability of a phonemically or phonetically labeled corpus of data. Recognizing the difficulties involved in arriving at an accurate *phonetic* transcription of speech, we are currently producing quasi-*phonemic* transcriptions of the languages in OGI_TS. This will allow us to pursue a phonemic approach to automatic language identification in future studies.

A new initiative in automatic language identification is now underway in the United States. Researchers at eight different sites across the United States are working on different approaches to automatic language identification using OGI_TS. The evaluation of these different approaches is being conducted by NIST. The tasks consist of identification of all ten languages and the language groups $L - Other$, English$-L'$, and English$-L' - Other$ described in Section 4.2.

Results of a HMM-based approach to automatic language identification using OGI_TS have been reported by Zissman [25]. Using continuous observation, ergodic hidden Markov models (HMMs) with tied Gaussian observation probability densities, he obtained 46.0% identification accuracy on the ten-language task (with short utterances). His results on the English$-L'$ and $L - Other$ tasks were also comparable to our results. He did not examine the English$-L-Other$ task.

This multi-site evaluation augurs well for the future of research in automatic language identification. By the collection, development and distribution of OGI_TS, we have achieved our twin objectives of providing a public-domain, multi-lingual speech corpus and fostering basic research in the field. We are confident that the interaction and exchange of ideas resulting from different sites working on different approaches to the problem using the same speech corpus will result in significant contributions to the area of automatic language identification.

Acknowledgments

This research was funded by NSF, ONR and the Center for Spoken Language Understanding. The authors thank Mark Fanty, Etienne Barnard and Todd Leen for their insightful comments and discussions, and Terri Durham for running the perceptual experiments.

REFERENCES

[1] E. Barnard and R. A. Cole. A neural-net training program based on conjugate-gradient optimization. Technical Report CSE 89-014, Department of Computer Science, Oregon Graduate Institute of Science and Technology, 1989.

[2] D. Burnett. Toward multi-language pitch tracking for telephone speech. Presented at the 1992 OGI CSE Student Research Symposium, May 1992.

[3] D. Cimarusti and R. B. Ives. Development of an automatic identification system of spoken languages: Phase 1. In *Proceedings IEEE International Conference on Acoustics, Speech, and Signal Processing 82*, Paris, France, May 1982.

[4] S. J. Eady. Differences in the F_0 patterns of speech: Tone language versus stress language. *Language and Speech*, 25(1):29–42, 1982.

[5] M. A. Fanty, R. A. Cole, and K. Roginski. English alphabet recognition with telephone speech. In J. E. Moody, S. J. Hanson, and R. P. Lippmann, editors, *Advances in Neural Information Processing Systems 4*, San Mateo, CA, 1992. Morgan Kaufmann Publishers.

[6] W. Fisher, G. R. Doddington, and K. Goudie-Marshall. The DARPA speech recognition research database: Specification and status. In *Proceedings DARPA Speech Recognition Workshop*, pages 93–100, February 1986.

[7] J. T. Foil. Language identification using noisy speech. In *Proceedings IEEE International Conference on Acoustics, Speech, and Signal Processing 86*, Tokyo, Japan, 1986.

[8] H. Hermansky. Perceptual linear predictive (PLP) analysis of speech. *Journal of the Acoustical Society of America*, 87:1738–1752, April 1990.

[9] A. S. House and E. P. Neuberg. Toward automatic identification of the language of an utterance. I. Preliminary methodological considerations. *Journal of the Acoustical Society of America*, 62(3):708–713, 1977.

[10] R. B. Ives. A minimal rule AI expert system for real-time classification of natural spoken languages. In *Proceedings 2nd Annual Artificial Intelligence and Advanced Computer Technology Conference*, Long Beach, CA, April-May 1986.

[11] S. C. Kwasny, B. L. Kalman, W. Wu, and A. M. Engebretson. Identifying language from speech: An example of high-level, statistically-based feature extraction. In *Proceedings 14th Annual Conference of the Cognitive Science Society*, 1992.

[12] R. G. Leonard and G. R. Doddington. Automatic language identification. Technical Report RADC-TR-74-200, Air Force Rome Air Development Center, August 1974.

[13] R. G. Leonard and G. R. Doddington. Automatic language identification. Technical Report RADC-TR-75-264, Air Force Rome Air Development Center, October 1975.

[14] R. G. Leonard and G. R. Doddington. Automatic language discrimination. Technical Report RADC-TR-78-5, Air Force Rome Air Development Center, January 1978.

[15] K.P. Li and T. J. Edwards. Statistical models for automatic language identification. In *Proceedings IEEE International Conference on Acoustics, Speech, and Signal Processing 80*, Denver, CO, April 1980.

[16] R. G. Leonard. Language recognition test and evaluation. Technical Report RADC-TR-80-83, Air Force Rome Air Development Center, March 1980.

[17] L. Lamel, R. Kassel, and S. Seneff. Speech database development: Design and analysis of the acoustic-phonetic corpus. In *Proceedings DARPA Speech Recognition Workshop*, pages 100–110, February 1986.

[18] Y. K. Muthusamy and R. A. Cole. A segment-based automatic language identification system. In J. E. Moody, S. J. Hanson, and R. P. Lippmann, editors, *Advances in Neural Information Processing Systems 4*, San Mateo, CA, 1992. Morgan Kaufmann Publishers.

[19] Y. K. Muthusamy and R. A. Cole. A Review of Research in Automatic Language Identification. Technical Report CSE 92-009, Center for Spoken Language Understanding, May 1992.

[20] Y. K. Muthusamy, R. A. Cole, and B. T. Oshika. The OGI Multi-language Telephone Speech Corpus. In *Proceedings International Conference on Spoken Language Processing 92*, Banff, Alberta, Canada, October 1992.

[21] Y. K. Muthusamy. A Segmental Approach to Automatic Language Identification. Doctoral Dissertation, Oregon Graduate Institute of Science & Technology, Beaverton, OR, July 1993.

[22] S. Nakagawa, Y. Ueda, and T. Seino. Speaker-independent, text-independent language identification by HMM. In *Proceedings International Conference on Spoken Language Processing 92*, Banff, Alberta, Canada, October 1992.

[23] M. Savic, E. Acosta, and S. K. Gupta. An automatic language identification system. In *Proceedings IEEE International Conference on Acoustics, Speech and Signal Processing 91*, Toronto, Canada, May 1991.

[24] M. Sugiyama. Automatic language recognition using acoustic features. Technical Report TR-I-0167, ATR Interpreting Telephony Research Laboratories, 1991.

[25] M. A. Zissman. Automatic Language Identification Using Gaussian Mixture and Hidden Markov Models. In *Proceedings IEEE International Conference on Acoustics, Speech and Signal Processing 93*, Minneapolis, MN, April 1993.

13

TEXT-INDEPENDENT TALKER VERIFICATION USING COHORT NORMALIZED SCORES

David J. Burr

Bellcore

1 INTRODUCTION

It is difficult to implement talker recognition on the telephone network because of normal variation in the channel characteristics. The primary component of variation is due to the different telephone handsets or microphone frequency characteristics (Rosenberg and Soong, 1992). Lack of availability of telephone speech databases has also contributed to slow progress in the solution of these problems, though clean speech databases such as TIMIT (Garofolo *et al.*, 1988) have been available. A telephone speech database suitable for talker identification research (Godfrey, 1992) was not generally available at the time of this research.

In place of a real database, a simulated telephone quality speech database (Yang, 1992) was used. Yang built a telephone simulator, which he then used to generate a telephone quality version of the TIMIT database, called STIMIT. With STIMIT it was possible to study several factors affecting the implementation of talker verification on the telephone.

A widely used algorithm for text-independent talker identification is based on vector quantization (Soong *et al.*, 1985). Though highly accurate on clean speech, the performance of this algorithm degrades appreciably for telephone quality speech. This requires ways to restore performance. Two classes of techniques were considered: those that (1) make the acoustic feature representation more robust to channel variations, and (2) improve the decision making process used to verify talkers. Two solutions in each category were studied. In the acoustic feature category dynamic information was added to the spectral representation since this is claimed to be more resistant to channel variability. Zero-crossing rate was also added, since this is related to pitch, which is important in talker recognition (Atal, 1976). Pitch is a feature that is little affected by channel variations. Decision making was improved by use of a self-adjusting dynamic threshold based on a cohort or reference set (Higgins *et al.*, 1991). A set of "cohorts" assigned to each talker generates an acceptance threshold which compensates for some kinds of trial-to-trial variations. The acceptance threshold is just the average score of all the cohort models. Finally, a multilayer perceptron (MLP) was used for the final decision output function since it is known to be powerful method for combining sensory data.

MLP's are useful for this problem because they can be effectively integrated with conventional methods. Both raw and normalized scores from the conventional method are combined using an MLP trained on example scores. The result is an improvement in recognition scores over a strictly conventional method based on setting a global acceptance threshold.

When analyzing the entire 630 talker set in the STIMIT database it is shown that over 50 percent of the error can be eliminated using the cohort normalization technique combined with MLP scoring. When the length of the training sentence is increased, performance increases to cut the error again in half. The final performance figure is 93.7 percent correct verifications on 1260 trials consisting of 630 true talkers and 630 randomly selected impostors. This figure is comparable to other reported results using cohort normalization for talker verification.

2 TALKER IDENTIFICATION BACKGROUND

Talker identification can be classified into text-dependent and text-independent paradigms. In text-dependent processing, acoustic word or phoneme models dedicated to each talker are used to score an utterance prompted by the system. In text-independent mode, an average acoustic model of the talker's speech is constructed, independent of the specific words spoken. In text-independent processing the talker is identified without prompting for a specific utterance. This is an advantage, as it allows a more user friendly interaction, but security can be broken with a recording of the true talker's speech. It also has the advantage that talker can be identified unobtrusively.

Talker identification can be further grouped into three additional categories: (1) talker recognition, which is a closed set determination; (2) talker verification, which is open set, and (3) talker change detection. In talker recognition N talker models are available and the system uses them to decide which of the N talkers is present. In talker verification an identity claim is made and the system decides to accept or reject the claim based on a stored model of the talker. In talker change detection, two or more talkers are present on a channel and the system must label the time intervals corresponding to each.

Accuracy for both text-dependent and text-independent talker identification increases with the length of the test utterance. However, for the same length utterance, text-dependent methods are generally more accurate. High accuracies approaching 100 percent have been reported for small talker populations using a single high bandwidth microphone for training and testing. However, substantial performance degradations have been reported for both methods when training and testing over different microphones.

Vector quantization (VQ), hidden Markov models (HMM), and multilayer perceptrons (MLP) have all been applied to talker identification. More recently hybrid approaches which combine one or more of the above methods have become an area of intense interest.

3 RELATED WORK

Matsui and Furui (1992) claim that VQ-based methods are superior to ergodic HMM's since parameters are fewer and hence more readily estimated. They say that a one-state model is sufficient since the transition information is not reliable. However Soong and Rosenberg (1992) prefer the HMM approach. They also claim that those features which are robust to channel variations (i.e. delta spectra) are not as good for talker discrimination as the absolute spectra, which degrade with additive channel variations.

Chang (1992) describes an augmented acoustic map method which tracks the derivative of phase spectrum (DPS) to enhance conventional energy-based acoustic features. He shows improvement over conventional methods when tested on 11 different telephone handsets, based on the ability to track formant frequency.

Non-telephone based talker identification generally yields higher accuracies since high quality microphones are used and handset variability is not usually an issue. Savic and Sorenson (1992) argue that not all speech is equally suited to talker verification. Their system first detects those frames containing the vowel *iy* and then uses these frames to discriminate between talkers. Equal-error rates of 7 percent were obtained using the LPC cepstrum. (The equal-error rate is that rate which balances the accept and reject errors). Hattori (1992) reports 92 percent text-independent recognition on 24 female talkers. He used an auditory front end with a predictive neural network to model spectral transitions. Data were obtained by selecting a subset of talkers from TIMIT whose spectra were very similar to each other.

Lund and Lee (1992) apply sequential decision strategy using Wald's sequential probability ratio test (SPRT) to a talker population of 6 males and 6 females. They obtain an equal error rate of 5% on clean speech using an average of 2.6 sentences per verification. Utterances were recoreded as digit pairs selected at random and the VQ distortion approach was used to model each talker.

4 COHORT NORMALIZATION

Higgins *et al.* (1991) describe a method for computing an acceptance threshold for talker verification. Each talker is assigned a distinct group of similar talkers, called the cohort set. They show that sets with five members are optimal (i.e. increasing the number does not improve accuracy). Verification based on scores for the talker model and its five cohorts are computed. If the score for the talker is lower than the average score of the five cohorts, the talker is accepted, otherwise it is rejected. This method is very sensitive and is claimed to be able to normalize out varying channel and microphone characteristics. Higgins *et al.* report an equal-error rate of 3.3% using two-number phrases such as "seventy-five, thirty-four, sixty-seven". However, their results were obtained using a clean speech database of 178 talkers.

Rosenberg *et al.* (1992) show that cohort normalization is theoretically sound since it is derived from a likelihood ratio. They applied cohort normalization to the scoring of an HMM-based text-dependent talker identification system. They trained on carbon button microphones and verified on both carbon button and electret microphones. A talker population of 19 males and 19 females was used. Features consisted of 12 LPC-derived cepstra and delta cepstra. Training and verification utterances consisted of continuous digit strings of five numbers. They demonstrated a fivefold reduction in equal-error rate when going from unnormalized scores to cohort normalized scores for the cross microphone condition. The best equal error rate for cross microphones was

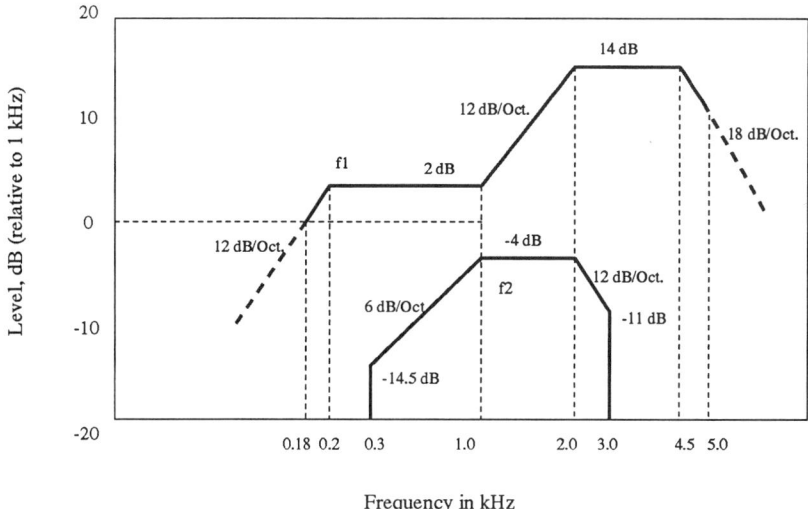

Figure 1 EIA transmit frequency response limits for an analog microphone

4.8% and for same microphones, 2.6%. Note that this does not model cordless vs. non-corded, mobile vs. wire-line phones, etc.

5 STIMIT

Yang (1992) designed a telephone network simulator which is intended to introduce telephone distortion and bandwidth characteristics into clean audio data. It consists of a combination of three modules: (1) additive noise, (2) microphone transfer function, and (3) μ-law compansion and expansion. Typical signal-to-noise ratios existing in the field were used in the additive noise module, one ratio randomly selected for each sentence. Microphone transfer characteristic curves were simulated by linear interpolation between the two curves specified in the EIA standard RS-470 (Figure 1.). A random interpolation value from 0 to 1 was selected for each sentence. Finally mu-law compansion and expansion was applied to simulate the telephone codec of eight bits resolution. Each sentence of TIMIT was modified with different values of random signal-to-noise ratio and microphone characteristic curves.

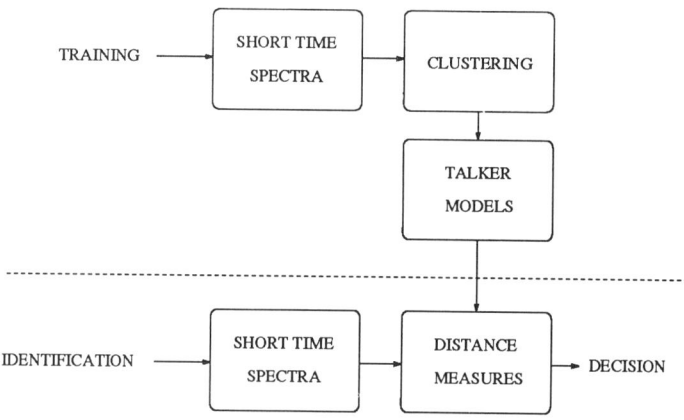

Figure 2 Block diagram of the talker identification system

6 TALKER IDENTIFICATION SYSTEM

A block diagram of a text-independent talker identification system is illustrated in Figure 2. A text-dependent system might be very similar. The diagram is divided into two phases, a training and an identification phase. In training, talker models are constructed from short time spectra of the speech signal. The spectra are encoded into cluster sets which reduce the number of parameters. A separate cluster set is formed for each talker.

In the identification phase, short time spectra are again computed. They are compared with the set of N talker models, and a score is generated for each. The model which generates the lowest score (distance) identifies the talker.

7 VECTOR QUANTIZATION

Soong *et al.* (1985) describe a technique for text-independent talker identification based on vector quantization. Vector quantization is a method for condensing or encoding the spectral space by replacing several similar spec-

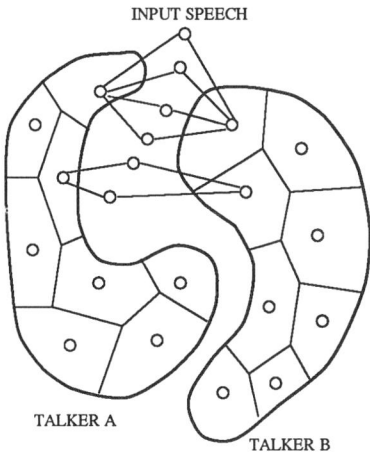

Figure 3 Illustration of VQ talker models and scoring of an input speech signal. The score for talker A is the sum of the distances from the input speech vectors to the talker A model.

tral vectors with a single vector (Figure 3.). Typically a spectral space is represented by a fixed number of these vectors independent of the number in the original data. This fixed set of vectors is referred to as a codebook. In VQ-based talker identification a separate codebook is computed for each talker using one or more training utterances. Identification of an unknown talker is done by computing the VQ distortion of the test utterance relative to each talker codebook. That codebook which produces the least distortion is assumed to be that of the unknown talker, and thus the talker is identified. Codebooks are generated using the conventional Linde-Buzo-Gray algorithm (Linde *et al.* 1980).

Neural networks are useful in the vector quantization model as they allow the combining of VQ scores in a somewhat optimal manner. Scores are combined by the nonlinear weighting of an MLP in such a way as to minimize the activation error.

8 EXPERIMENTS ON 90 TALKERS

STIMIT speech files quantized at 8 kHz were used in the experiments along with clean TIMIT files quantized at 16 kHz for comparison purposes. Sixty-Four element talker codebooks were generated using the LBG algorithm on 16-dimensional cepstral vectors. Cepstral coefficients were computed using the FFT spectrum on 32 msec Hamming windowed speech frames shifted every 16 msec. The first 16 non-energy cepstral coefficients were used to represent each frame of speech.

STIMIT contains ten sentences spoken by each talker. Of these ten sentences, eight are different for each speaker, and two are the same – referred to as *dialect normalization* sentences. Of the eight variable sentences, five were obtained at MIT, and three were obtained at Texas Instruments (TI). The two dialect normalization sentences were used to build each talker codebook. These sentences are the same for each talker. The remaining eight sentences (different for each talker) were combined into one long utterance for testing purposes. This produced test utterances whose average lengths were 25 seconds. A squared Euclidean distortion score was used to compute VQ distortion relative to each talker codebook.

Preliminary results on the telephone simulated speech were reported in Burr (1992). In a talker recognition experiment using 45 male talkers in three groups of 15, seven errors were made in all, for an accuracy of 84% (see Table 1). The talker was recognized as the one whose codebook generated the minimum score out of the 15. In a similar experiment on 45 females, nine errors were made, for an accuracy of 80%. Separate codebooks were generated for cepstra and delta cepstra, but the zero crossing rate was added as the 17th component of the cepstral vector.

	Telephone	Clean (16 kHz)
Male:	84%	100%
Female:	80%	100%

Table 1 Talker Recognition Accuracy.

The cepstral vector $C(j)$ was augmented with a delta cepstrum $D(j)$ computed by regression on twelve consecutive frames of speech (Furui, 1986).

$$D(j) = \sum_{i=-6}^{6} i * C(j+i) \qquad (1)$$

Addition of delta cepstra to the input features improved performance to 87% for both males and females. When cepstra were augmented with the zero crossing rate (sum of positive and negative zero crossings in the Hamming window interval) performance increased to 91% for males. When all three features were used, performance increased again to 98% for males and 93% for females.

	CEP	CEP+DCEP	CEP+ZCR	CEP+ZCR+DCEP
Male:	84%	87%	91%	98%
Female:	80%	87%	89%	93%
Both:	82%	87%	90%	96%

Table 2 Telephone Quality Speech.

As can be seen from the combined male and female scores, the greatest improvement per additional parameter occurred with the addition of zero crossing information. This increased the total parameters to 17 from 16 and improved performance by 8%, while delta cepstra doubled the number of parameters from 16 to 32, but improved performance by only 5%. Therefore in the following experiments on larger number of talkers, only the 17 dimensional pitch enhanced features set was used for computational efficiency.

9 EXPERIMENTS ON ALL 630 TALKERS

The entire STIMIT database of 630 talkers was analyzed for an experiment in talker verification using cohort normalization. The same parameters were used to compute FFT-derived cepstra with the exception that codebook size was limited to 32 for computational reasons. The distance measure employed a 17-dimensional weighted variance with intra-class variance computed over the entire training set.

	Global Threshold	Cohort Score	Cohort Score + MLP
Male:	70.3%	87.1%	87.3%
Female:	84.6%	79.4%	91.1%
Both:	74.2%	84.8%	88.1%

Table 3 Full STIMIT Database (CEP+ZCR).

First the cohort was identified for each of the 630 talkers by finding those five talkers whose VQ codebooks produced the lowest distortion scores over the entire database relative to the test talker VQ codebook. Then the three TI sentences were concatenated for each talker and used as the test sentence for verification purposes. To obtain impostor trials, each talker model was challenged by an impostor selected at random from the full database. This technique obtained 630 true talker and 630 random impostor trials.

When the cohort, or dynamic, threshold was used, the equal error rate was 15.2%. This can be compared to the standard verification scheme in which a global threshold is computed for the entire population of talkers. The global threshold is obtained by determining that threshold value which equalized the accept and reject errors. The equal error rate for the global threshold was 25.8%. The cohort technique resulted in a decrease of over 50% of the error.

10 MULTILAYER PERCEPTRON

A multilayer perceptron (MLP) was suggested by Higgins *et al.* (1991) to map the scores produced by the cohort normalization into accept/reject categories. Inputs to the MLP were the raw VQ score of the claimed talker model, and the difference between the raw score and the cohort average score. The number of hidden units in each case was three. Two output neurons corresponded to true talker and to impostor categories, respectively.

Training was accomplished by splitting the 630 true talker and 630 impostor scores into two equal sets. The first half was used for training and the last half, for cross validation. The testing and cross validation sets were then interchanged and the training was repeated. The reported accuracy was the average accuracy of the two trials.

When the neural network was used for scoring, the performance improved overall from 84.8% to 88.1%, an increase of 3.3%. The increase was greater for females than for males, but performance in all cases was higher than either the global threshold or cohort scores used alone. It is interesting to note that the cohort score used alone actually reduced the performance for females, but the MLP successfully combined the two scores into a higher composite score.

11 FIXED VERSUS VARIABLE TRAINING SENTENCES

Another issue is whether there is any advantage to using a fixed training utterance instead of a variable-text training utterance. All the results reported so far have used the two fixed STIMIT sentences to train each talker model. For comparison, training was done on a portion of the SX sentences, which varied in text from talker to talker. The length of the SX sentence was the average length of all the SA sentences over the entire talker set. Classification accuracy of VQ models built from the variable sentences was compared to that of VQ models built from the fixed sentences. In both cases testing was done on the concatenated SI sentences. All system parameters were otherwise the same as the previous experiment. Table 4 shows the results. Improved performance was expected for the fixed sentence training since this provided an additional dimension of normalization among the talkers. However within the margin of experimental error the results are equivalent. Thus there appears to be no performance advantage in training with a fixed utterance.

Fixed Text	Free Text
84.8%	84.6%

Table 4 Type of Training (CEP+ZCR).

Finally, the fixed-text and free-text training utterances were combined into one long utterance of average length 14 seconds. These were used to build VQ models for each talker. Performance scores were obtained for these long training utterances using the same verification utterances in the previous experiment. When cohort normalization was used for verification, the accuracy was 91.6%, or an equal error rate of 8.4%. When the MLP was trained to verify the talker using the cohort scores as input, accuracy increased to 93.7%. This performance corresponds to a true talker acceptance accuracy of 91.7% and an impostor reject accuracy of 95.7%. As in the previous experiment these figures were obtained by splitting the 1260 utterances into two equal sets, training on one and validating on the other, then reversing the two.

Cohort Score	Cohort Score + MLP	True Accept	Impostor Reject
91.6%	93.7%	91.7%	95.7%

Table 5 Training on One Long Utterance.

12 PRELIMINARY EXPERIMENT ON REAL TELEPHONE SPEECH

Recently a database was obtained from the Oregon Graduate Institute (OGI white pages database) containing utterances on dialed-up telephone lines. A common isolated alphabet utterance was used to train each caller's VQ model. For testing, three utterances were concatenated consisting of the caller's hometown, calling location, and last name. Preliminary results are based on an analysis of 75 of the callers. A global threshold gave 72.6% accuracy and the cohort-based dynamic threshold yielded 81.3% accuracy. These compare best with the third line of Table 1, which contains the shorter training utterance. Note that the real telephone speech had isolated alphabet as opposed to continuous speech training and the test utterances were shorter than those from TIMIT. These together probably account for the decreased performance on the real telephone data. Another possible factor is the presence of nonlinearities in real microphones (especially carbon button), which are not modeled in the telephone simulator.

Another experiment was run on a different database from OGI, consisting of the first 25 seconds of spontaneous stories collected from different telephone callers (OGI 300 stories database). Thirty stories from this database were analyzed using the cohort scoring algorithm. The first 20 seconds of each conversation was used for training and the last 5 seconds was used for testing. So training and testing utterances were obtained from the same telephone. It is not certain that all conversations were from different talkers, but we understand that they are likely to be from different talkers. The result using a global threshold was 66.6% on a total of 60 (30 true plus 30 impostor) trials.

When cohort scoring was used, the accuracy increased to 91.6%. This confirms that high accuracy is still obtainable with real telephone speech.

	Global Threshold	Cohort Score
OGI white pages (75 callers)	72.6%	81.3%
OGI 300 stories (30 callers)	66.6%	91.6%

Table 6   Real Telephone Speech.

Notice that training on the 20 second real telephone speech (91.6%) gave a cohort score comparable to training on the 14 second simulated telephone quality speech (91.6%). This is very encouraging since it indicates not too great a loss of accuracy on real telephones. Keep in mind, though, that the telephone result is based on training and testing on the same handset, whereas the simulated speech results are based on different simulated handsets for training and testing. The final experiment, training and testing on different telephone handsets needs yet to be performed.

13 DISCUSSION

These results can be compared to that of Rosenberg *et al.* (1992) who studied cohort normalization in a text-dependent talker recognizer using cepstra and delta cepstra as features. The text-dependent approach would be expected to yield higher performance than our text-independent system using the same length utterances. Our utterance length is four sentences for training and three for testing (average 14 sec and 10 sec, respectively). Rosenberg *et al.* used five-digit strings, so their utterances were about one-fourth our length. Research shows that text-independent utterances must be longer than text-dependent utterances to compete in performance. Therefore, our result of 93.7% compares favorably to their result of 95.2%. Furthermore, they left out of the impostor trials those talkers who were in the talker's cohort set. This would result in a higher accuracy since the "most similar" impostors were not allowed to compete. We did not arbitrarily exclude an impostor who happened to be in a claimed talker's cohort set. Therefore our task was a more difficult one.

14 CONCLUSION

An accurate method for text-independent talker verification has been demonstrated for simulated telephone quality speech using a large database of talkers. Its accuracy of 93.7% compares well with other work, in particular to that of Rosenberg *et al.*, who used cohort normalization for text-dependent talker verification on real telephone speech.

Though delta cepstrum information was useful in reducing the error rate for the reduced talker set, for computational reasons it was not included in the full database. The delta cepstrum should be tried on the full database. Plans are to use the SWITCHBOARD database (Godfrey, 1992) when it is available, since it would permit longer training and test utterances with real telephone speech.

Our results on simulated telephone quality speech confirm the notion that the standard VQ method deteriorates with telephone channel variations. This can be at least partially augmented by addition of either zero crossing or dynamic spectral features. However it is best dealt with by using a combination of robust features and sophisticated decision making.

Zero crossing features work together with the cohort normalization and MLP to make decisions which are robust to channel mismatch between training and testing. The role of the MLP in this process is to *combine* the raw and normalized scores into a single score which minimizes the accept and reject error. In this sense the MLP is performing a kind of sensory data fusion.

REFERENCES

[1] B. S. Atal, "Automatic Recognition of Speakers from their Voices", *Proc. IEEE*, **643**, 460-475, April, 1976.

[2] D. J. Burr, "On Text-Independent Talker Identification using Simulated Telephone Quality Speech", *Proc. IEEE International Workshop on Interactive Voice Technology for Telecommunications Applications*, Piscataway, NJ, Oct. 19-20, 1992.

[3] H. Chang, "Augmented Phonetic Map for Voice Verification", *Proc. ICASSP*, II 169-172, San Francisco, CA, 1992.

[4] S. Furui, "Speaker-Independent Isolated Word Recognition Using Dynamic Features of the Speech Spectrum", *IEEE Trans. on ASSP*, Vol. ASSP-34, No. 1, pp. 52-59, February, 1986.

[5] J. S. Garofolo, "Getting Started with the DARPA TIMIT CD-ROM: An Acoustic Phonetic Continuous Speech Database," National Institute of Standards and Technology (NIST), Gaithersburg, MD, 1988.

[6] J. Godfrey, "The SWITCHBOARD Corpus for Telecom Speech Research," *IEEE Workshop on Interactive Voice Technology for Telecommunications Applications*, Oct., 1992.

[7] H. Hattori, "Text-Independent Speaker Recognition Using Neural Networks", *Proc. ICASSP*, II 153-159, San Francisco, CA, 1992.

[8] A. Higgins, L. Bahler, and J. Porter, "Speaker Verification Using Randomized Phrase Prompting", *Digital Signal Processing*, 1, 89-106, 1991.

[9] Y. Linde, A. Buzo, and R. Gray, "An Algorithm for Vector Quantizer Design," *IEEE Trans. on Communications*, COM-28, 84-95, Jan. 1980.

[10] M. Lund and C. C. Lee, "Wald's SPRT Applied to Speaker Verification", 30th Annual Allerton Conference on Communication, Control, and Computing, Sept. 30 - Oct. 2, 1992.

[11] T. Matsui and S. Furui, Comparison of Text-Independent Speaker Recognition Methods Using VQ Distortion and Discrete/Continuous HMM's," *Proc. ICASSP*, II 157-160, San Francisco, 1992.

[12] A. E. Rosenberg and F. K. Soong, "Recent Research in Automatic Speaker Recognition", in *Advances in Speech Signal Processing*, S. Furui and M. M. Sondhi (ed.), Marcel Dekker, New York, 701-738, 1992.

[13] A. E. Rosenberg, J. DeLong, C. H. Lee, B. H. Juang, and F. K. Soong, "The Use of Cohort Normalized Scores for Speaker Verification", Proceedings International Conference on Spoken Language Processing, 599-602, Banff, Alberta, CANADA, October, 1992.

[14] M. Savic and J. Sorensen, "Phoneme Based Speaker Verification", *Proc. ICASSP*, II 165-169, San Francisco, CA, 1992.

[15] F. K. Soong, A. E. Rosenberg, L. R. Rabiner, and B. H. Juang, "A Vector Quantization Approach to Speaker Recognition," *Proc. ICASSP*, pp. 387-390, 1985.

[16] K. M. Yang, "A Network Simulator Design for Telephone Speech Recognition Applications," *IEEE Workshop on Interactive Voice Technology for Telecommunications Applications*, Oct., 1992.

14

NEURAL NETWORK APPLICATIONS IN CHARACTER RECOGNITION AND DOCUMENT ANALYSIS

L.D. Jackel, M.Y. Battista, J. Ben, J. Bromley,
C.J.C. Burges, H.S. Baird, E. Cosatto,
J.S. Denker, H.P. Graf, H.P. Katseff,
Y. LeCun, C.R. Nohl, E. Sackinger,
J.H. Shamilian, T. Shoemaker,
C.E. Stenard, B.I. Strom, R. Ting, T. Wood,
and C.R. Zuraw

AT&T Bell Laboratories

1 INTRODUCTION

Character Recognition has served as one of the principal proving grounds for neural-net methods and has emerged as one of the most successful applications of this technology. This chapter outlines optical character recognition / document analysis systems developed at AT&T Bell Labs that combine the strengths of machine-learning algorithms with high-speed, fine-grained parallel hardware. From our point of view, the most significant aspect of this work has been the efficient integration of diverse methods into end-to-end systems. In this paper we use the task of locating and reading ZIP codes on US mail pieces as an illustration of the character recognition / document analysis process. We will also describe other applications of the technology, including interpretation of faxed forms and bit-mapped text to ASCII conversion.

2 The Character Recognition Process

Figure 1. shows the "typical" character recognition process, which starts with an optical image and ultimately produces a symbolic interpretation. The process is divided into a series of tasks that are usually executed independently. It begins with image capture in which an optical image is converted to a bit-map. Next the region of interest, in this example the address block, is located. Then the desired field, the ZIP code, is found. This field is then usually size-normalized and sometimes (not shown here) de-slanted. Finally, the characters are segmented and recognized. Note that the recognition phase is only one step in a long process. In our systems we modify this model, using feedback from down-stream stages to influence up-stream decisions. We also apply neural-net hardware and algorithms where they are advantageous.

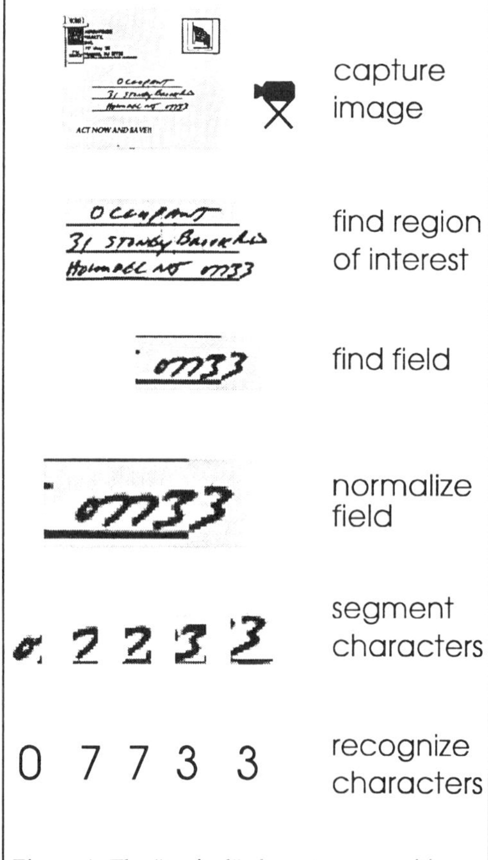

Figure 1. The "typical" character recognition process.

For a user, the end-to-end system performance is what counts. It matters little if the recognizer module is fast or accurate if most of the time budget and most of the errors arise from other modules. It has been our goal to design a system that has no "weak links" in either accuracy or speed. The technology driver for us has been a program sponsored by the US Postal Service whose goal is to produce an automatic address reading system that starts with bit-mapped images of envelopes and produces interpreted ZIP codes. Both speed and accuracy are key issues. Most of this paper describes technology developed for this application. This kind of problem has also been addressed by other workers [1].

3 The Basic Recognizer: LeNet

At the core of our recognition system is an isolated character recognizer that we now call LeNet [2]. The LeNet architecture is shown in Figure 2. LeNet takes a 20 x 20 pixel field as input and returns a rank-ordered list of possible single-character interpretations of the input image, along with confidence scores for each interpretation.

LeNet is an example of a highly structured neural-net in which the structure seeks to incorporate *a priori* knowledge about the task domain. For the OCR task, this knowledge includes the local two-dimensional geometric relationships that exist in images. LeNet is designed to extract local geometric features from the input field in a way that preserves the approximate relative locations of these features. This is done by creating feature maps that are formed by convolving the image with local feature-extraction kernels. (An important feature of LeNet is that the feature extraction kernels are learned as opposed to being hand-crafted.) These maps are then spatially smoothed and sub-sampled; this latter step builds in invariance to small distortions of the input image [3,4]. In the same way, higher-level feature maps are extracted from the sub-sampled first-level maps. The higher level maps then provide input to a linear classification layer. Although the network has over 100,000 connections, the network structure imposes constraints so that only about 3,000 different weight values have to be learned.

LeNet has several advantages that make it attractive for recognizing characters when high variability (like we see in images of mailed envelopes) is expected. First, LeNet has state-of-the-art accuracy as illustrated by its strong performance in a competition sponsored by NIST, the US. National Institute of Standards and Technology [5]. Second, LeNet runs at reasonable speeds on standard hardware (~10 characters/sec on a workstation) and high speed (~1000 characters/sec) on specialized hardware. LeNet can also be readily trained to recognize new character styles and fonts. It works well for both handwritten and machine printed characters.

In general, in machine-learning tasks, best performance on test data is obtained by controlling the capacity of the learning machine to match the available training data. With this idea in mind, we have modified the architecture of LeNet, hence controlling capacity, depending on the amount of training

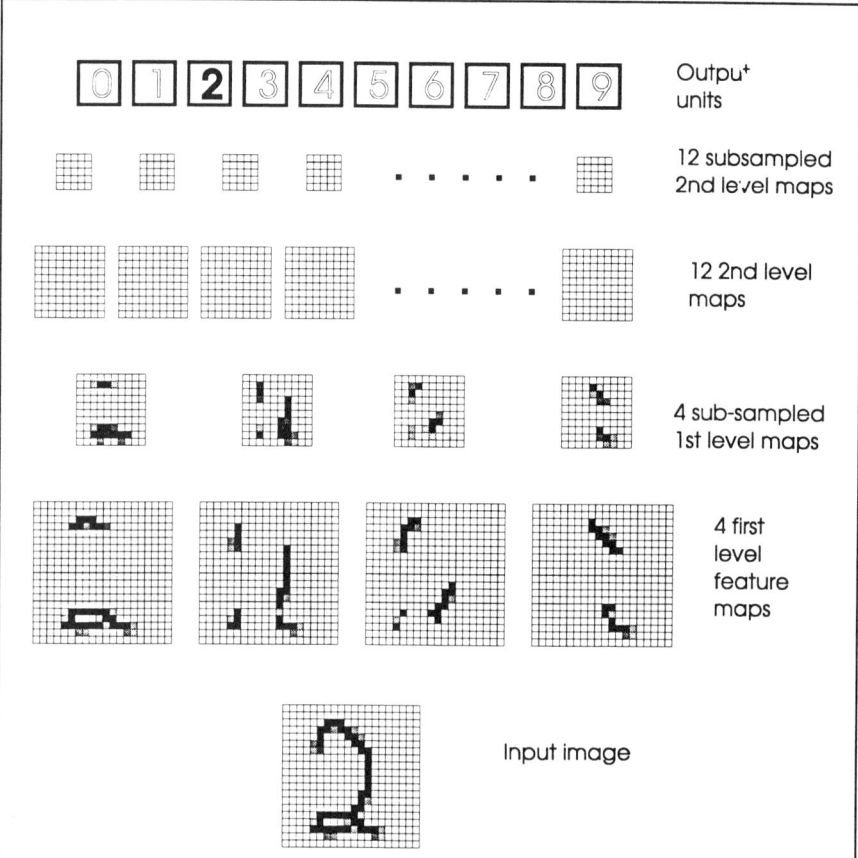

Figure 2. Architecture of LeNet. Each small box represents a neural-net "unit" or "neuron". The weighted connections between units (not shown in this figure) are highly structured. The maps are generated by convolutions with feature extraction kernels. Input images are sized to fit in a 20 x 20 pixel field, but enough blank pixels are added around the border of this field to avoid edge effects in the convolution calculations. In this figure, the activation of the input and the first two layers of the network are indicted: darker shading indicates greater activity.

data available. The version shown in Figure 2 was optimized for a training set of 7000 ZIP code digits. We found that a version with more hidden units and about twice as many weight values provided better results when we switched to a database of 50,000 digits. Even larger nets will be effective for bigger training data bases.

We note that there are other methods for OCR that may be more appropriate than LeNet for some applications. In particular, when our task is to read cleanly printed text with a limited range of fonts, a much simpler network may give adequate performance. In this case, acceptable accuracy at very high recognition speeds can often be attained by simple template matching. The problems we address in this paper are those in which recognition accuracy is most important and where the quality of the input images or characters may be poor. It is in this regime that LeNet excels. We also note that recognition accuracy equaling or exceeding LeNet has been obtained with a sophisticated pattern matching technique that uses a special metric, known as "tangent distance", for comparing patterns [6]. Currently, this method lags LeNet in speed, but this situation may change as this new method evolves.

4 Segmentation

LeNet recognizes one character at time. If our objective is to recognize a string of characters, the string has to be cut up into individual characters, a process called segmentation. The difficulty of the segmentation task strongly depends on the quality and type of the string. If we have fixed-pitch machine-printed fonts `like this`, and if we can detect this condition, the task is straightforward. For cleanly-printed, variable-pitch machine fonts, the task is more difficult, although a connected components analysis can almost always identify individual characters. For machine printing of poor quality or for handwriting, where different characters might touch and single characters might be broken into several pieces, the task is very difficult. For these difficult cases, segmentation is like a "chicken and egg" problem: in order to recognize we need to segment, but in order to segment we may have to recognize [7,8,9,10,11].

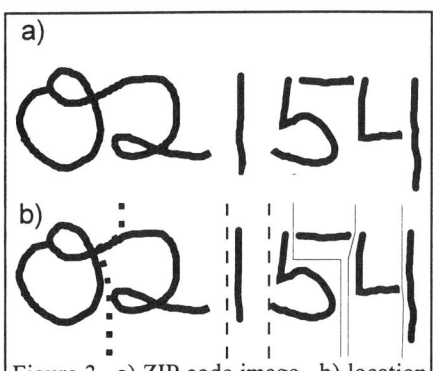

Figure 3. a) ZIP code image. b) location of definite cuts (dashed lines) and tentative cuts (dotted line and thin solid line).

Our current systems [7] take a hierarchical approach to a combined segmentation / recognition process. Our strategy is to find probable segmentation points or "cuts", snip out the "inked" regions between these points, and then see if LeNet can recognize these segments with high confidence, either in isolation, or in combination with neighboring segments. We then choose the set of segment combinations that gives highest overall confidence while accounting for all the ink in the image.

We proceed in the following way: given a string, we first find the "definite" cuts between characters. These are places where there are substantial horizontal gaps in the "inked" image. For the ZIP code image shown in Figure 3, definite cuts occur between the "2" and the "1" and between the "1" and the "5". We denote them by the vertical dashed lines shown in Figure 3b. Then we consider "tentative cuts" where a connected components analysis locates gaps between blobs of ink. In Figure 3, such cuts are within the "5" and the "4" and between the "5" and "4". These tentative cuts are shown as thin, solid lines in Figure 3b. Finally, using heuristic rules, we identify additional tentative cuts at places where characters are likely to be joined or touching. Such a tentative cut is shown as the dotted line in Figure 3b.

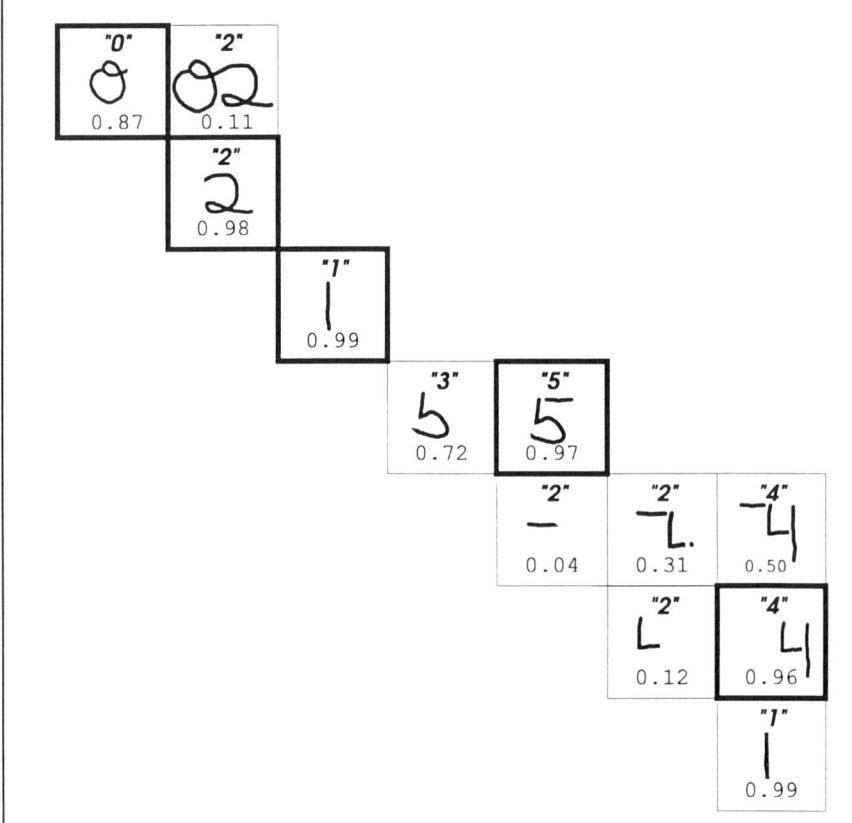

Figure 4. Blocks of segments from Figure 3 that are passed to LeNet. The number near the top of each box is the most likely classification of the "ink" in the box. Blocks in the same row are bounded on the left by the same cut; blocks in the same column are bounded on the right by the same cut. The number near the bottom is LeNet's relative confidence in the classification. In order to choose a consistent segmentation of the image we must ensure that all the "ink" is accounted for, and that it is only used once. Using this rule, the highest overall confidence classification / segmentation is indicated by the boxes with the heavy borders.

The next step in our segmentation process is to pass segments and possible segment combinations to LeNet for scoring as possible characters. Figure 4. shows these segments and segment combinations along with their top scoring classification and confidence level. The number near the top of each box is the most likely classification of the "ink" in the box. Blocks in the same row are bounded on the left by the same cut; blocks in the same column are bounded on the right by the same cut. The number near the bottom is LeNet's relative confidence in the classification. In order to choose a consistent segmentation of the image we must ensure that all the "ink" is accounted for, and that it is only used once. Using this rule, the highest overall confidence classification / segmentation is indicted by the boxes with the heavy borders.

Notice that for the example in Figure 3, in order to recognize and segment a 5 digit ZIP code we had to make 12 calls to LeNet. For 5 digit ZIP code images with no large blobs of extraneous ink, we have to make an average of 7.5 calls. In actual mail streams we expect more calls will be necessary. This places additional demands on the required speed of the recognizer engine and further motivates the use of special purpose hardware to implement LeNet.

A hardware system that has been effective in speeding the recognition / segmentation process is one based on the ANNA neural-net chip [12]. This chip, which mixes analog and digital processing, was specifically designed to speed evaluations of networks like LeNet. A key feature of ANNA's design is the provision for parallel evaluation of non-linear 2-dimensional convolutions, which represent the bulk of the computing required by LeNet. Because LeNet is large (over 100,000 connections), it cannot be evaluated entirely in parallel by ANNA. Instead, sections of the input image are sequentially evaluated, with the corresponding sections of the feature maps being evaluated in parallel. In order for ANNA, (or most any neural-net chip) to run efficiently, heavy system demands are placed in the management of data on and off the chip. The current board-level ANNA system has special hardware to speed I/O operations along with a custom sequencer to control ANNA's instruction execution. The latest versions of this system can evaluate LeNet in about 1 msec, so that, accounting for multiple calls to LeNet during the recognition/segmentation process, a throughput of over 400 character recognitions/sec can be sustained. This speed is about 25 times faster than a state-of-the-art workstation.

Figure 5. NET32K board-level system. The board contains two NET32K chips, along with a sequencer and circuitry to provide high-speed data input/output paths to the NET32K chips. In image processing applications this board sustains over 100 billion multiply/add operations per second.

5 Normalization

Recognition accuracy can be increased if we know that characters presented to the recognizer engine are limited in their range of size and orientation. This can be accomplished by normalizing the character strings with respect to size and slant angle. We found that while size normalization could be done quickly on a standard work station, slant normalization could not. In this de-slanting process, the overall slant in a string must be detected. Then the image bit-map of the string is modified so that the overall slant is set to zero. Here the most computationally intensive step is the measurement of the string slant angle. We have found that this potential speed bottleneck can be eliminated by using a second neural-net board-level system. This system is based on Hans Peter Graf's NET32K chip [13], which like ANNA mixes analog and digital processing, but unlike ANNA, NET32K has more stored weights (up to 32K vs. 4K for ANNA) at the expense of decreased accuracy (1 bit vs. 6 bits). NET32K excels at scanning relatively large images with large kernels (up to 16 x 16). A working board-level NET32K system, now in use, contains two NET32K chips, as well as custom sequencers and on-board memory. This system, which is shown in Figure 5, is designed to support a high I/O rate for the NET32K chips. In the applications described below, this system achieves a sustained rate of 100 billion multiply-adds per second at 1.5-bit precision. To our knowledge this is the highest processing rate yet attained in any single-board image processing system.

In order to measure average image slant, the NET32K system scans the image field with a set of oriented edge detector kernels, with each kernel tuned for a particular edge orientation. Examples of these kernels and detected feature maps are shown in Figure 6. After the scan is completed, we count how many places each kernel matched in a section of the image. The kernel that scored the most matches indicates the dominant slant angle in the image. With NET32K, we can find the slant angle of a ZIP code string in about 20 msecs.

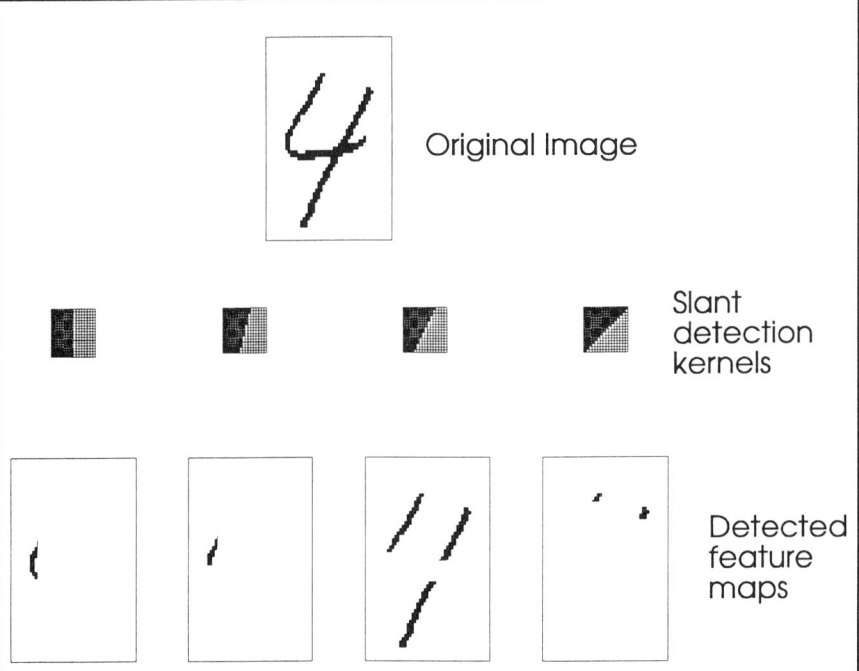

Figure 6. Detection of slant angle using a set of slant detection kernels. A set a 16 kernels (only 4 are shown in this figure) are scanned across the image. For the kernels shown here, a dark shading indicates a region where the required match is white space, and a lighter shading indicates a region where the required match is inked. Feature maps for each kernel record where matches are found in the image. The kernel that scores the most matches across the image indicates the predominant slant.

6 Finding the Region of Interest

Locating the region in an image that contains the desired text can be very challenging, especially if the image is cluttered with extraneous text, graphics, and/or background noise, e.g. Figure 7. Because the images typically contain millions of pixels and because techniques for image analysis usually require many operations per pixel, field location is a computationally intensive task. We have used custom hardware to find address blocks at a rate of more than 10 images/sec. Here again, NET32K shows its effectiveness. The system scans the input image with feature detection kernels that are tuned to the characteristics of text lines. Using the resultant feature maps and our *a priori* knowledge about where address blocks are likely to occur on an envelope, we can find likely candidate fields at the required rate. All the processes described above are now being integrated into a complete end-to-end system.

Figure 7. An example of address block location performed by the NET32K system on a particularly challenging mail piece. Two candidate address blocks were found; rules about the likely location of address blocks makes the candidate on the lower right the first choice. The NET32K system can sustain a processing rate of more than 10 images/sec.

7 Additional Applications

The technology developed to solve postal tasks has been successfully built upon and used in telecommunication applications. In this section several of these applications are discussed.

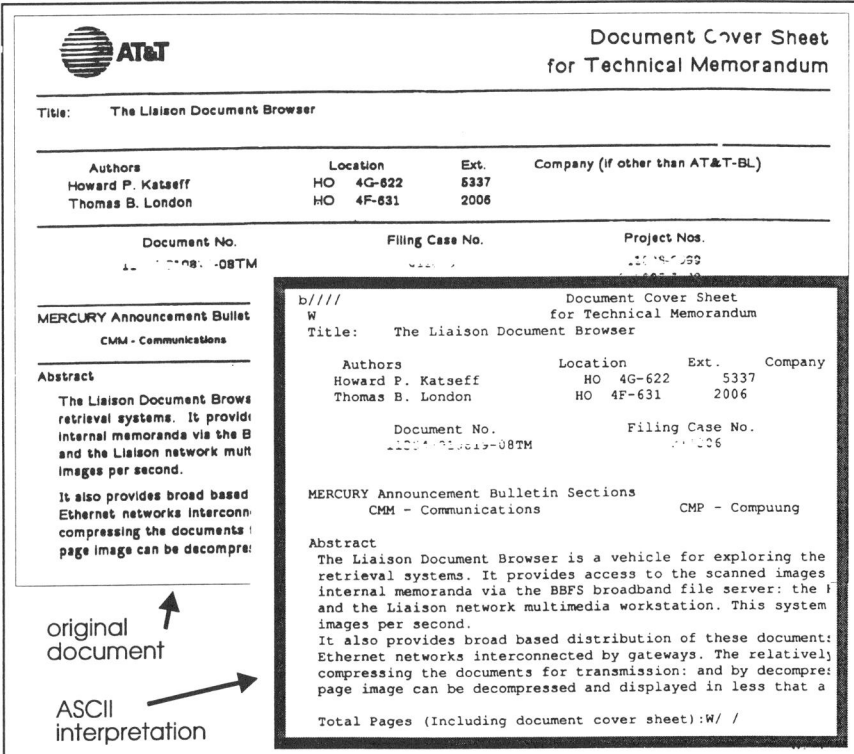

Figure 8. Conversion of bit-mapped documents to ASCII. Using a combination of page-layout analysis and LeNet, images from an internal AT&T document retrieval system have been converted to ASCII text. Part of a bit-mapped document is shown in the background above. The overlay is part of that document's ASCII interpretation. The ASCII text allows word or phrase searching as well as lifting of the text for further word processing.

Bit-Map to ASCII Conversion for Document Retrieval

A document retrieval system that allows users to browse internal technical publications has been in service at AT&T Bell Laboratories for more than a year [14]. The system displays bit-map images of document pages on users' workstation screens. It is now being upgraded to provide an ASCII version of the text as a companion of the bit-map. This allows users to search for words or phrases and to lift sections of text for inclusion in other documents. The bit-map images include many fonts and vary in their image quality.

To obtain the ASCII version of a document, page-layout analysis is first performed using a software package developed by H. S. Baird [15]. This package finds text blocks, and then segments out individual characters. These clipped character images are then recognized using a version of LeNet that was trained on numerous printed fonts and sizes, including character examples that were corrupted with synthetic noise with characteristics similar to those encountered in actual documents [16]. Overall OCR accuracy for this system typically exceeds 99%. An example of an original document and its ASCII version are shown in Figure 8.

Processing of Faxed Forms

As a further example of an application of recognition technology, we describe a system, deployed internally in AT&T, that automates processing of new service orders for parts of the AT&T network. A block diagram of the system is shown in Figure 9. In this system, a client requesting a new service faxes a form to a central AT&T facility. There, the bit-map is first used to identify the type of form. Next, registration marks on the image are located and the image is adjusted to compensate for distortions generated by the fax-scanning process. The image fields that specify the order requisitioner and the ordering information are then clipped out, normalized and passed to LeNet. The forms are designed so that the characters to be processed are written in boxes, eliminating the need for segmentation.

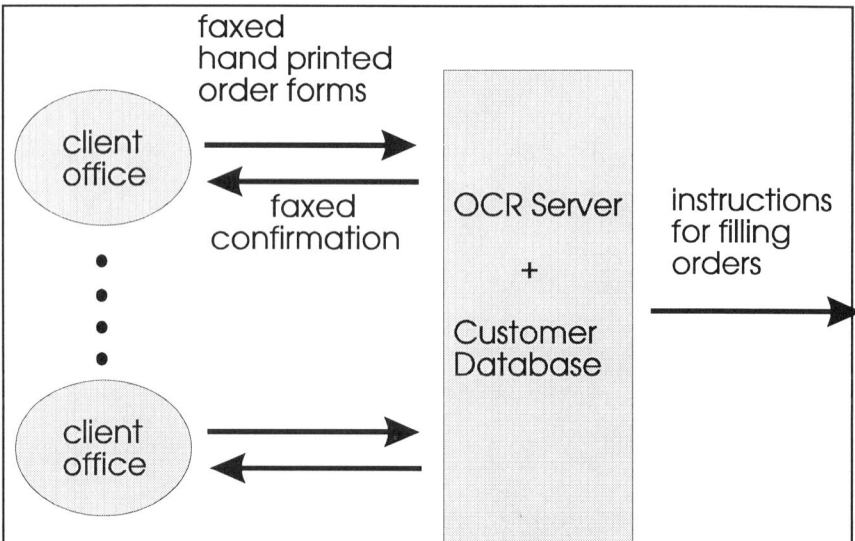

Figure 9. Block diagram of automatic order processing system. Clients fax order forms to a central facility where OCR is performed and the instructions for filling the order are then generated. A confirmation notice is faxed back to the client.

For this task the considerable contextual information available is used to maintain high recognition accuracy. As examples, the requisitioner's name and organization can be cross-checked against a database and the service order can be matched against allowable service codes After this cross-checking, if LeNet still has low confidence in a particular field, that field is passed to a human correction station operator who makes the final decision. This system was successfully deployed in 1992 and is now in everyday use.

Processing of Tabular Text

Another application of the document processing technology is being used by an AT&T Data Center in Kansas City. For this application very high recognition accuracy is essential. First introduced in 1992, this application translates large volumes of densely printed tabular text from scanned documents into structured ASCII format. Typically, the printed text is one of a large variety of small machine printed fonts, and may be poorly registered on the document. The document pages (either original or copy) are scanned in by document type and passed to the AT&T document analysis system, which uses Baird's page-layout analysis. The system locates text strings according to a user-defined template that describes fields within text lines while ignoring other text. Then using fast and accurate algorithms invented by David Itner [17] for fixed-pitch printed text , such as from impact printers, each line is parsed into fields according to the user-defined template. The fields are then passed on to the neural-net character recognizer. After performing a contextual analysis, 99.9% accuracy is achieved. Any low-confidence characters are marked for review by human operators.

8 Conclusions

In this paper we have described some examples of applications of neural-net character recognition and document analysis that have been developed at AT&T Bell Labs. We have concentrated on a system designed to find and read ZIP codes on envelopes for the US Postal Service. In order to meet real-time requirements, this system includes special purpose hardware with neural-net chips. The system also uses a combined approach to the interdependent problems of recognition and segmentation. We have also sketched applications to document retrieval and to automatic processing of faxed forms.

We gratefully acknowledge support of this work in part by the U.S. Postal Service.

References

1. For example, see G. Martin, M. Rashid, D. Chapman, J. Pittman, "Learning to See Where and What: Training a Net to Make Saccades and Recognize Handwritten Characters," in *Advances in Neural Information Processing* **5**, (NIPS 92) S.J. Hanson, J.D. Cowan, and C.L. Giles eds., pages 441-447, Morgan Kaufmann (1993).

2. Y. LeCun, B. Boser, J. S. Denker, D. Henderson, R. E. Howard, W. Hubbard, L. D. Jackel, "Handwritten Digit Recognition with a Back-propagation Network," in *Advances in Neural Information Processing* **2**, (NIPS 89) D. S. Touretzky ed., pages 396-404, Morgan Kaufmann (1990).

3. K. Fukushima "Neocognitron: A Self-Organizing Neural Network Model For a Mechanism of Pattern Recognition Unaffected by a Shift in Position," *Biol. Cybernetics* **36**, 193-202 (1980).

4. M. C. Mozer, "Early Parallel Processing in Reading: A Connectionist Approach," in *Attention and Performance, XII: The Psychology of Reading*, M. Coltheart, ed., **Vol. XII**, pages 83-104, Erlbaum, Hillsdale, NY (1987).

5. See Proc. of the First Census Optical Character Recognition System Conf., **NISTIR 4912,** August 1992.

6. P. Simard, Y. LeCun, and J. S. Denker, "Efficient Pattern Recognition Using a New Transformation Distance," in *Advances in Neural Information Processing* **5**, (NIPS 92) S.J. Hanson, J.D. Cowan, and C.L. Giles eds., pages 50-58, Morgan Kaufmann (1993).

7. C. J. C. Burges, O. Matan, Y. LeCun, J. S. Denker, L. D. Jackel, C. E. Stenard, C. R. Nohl, J. I. Ben, "Shortest Path Segmentation: A Method for Training a Neural Network to Recognize Character Strings." *in Proc. of IJCNN, International Joint Conference on Neural Networks*, Baltimore MD, **Vol. III**, pages 165- 170 (1992).

8. J. Keeler, D. Rumelhart, and W. K. Leow," Integrated Segmentation and Recognition of Handprinted Numerals", in *Advances in Neural Information Processing* **3**, (NIPS 90) R. P. Lippmann, J. E. Moody, and D. Touretzky eds., pages 557-563, Morgan Kaufmann (1991).

9. O. Matan, C.J.C. Burges, Y. LeCun, and J. S. Denker ," Multi-Digit Recognition Using a Space Displacement Neural Network," in *Advances in Neural Information Processing* 4, (NIPS 91) J. E. Moody, S.J. Hanson, and R. P. Lippmann eds., pages 488-495, Morgan Kaufmann (1992).

10. G. L. Martin and M. Rashid, "Recognizing Overlapping Hand-Printed Characters by Centered-Object Integrated Segmentation and Recognition," in *Advances in Neural Information Processing* 4, (NIPS 91) J. E. Moody, S.J. Hanson, and R. P. Lippmann eds., pages 504-511, Morgan Kaufmann (1992).

11. J. Keeler and D.E. Rumelhart, "A Self-Organizing Integrated Segmentation and Recognition Neural-Net," in *Advances in Neural Information Processing* 4, (NIPS 91) J. E. Moody, S.J. Hanson, and R. P. Lippmann eds., pages 496-503, Morgan Kaufmann (1992).

12. E. Sackinger, B. E. Boser, and L. D. Jackel, "A Neurocomputer Board Based on the ANNA Neural Network Chip," in *Advances in Neural Information Processing* 4, (NIPS 91) J. E. Moody, S.J. Hanson, and R. P. Lippmann eds., pages 773-780, Morgan Kaufmann (1992).

13. H. P. Graf, C. R. Nohl, and J. Ben, "Image Recognition with an Analog Neural Net Chip," *in Proc. 11th IAPR Int. Conf. Pattern Recognition,* **4**, pages 11-15 (1992).

14. H. P. Katseff and T. B London, "The Ferret Document Browser", to appear in *Proc. of USENIX Summer 1993 Tech Conf.* Cincinnati, June 1993. The Ferret browser is now being combined with the "Right Pages " electronic library system; see G. A. Story, L. O'Gorman, D. Fox, L. Schaper, and H. V. Jagadish, "The Right Pages Image-Based Electronic Library for Alerting and Browsing," IEEE Computer, pages 17-26, Sept. 1992.

15. H. S. Baird, "Global-to-Local Layout Analysis," *in Proc. IAPR Workshop on Syntactic and Structural Pattern Recognition,* Pont-a-Mousson, France 12-14 September 1988.

16. H. S. Baird, "Document Image Defect Models" *in Structured Document Image Analysis,* H. S. Baird, H. Bunke, & K. Yamamoto eds. Springer-Verlag, New York (1992).

17. D.J. Itner and H. S. Baird, "Language-Free Layout Analysis," *in Proc. of 1993 International Conf. on Document Analysis and Recognition,* Tsukuba Science City, Japan, October 1993.

15

IMAGE VECTOR QUANTIZATION BY NEURAL NETWORKS

Rosa Lancini

CEFRIEL, Italy

1 INTRODUCTION

Image compression constitutes an essential tool for applications such as broadcast television, remote sensing via satellite, teleconferencing, computer communication and facsimile transmission. Compression techniques become necessary to reduce the amount of data to describe a still image or an image sequence correctly. Vector Quantization (VQ) is already known as a very efficient compression method when used in image coding scheme. In fact, using Vector Quantization in place of Scalar Quantization it is possible to reduce the bit-rate of a data compression system at the same quality or viceversa to improve the reconstructed quality image at the same bit-rate. The neural network paradigm represents, for VQ problems, an interesting alternative to traditional algorithms. Neural Networks provide good performance both in quality image reproduction and in computational effort, allowing the use of adaptation schemes to follow the statistics of the incoming images during the coding process. In this chapter, *Vector Quantization* and *Adaptive Vector Quantization* applications, solved by using neural net approach, will be presented. Section 2, after a brief review on the theory of Vector Quantization, presents the application of neural network algorithms for two particular cases: *colormap design* and *interframe coding scheme for videoconference sequences*. Section 3 proposes a solution for adaptive vector quantization problem with a neural network method. Conclusions are given in Section 4.

2 NEURAL NETWORKS FOR VECTOR QUANTIZATION

In recent years, there has been much interest focused on Vector Quantization [1] for various signal sources, especially speech and image signal. It has been proved by Shannon's rate distortion theory [2], a better performance is usually achievable by coding a block of signal instead of coding each signal individually (Scalar Quantization). In most data compression systems, Vector Quantization (VQ) plays an important role in generating the necessary information to describe the input image with a certain quality. VQ [1] has proved to be a very powerful coding technique due to its inherent theoretical superiority over scalar quantization. In fact it is possible to obtain an effective bit-rate/quality ratio; a good reconstructed quality image is achieved at a very low bit rate. The introduction of neural network approach is fundamental in vector quantizer design to assure good performance both in reconstruction quality image and computational effort. Most adaptation schemes employed in neural network methods can be classified as either *Supervised* or *Unsupervised*. Among various unsupervised learning paradigms, Competitive Learning [3] algorithm is useful for VQ applications.

2.1 Vector Quantization

A Vector Quantization can be defined as a mapping from K-dimensional Euclidean space R^k to a finite subset C of R^k. This finite set $C=\{c_i : i = 1,..., M\}$ is called *codebook*, where M is the size of the codebook, and each $c_i = (c_1,..., c_k)$ in C is called *codeword*. A VQ codec has two parts, an *encoder* and a *decoder* (see Fig. 1), both equipped with the same codebook. The encoder assigns the input vector $x = (x_1,..., x_k)$ in R^k to an index i, that points to the closest codeword c_i in the codebook. The decoder uses the index i to address the codeword c_i in the codebook. By choosing the size of the codebook, we can control the transmission rate of a VQ process. The distortion between the input vector x and its corresponding codeword c_i is measured, in general, by the squared Euclidean distortion $d(x,c_i) = |x-c_i|^2$. The quality of the codebook is judged by the reproduced image fidelity, usually quantified by the Signal to Noise Ratio (*SNR*):

$$SNR = -10 log_{10}(\sum_i [(x_i - c_i)/x_i^2]\ dB$$

where $\mathbf{x}_i$ is the input signal and $\mathbf{c}_i$ the reproduced output (chosen codeword). The compression obtained by VQ depends on the dimension of the codebook C and the number of elements of the codeword. If M is the dimension of C and

Image Vector Quantization by Neural Networks

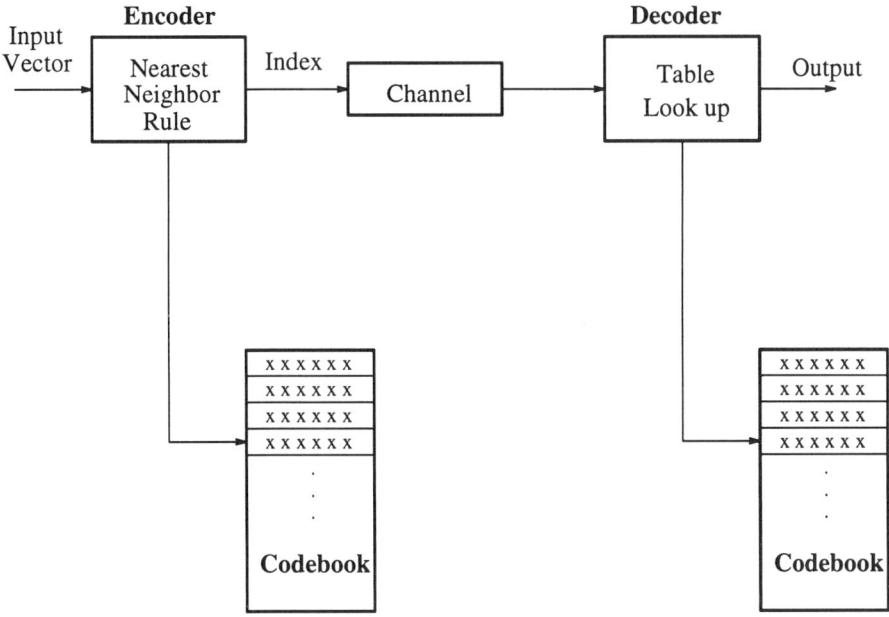

Figure 1 Block diagram of a simple Vector Quantization

each codeword has m elements, then the compression rate r in bits/pixel (bpp) is:

$$r = R/k$$

where $R = \log_2 m$ and k is the vector size in pixel/vector.

2.2 Neural Network Algorithms for Vector Quantization

There are several unsupervised learning paradigms; a brief review is given hereafter. They are the simple Competitive Learning (*CL*), the Self Organizing Feature Maps (*SOFM*), the Frequency Sensitive Competitive Learning (*FSCL*) and the Competitive Learning with initialization from the Training Set (*CL-TS*).

Competitive Learning (CL)

In competitive learning, a vector w_i is associated to each node of the network. Vector dimension equals that of input vector and its components are called the node weights. At the end of the codebook generation process, w_i will represent the *i-th* codeword. At the beginning w_i's are initialized to small random values (from 0 to 1 [4]-[5]). Then, the algorithm iterates through the training data a number of times in order to reach a final representation for the codewords. The learning phase consists of two steps: the *Similarity Matching* and the *Updating Phase*. In the similarity matching, for each training vector **x** and for all output neural units, the distortion $D_i=d(x,w_i)$ (a simple Euclidean distance) is found and the output unit w_i^* with the lowest distortion is selected. W_i^* is called the winner weight. Then in the updating phase, w_i^* (the winner weight) is updated as follows:

$$w_i^*(n+1) = w_i^*(n) + e(n)[x(n) - w_i^*(n)]$$

where *n* is the training time index, and $0 < e(n) < 1$ represents a "learning" factor that decreases during the training. A major problem with the *CL* algorithm is that usually a fraction of the units is not updated in the learning process [5]; that means that these units are not representative of any input data.

Kohonen Self Organizing Feature Maps (SOFM)

The Kohonen Self-Organizing Feature Maps (*SOFM*) [6]-[7] is a *CL* network that was proposed by Kohonen. In *SOFM*, the problem of underutilized nodes is overcome by associating each neural unit to a topological neighborhood of other neural units. For each training vector the closest unit is selected (using an Euclidean distance). This unit is updated to reduce the distortion. Furthermore, all its neighbors are adapted in order to allow the network self-organize easier during the learning process. In this way, during the learning process, the winner weight as well as its neighboring units are updated. At each iteration step the neighborhood domain is reduced, so that only the selected unit is updated at the end of the learning process. The learning algorithm works in the following way. Initially N_c (radius of neighborhood region) is chosen. The similarity matching is the same as in *CL*: for each input vector **x** and each output unit a distortion measure $D_i(x,w_i)$ is computed; the output unit w_i^* with the smallest distortion is selected. The selected vector (winner weight) is adjusted as follows:

$$w_i^*(n+1) = w_i^*(n) + e(n,0)[x(n) - w_i^*(n)].$$

All neighboring units whose distance from w_i^* is less than N_c are updated as follows:

$$w_i(n+1) = w_i(n) + e(n,N)[x(n) - w_i(n)]$$

where *n* is the iteration index and $0 < e(n,N) < 1$. Periodically the extent of the neighborhood is reduced by decreasing N_c until $N_c = 0$. Also the amplitude of the gain term $e(n,N)$ decreases with time. Good results are normally obtained by using a gain factor shaped like a cardinal sine ("Mexican hat"where around the receiver neuron there exists a strong excitatory area and, externally, an inhibitory area of less strength). In this case the amplitude of *e(n,N)* decreases during the learning process, but the dimension of the neighborhood region is always the same. A detailed description of this method can be found in [7].

Frequency Sensitive Competitive Learning Network (FSCL)

FSCL [5] directly addresses the underutilized neural unit problem, typical of the *CL* algorithm, by using a different approach to search for the winner node. This algorithm tries to overcome the drawback of *CL* by modifying the similarity matching while preserving the same updating phase. Each neural unit counts the number of times u_i, that it was selected as a winner. The distortion measure, used to select the winner node, is redefined as $D = f(u_i) \cdot d(x, w_i)$, where $f(u_i)$ is a monotonically increasing function called *"fairness function."* The *"fairness function"* introduces a count-dependent weight to the distortion measure. Its effect is to increase the probability of underutilized neural units being selected as a winner, thus preventing dominant winners. In this way a neural unit that "wins" frequently, though its count, u_i, increases, its relative modified distortion measure, *D*, increases, thus giving other units an opportunity to win the competition. The *"fairness function"* can simply be *f(*u_i*)* = u_i, but it is also possible to make it iteration dependent, as in [5] where $f(u_i) = u_i \cdot exp(a \cdot e^{-n/T})$ was chosen, *n* is the iteration index and *a*, *T* are constants. In general, the choice of *f(*u_i*)* depends on the input data, and it affects the performance quality.

Competitive Learning with initialization from the Training Set (CL-TS)

In comparison to *SOFM* and *FSCL*, *CL* offers a saving in time because neither the updating of the neighboring units as in *SOFM* nor the calculation of the *"fairness function"* as in *FSCL* is required. On the other hand, *CL* poses the drawback of the underutilized nodes; this means a fraction of units is never updated in the learning process and, as a consequence, this fraction is never used, not even during the mapping phase. To improve the performance of the neural networks, we can intervene at three different stages: at the similarity matching (as *FSCL* does) or at the updating phase (as *SOFM* does) or during the initialization phase. Considering the simplicity of the *CL* algorithm, a possible solution for the problem could be a "different" initialization of the net in the training phase. Generally speaking most neural networks are initialized with small random values (values that go from 0 to 1). If the input data have a different dynamic range, this initialization can generate bias in the neural net. In fact during the learning process, in this latter case, the first updated data will have a higher probability to win (because of their closest distance on similarity matching). Therefore it is likely that better results were provided by initially choosing data from the training set at random. In this case all units of the net have the same probability to win, avoiding the bias given by small random values. In this way, Competitive Learning with Initialization from the Training Set (*CL-TS*) works like the simple *CL* (same similarity matching and updating phase) with a different initialization.

2.3 Applications and Results

To test the efficiency of the codebooks designed by neural networks, two different simulations are presented. The first one is related to the codebook design for displaying full color images (24 bits/pixel) on screens with 4, 6 and 8 color planes (*colormap design*). The second application is related to the *interframe coding of videoconference sequence* using vector quantization. All these results are compared (in term of reproduced quality) to those obtained by using the *LBG* algorithm [8], that represents a classical and widely known codebook design technique.

Colormap Design

Most image display systems limit the number of colors which can be simultaneously viewed to a small set called *colormap*. The colormap is a table usually containing 16, 64, 256 triplets of (Red, Green, Blue) colors which should be the best representative of the original image colors. Various VQ methods ([9]-[10]-[11]) have used for the colormap design. Here, results obtained by the neural nets are compared to those obtained by *LBG*. The comparative study has been conducted on a variety of 4:2:2 CCIR 601 [12] color images ("Barbara," "Zelda," "Boat" and "Black"). These images are full color and their dimension are 720 by 576 pixels. For a colormap with a size of 256 (see Fig. 2) *SOFM* and *CL-TS* offer the best performance in terms of SNR closely followed by *LBG*. *CL* presents the worst performance due to the fact that a considerable number of units in the neural nets (about 39%) do not represent any input data. The results obtained by *FSCL* are very dependent on the applied "*fairness function*" [11]. *CL-TS* is computationally less intensive than SOFM, and it provides a considerable saving over LBG. Similar comparative results are obtained for a colormap of 16 and 64 elements.

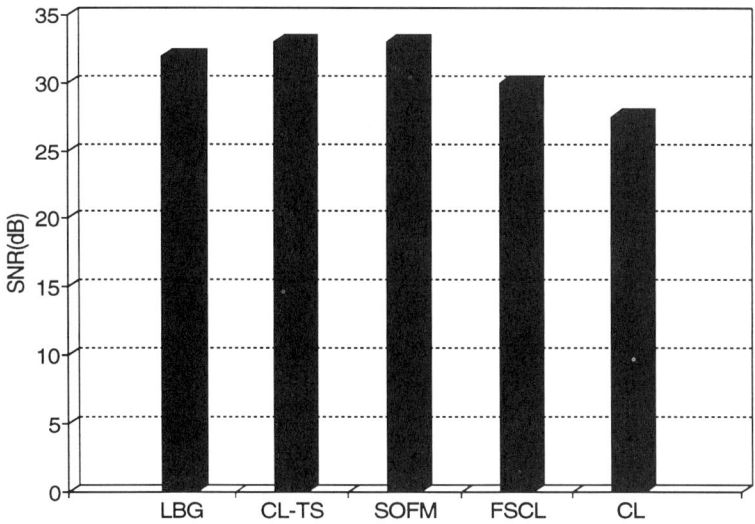

Figure 2 Average representation quality image for 256 colormap size.

Interframe coding using vector quantization

The proposed codebook design algorithm is also used for image sequence coding based on hybrid structure. The utilized coding scheme [13] is shown in Fig. 3. In interframe coding the difference between current frame and the previously coded frame is calculated and then coded (quantized). Much improved performance can be achieved by compensating for the inherent motion among frame, in this case the information calculated and then coded is the Motion Compensated Luminance Difference $MCLD$ (16 by 16 pixels). For each MCLD the estimated motion vector and the mean value of the Motion Compensated Luminance Differences (MCLDs) are transmitted to the receiver. The image representing the MCLDs is broken into blocks of 8 by 8 pels. These blocks are divided into four classes on the basis of their energy (that means to classify the blocks for their importance in the image description). After these operation the Discrete Cosine Transform (DCT) is performed over each MCLD block. Since the variances of DCT coefficients vary widely, the quantization phase has to adapt itself to their dynamic range ([14]-[15]-[16]). For each class, the transformed block is partitioned into a number of subvectors of small dimensions according to their energy correlation. The partitions have

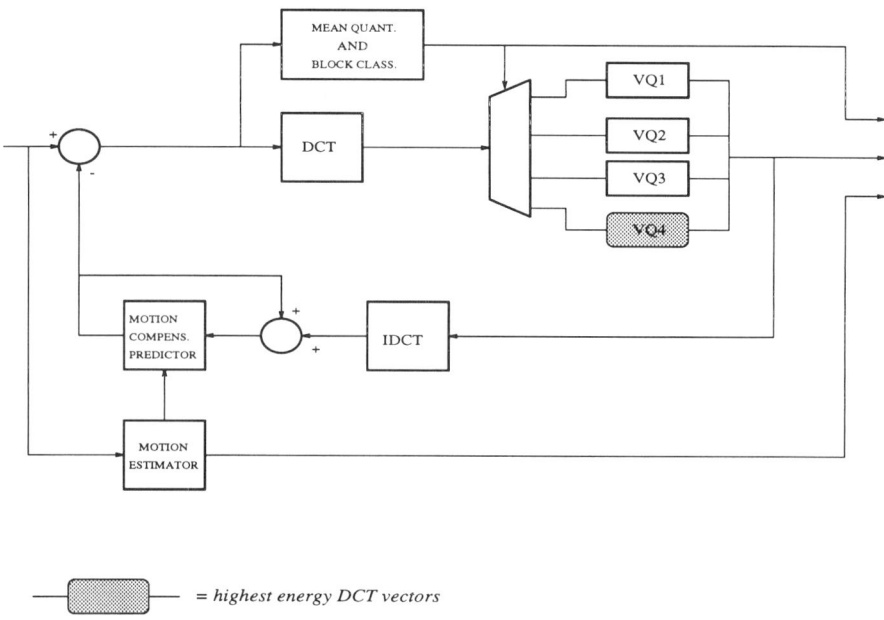

= highest energy DCT vectors

Figure 3 Block diagram of the coder using DCT and VQ.

different statistics as well as different dimensions, and for this reason different codebooks must be used to code different subvectors. The partitions with large variances contribute more significantly than the ones with smaller variances, and therefore a large number of bits are allocated to the partitions with large variances. The low frequency coefficients of the most active class (VQ4 in Fig. 3) have very high variances. The coding scheme, described above, has been applied to encode monochrome images obtained from the odd frames of CIF sequences ("Trevor White" and "Claire"). The dimension of the images is 256 by 256 pixels. A time subsampling factor of 2:1 is used. Therefore, for each test sequence, two subsampled sequences are considered. One is employed in the training phase, and the other is used for the tests. Because the vector quantizer works in a motion compensated interframe loop, the data, actually used to design the codebooks, are obtained from the motion compensated differences between couples of consecutive original images. All the codebooks have been designed using either a *LBG* algorithm or the *CL-TS* algorithm with the proposed initialization. A full search procedure is used during the coding process to select the codewords that best match the incoming vectors. Fig. 4 shows the peak signal to noise ratio of the coded images obtained using the two different sets of codebooks. The average bit rate is about 0.4 bit/pixel.

Image Vector Quantization by Neural Networks

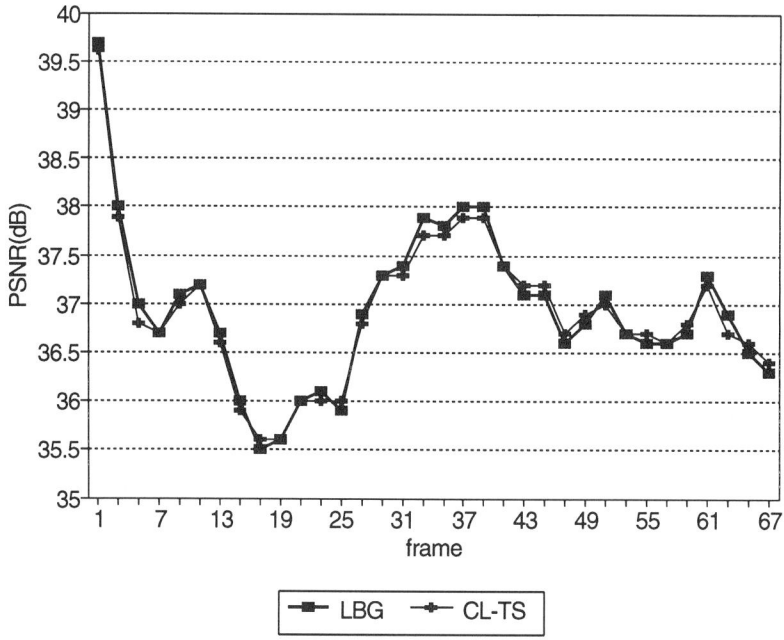

Figure 4 Reproduced quality for "*Trevor White*" sequence.

The two curves are similar. This indicates that the *CL-TS* clustering algorithm achieves (with lower computational cost) comparable performance with the *LBG* algorithm.

3 NEURAL NETWORKS FOR ADAPTIVE VECTOR QUANTIZATION

In general, image sequences can show significant variations in their frames' or scenes' statistics. If data with different statistics are presented to a fixed VQ, its performance cannot be maintained over the whole sequence. For this reason, it is important to introduce some adaptive mechanisms to improve the quality performance. Two main approaches ([17]-[18]-[19]) can be used to fix the problem. The first propose to design a codebook of a bigger dimension; in this way, the codebook contains more information. This approach affects the bit-rate (the index to address the codeword in the codebook needs more bits because the codebook dimension is bigger), and increases the computation spent in searching for the codeword closest to the input vector. The other proposes that each frame has its own codebook (smaller than the general one). This approach heavily affects the bit-rate, because all the codewords of the new codebook must be sent to the receiver. Here, a different approach will be considered: the *codebook replenishment method* [20], that represents a trade-off between the approaches presented earlier.

3.1 Codebook replenishment

Goldberg and Sun [20] presented an image coding scheme that uses a VQ technique in which the codebook is gradually modified in order to track the local (frame) statistics by updating only a part of it. The new codebook follows the local characteristics better while preserving the original structure of the codebook. However, it requires the transmission of additional information, that is the updated codewords that replace the ones already present in the codebook. This technique is called *codebook replenishment*. Let's define $C(n-1) = \{w_i(n-1); i = 1, ..., M\}$ the codebook obtained until the time $(n-1)$ *frame(n-1)*, where $w_i(n-1)$ is the codeword and M is its dimension. $LC(n)$ is the local codebook obtained at *frame(n)*. To avoid possible redundancies, only a part of $C(n-1)$ is changed, thus N codewords of $LC(n)$ substitute N codewords of $C(n-1)$ to obtain the updated codebook $C(n)$. $C(n)$ is the codebook actually used to map the *frame(n) data*. Both the N new codewords and their relative index are sent to the receiver. In [20], the codebook design of $LC(n)$ is carried out by using the clustering method of *LBG*. Because *LBG* algorithm requires a heavy computational time, in order to obtain a rapid convergence, the previous codewords of codebook $C(n-1)$ are taken as seed for generating the codebook $LC(n)$. Note that in this way, the codebook dimension of $C(n-1)$ must be the same as that of $LC(n)$.

3.2 Codebook replenishment with Neural Network approach

The scheme proposed by Goldber and Sun [20] poses some limits. They principally concern the complexity of the implementation in real-time of the classical *LBG* algorithm necessary to introduce in a vector quantization some adaptive capabilities. Neural nets offer an interesting alternative to overcome this problem. Transputer and VLSI hardware allow achievable speedup with easy implementation for these algorithms. The codebook design of $LC(n)$ and

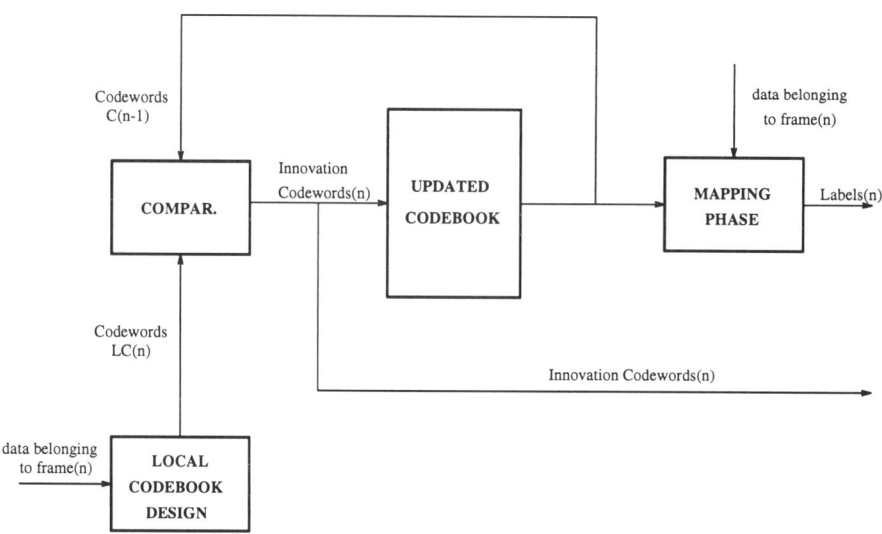

Figure 5 Codebook replenishment scheme.

the initial *C(n-1)* is carried out by applying the *CL-TS* method. By using the neural net approach, it is not necessary to generate *LC(n)* starting from *C(n-1)*. The design of *LC(n)* design is completely independent from the information on the coding of the previous frames. Also, the dimension of *LC(n)* is greatly reduced with respect to *C(n-1)* because the local codebook *LC(n)* represents only the most significant data configuration on the current frame. For this reason, the *LC(n)* design results even faster than in the *LBG* approach case. The codebook *C(n)* used for quantizing the *frame(n)* is obtained by adding to the old codebook *C(n-1)* the most "significant" codewords selected from *LC(n)*. This codebook replenishment scheme is depicted in Fig. 5.

Codeword selection

The N new codewords of *LC(n)* to be inserted in *C(n-1)* are selected in the following way. The current representative vectors in *LC(n)* are compared to the previous ones in *C(n-1)* (that is each codeword is compared to the closest one in the previous codebook). The N codewords from *LC(n)* that have the greatest differences are chosen. Then, the newly selected codewords from *LC(n)* are added to *C(n-1)* to form an expanded codebook *C(n)*. To retain identical dimension of the codebook, *C(n)*, N less used codewords in the expanded codebook are discarded. This procedure is applied by taking into account not only the quantization of the last frame, but also results from most previous frames. The discarding procedure is obviously known to both the transmitter and the receiver, therefore it is not necessary to send any information regarding this discharge procedure over the communication channel. Note that only the new N codewords have to be sent to the receiver side for updating the codebook. The trade-off is a slight increase of the bit-rate if compared to the fixed VQ. The total bit-rate for the fixed VQ is R_t is $R_t = R_l = (log_2 M)/L$, where R_l is the bit-rate for transmitting the index due to the quantization, M is the dimension of the codebook and L is the vector dimension, while the total bit-rate for the adaptive VQ is $R_t = R_l + R_{nc}$, where R_{nc} is the additional overhead for updating the codebook.

3.3 Applications and Results

To test the described adaptive vector quantization algorithm, it was applied to the motion compensated interframe image coding scheme, previously proposed (see Fig. 3). The performance of the entire coder is heavily determined by the quality of the codebooks used to quantize the vectors belonging to the most active class, that is VQ4 in Fig. 3. It has proved that these vectors are the most important to the image reproduction. Thus, to obtain good performances the codewords included in these codebooks must represent well all the significant values of the input vectors. This fact implies that the vector quantization must include an adaptive mechanism to track their local behavior better. For this reason the adaptive vector quantization has been applied in VQ4 to code the low frequency coefficients of DCT (the most energetic). The tests have been carried out by using the same CIF sequences ("Claire" and "Trevor White"). In the *non-adaptive* method the low frequency coefficients of transformed MCLD blocks belonging to the most active class are quantized by a codebook of 512 codewords [21]. In the *adaptive scheme*, the same coefficients are quantized with a codebook of 256 codewords. Frame by frame a small net $LC(n)$ (32 codewords) is trained with the incoming vectors (about 400 for each frame). These vectors are used, at first time to train the small net, employed to select the codewords to adapt the codebook. The number of training vectors practically limits the dimension of the local net, in fact a good empirical rule requires that the dimension of training set is about 10 times the dimension of the net. The codewords, that are most significant in the description of the vectors relative to the current frame (usually half of the entire net $LC(n)$), replace the less used codewords in the codebooks. In this way only the adapted codewords must be sent to the receiver and not their positions, because the information relative to the use of the various codewords is also available at the receiver side. In Fig.

Image Vector Quantization by Neural Networks

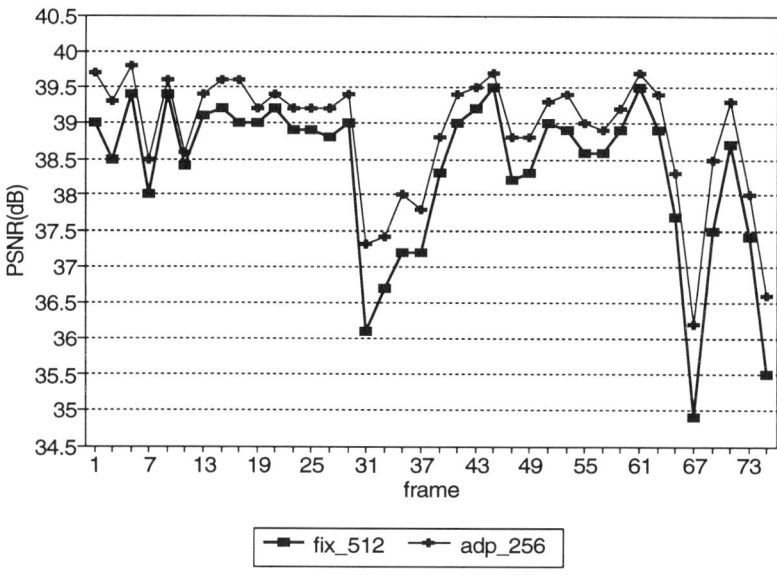

Figure 6 Reproduced quality for *"Claire"* sequence using fixed and adaptive vector quantization.

6 the results of the simulations (for the sequence *"Claire"*) are compared with those obtained with the standard coding algorithm. The performance of the adaptive coding scheme has been carried out using the odd frames of "Claire" and "Trevor White". The comparison between the non-adaptive case and the adaptive one is carried out at the same bit rate (0.47 for "Trevor White" and 0.26 for the less active sequence "Claire"). For the "Claire" sequence the starting codebooks have been designed using "Trevor White" sequence and viceversa. The performances of the coder which uses the adaptive quantization are always better (about 1-2 dB) than those obtained by using the non-adaptive one. Furthermore, using the adaptive scheme, the quality of the coded images appears more uniform all through the test sequences. Moreover when rapid motion is present in the scene, the subjective quality of the coded images is much better than that obtained with the standard vector quantization (see Fig. 7).

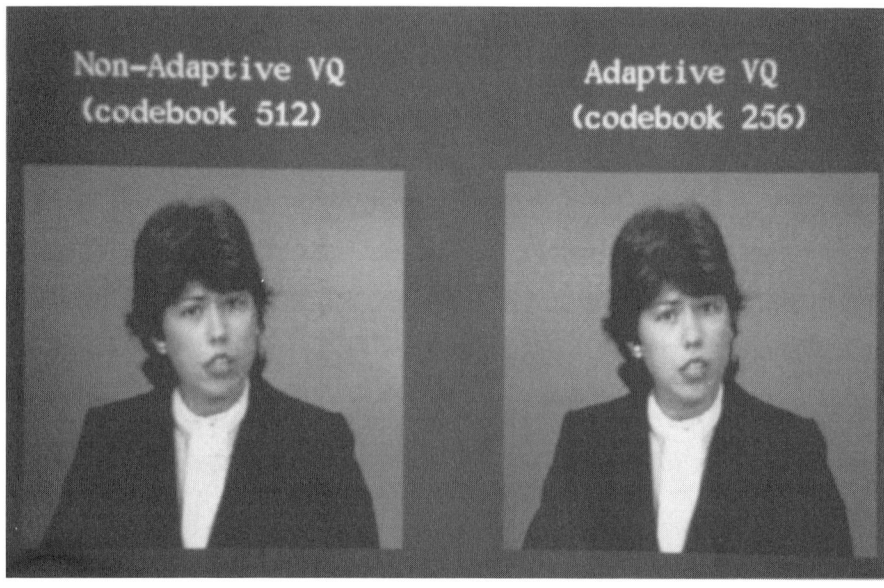

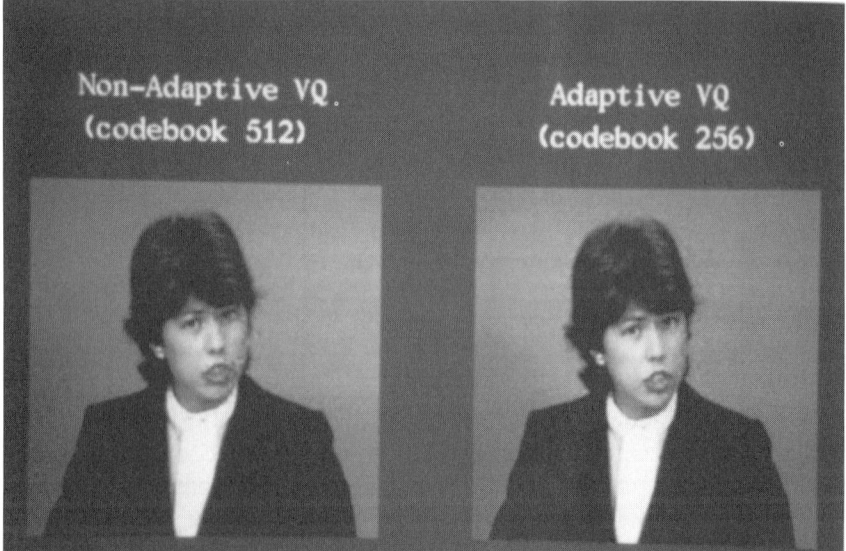

Figure 7 Reproduced quality image for sequence *"Claire"* using fixed and adaptive vector quantization.

4 CONCLUSIONS

In this chapter Vector Quantization (VQ) and Adaptive VQ, that uses a Competitive Learning (*CL*) strategy for codebook design, have been discussed. The performance of the presented algorithms for the generation of fixed codebooks to be used in two different applications has been analyzed. The first regards the design of reduced color palettes for full color image (24 bit/pixel) display on systems with a limited number of colors (4, 6 and 8 color plane). The second concerns the generation of codebooks in an interframe coding scheme. The results obtained with codebooks generated by using the Competitive Learning with initialization from the Training Set algorithm (*CL-TS*) performed better than those obtained by using *FSCL*, *SOFM* or a classical *LBG* approach, by offering a considerable saving of time in the codebook generation process. These encouraging results prompted the development of an adaptive vector quantization algorithm and its introduction in an interframe coding scheme. The results of the simulations are very promising. In fact the coder performance, compared with that using a fixed VQ, is considerably improved and the subjective quality of the coded images is much better than those obtained with the standard vector quantization, especially when rapid motion is present in the scene. As discussed before, a fundamental phase in the adaptive procedure consists in the selection of the new codewords used to update the codebook. Here, it has been decided to select, at each frame, a fixed number of codewords from the local codebook *LC(n)*. This is not an optimal choice. In fact, it is likely that a careful selection, which uses fewer codewords, can allow an adapting phase improvement while decreasing the bit-rate. For example, it could be useful to select few codewords when slow movements are present on the scene and more codewords for scenes with rapid motion. This implies that a variable codeword transmission, dependent on statistical changes, should be considered. For this reason, current research is oriented in two directions: further simplification of the algorithm to make easier the hardware implementation and a variable transmission of the selected codewords dependent on the statistical changes.

REFERENCES

[1] R. Gray, "Vector Quantization," IEEE ASSP Magazine, April 1984, pp. 4-29.

[2] Shannon, "A Mathematical Theory of communication," Bell. Syst. Tech., vol. 27, pp. 379-423, 623-656, 1948.

[3] A.K. Krishnamurthy, S.C. Ahalt, D.E. Melton and P. Chen, "Competitive Learning Algorithms for Vector Quantization," Neural Network, vol. 3, no. 3, pp. 277-290, 1990.

[4] E. Yair, K. Zeger and A. Gersho, "Competitive Learning and Soft Competition for Vector Quantizer Design," IEEE Trans. on Signal Processing, Vol. 40, No.2, February 1992.

[5] A.K. Krishnamurthy, S.C. Ahalt, D.E. Melton and P. Chen, "Neural network for vector quantization of speech and images," IEEE Journal on selected Areas in Communication, vol. 8, No. 8, October 1990.

[6] N.M. Nasrabadi and Y. Feng, "Vector Quantization of Images Based upon the Kohonen Self Organizing Feature Maps," in Proceeding IEEE Int. Conf. Neural Networks, 1988, pp. 1101-1108.

[7] T. Kohonen, *Self-Organization and Associative Memory.* New York: Springer-Verlag, 1984.

[8] Y.Linde, A. Buzo and R.M. Gray, "An algorithm for vector quantizer design," IEEE Trans. Commun., vol. COM-28, pp.84-95, Jan. 1980.

[9] R. Lancini and P. Migliorati, "Digitized Color Image Display with Limited Colormap," Proc. of PCS 91, Tokyo.

[10] P. Heckbert, "Color Image Quantization for Frame Buffer Display," Computer Graphics, vol. 16, pp.297-307, July 1982.

[11] R. Lancini, F. Perego, S. Tubaro, "Neural network Approach for Adaptive Vector Quantization of Images," Proc. GLOBECOM 91, Phoenix, December 2-5, 1991, pp. 4.6.1-4.6.5.

[12] CCIR Recommendation 601: *Encoding parameters of digital television for studios,* Geneva 1982.

[13] P. Migliorati, L. Ponte and S. Tubaro, "Application of Vector Quantization to a Motion Compensated Interframe Image Coder with and without DCT," Proc. PCS 90, Boston, March 1990.

[14] J.W. Kim and S. U. Lee, "Discrete Cosine Transform - Classified VQ Technique for Image Coding," Proc. ICASSP 89, pp. 1831-1834.

[15] Y.S. Ho and A. Gersho, "Classified Transform Coding of Images Using Vector Quantizer," Proc. ICASSP 89, pp 1980-1983.

[16] R.C. Reininger, J.D. Gibson, "Distribution of the Two Dimensional DCT coefficients of Images," IEEE Trans. on COMM., Vol. COM-31, N.6-June 1983.

[17] R.L. Baker and R.M. Gray, "Image compression using non-adaptive spatial vector quantization," Proc. Sixteenth Asilomar Conf. Circuits, Syst., Comput., Oct. 1982, pp. 55-61.

[18] Y.S. Ho and A. Gersho, "Variable-Rate Multi Vector Quantization for Image Coding." Proc. ICASSP 88, pp. 1156-1159.

[19] A. Gersho and M. Yano, "Adaptive Vector Quantization by Progressive Codebook Replacement," Proc. ICASSP '85, FL, 1985, pp.4.6.1-4.6.4.

[20] M. Goldberg and H. Sun, "Frame Adaptive Vector Quantization for Image Sequence Coding," IEEE *Trans. Commun.*, vol. 36, No. 5, pp. 629-635, May 1988.

[21] R. Lancini and F. Perego, "Frame Adaptive Vector Quantization with Neural Networks," Proc. GLOBECOM 92, Orlando, December 6-9, 1992, pp. 1310-1314

16

MANAGING THE INFOGLUT: INFORMATION FILTERING USING NEURAL NETWORKS

Thomas John

Southwestern Bell Technology Resources

1 INTRODUCTION

The promise of the information revolution is the easy availability of vast amounts of knowledge. The reality is that once the gates are opened to these pools of information, the flow quickly turns into a flood. Studies have pointed out poor retrieval rates even in commercial systems designed for retrieving information [1]. In addition to standardized sources of new information, electronic publishing has resulted in an exponential growth in the amount of computer readable text, with amount of electronic text doubling in volume in just three years [2]. Recent developments in communications imply that the problem will only get worse. The electronic superhighway will ensure that access to information every living room in the form of cable systems which offer hundreds of channels [3].

While useful information is the essentially valuable commodity, it is difficult to extract from irrelevant information. Tools are usually designed to filter specific pieces of information. These tools may take the form of some cryptic commands as in the *grep* family of search tools which select a few pieces of information of interest to a particular individual, discarding the rest. By contrast, library classification attempts to meet needs of many individuals by preclassifying information. But effective use of this information depends on experts who can match individual interests with these preclassified collections.

Neural networks are presumably models of human information processing. Thus it should be possible with the right architecture to approximate the filtering capability of an expert. This is the motivation behind neural information filtering models. Since non-neural filtering algorithms came before the neural ones, sometimes it is useful to consider the neural algorithms in comparison with the traditional algorithms. We will also consider properties of the neural filters from the perspective of real-time performance and adaptability. We will propose an experimental set up for comparing these characteristics and will examine some data on this topic. Finally, we will consider the opportunities presented by neural information filtering models.

Traditional filters offer various capabilities which make them perfectly adequate for many information processing tasks, though we will argue that most of them are lacking in terms of adaptiveness. Some of these traditional filters are extensively used in commercial products, thus they have undergone refinements guided by experience. Most neural filters we consider are experimental and have had only limited use. In many cases, interesting results may be obtained by combining some of the features of neural and non-neural filters.

2 PREVIOUS WORK ON FILTERING

Some authors make a clear distinction between information *filtering* and information *retrieval*. Information retrieval came before filtering; it refers to the activity of storing and retrieving information. The fact that information may consist of free form text and other disorganized data is what distinguished information retrieval from databases. Information filtering applies more towards streams of online data. While there has been lot of work on the topic of information retrieval, there is relatively little work associated specifically with information filtering. There are however several information filtering systems in use commercially, and in most cases, their underlying algorithms are well known. The algorithms used tend to be those developed for information retrieval. Although most work on information retrieval has concentrated on text processing, filtering is not limited to textual information. This is due to the fact that even non-textual information has to be indexed with some textual information in order to make them available on most computer systems.

A very detailed model of information filtering in comparison with information retrieval is considered in [4]. In broad terms, the difference between filtering and retrieval can be thought of as the difference between reading a newspaper and seeking information at a library. Information retrieval algorithms can be converted to filtering algorithms by querying information feeds against a set of queries. With a set of queries $\{q_1, \ldots, q_n\}$, the filtering algorithm F' works on all documents d as follows: (Note that $F'(d)$ may have many values. This is

Filter	Motivation	Author
FRUMP	Natural Language	DeJong
FERRET	Genetic Algorithms	Mauldin
SMART	Vector Space	Salton
RUBRIC	Probabilistic Logic	Cooper
LSI	Vector Space	Dumais

Table 1 Some Non-neural Filtering Methods

sometimes considered a desirable feature, i.e. the same document may appear in different categories.)

$$F'(d) = \begin{cases} i & \text{if } i \leq n \text{ and } d \in F(q_i) \\ 0 & \text{otherwise} \end{cases}$$

While there are differences between filtering and retrieval operations, at a fundamental, conceptual matching level these differences fade. Traditional retrieval systems have dealt with relatively stable collections, but with the information growing at current rates, these collections cannot be considered static. Even though we may retrieve the answer to some specific query at one time, it would be nice if the system remembers what we are interested in. Even though we may filter out information, that stuff we filtered out may be useful for somebody else. In discussing prior work on filtering, we will therefore cite various algorithms presumably designed for retrieval. In this section we will discuss the non-neural algorithms. While there are several algorithms which are not considered here, we will attempt, albeit briefly, to cover all the well-known algorithms. Table 1 lists the different filters we will describe.

We will first skim over the methods inspired by work in natural language processing, then we will consider some popular methods which ignore the linguistic structure of free form text. These latter methods are Vector space methods [5], Latent semantic indexing [6], and Belief networks [7], [8]. Other than simple key-word and boolean logic systems, these three systems seem to be those represented in non-neural commercial filtering efforts. These systems use various measurements of word and phrase occurrences within text and generally do not possess any understanding of the content. There are however several retrieval systems which include natural language capabilities.

2.1 Filtering with language awareness

Filtering using natural language is somewhat easier than traditional natural language interaction. This is due to the fact that filtering with natural language knowledge does not require the user to use long and cumbersome queries. One of the earliest filtering systems which had a good degree of natural language understanding was FRUMP [11]. This system used a form of skimming through news stories, looking for *sketchy scripts*, developing an understanding of the news story based on predicting outcomes and seeing whether the predictions are substantiated. FRUMP was able to understand many news stories it had not seen before and actually form summaries of their content.

Another notable system with a degree of natural language understanding is SCISSOR [12]. This system was designed especially to find news stories dealing with mergers and acquisitions of companies. After reading the stories, SCISSOR can also answer questions about these stories. In some cases, natural language understanding has been used to complement other methods of classifying information. A good example of this is FERRET [13]. This system combines extensions of script understanding as in FRUMP with a genetic learning component. The resulting system was used to retrieve Usenet articles from an Astronomy database.

2.2 Vector space methods

Outside of naive key-word and logic-based retrieval, the most popular text retrieval and filtering method appears to be Salton's vector space model [5]. It is the retrieval method used in a number of commercial products such as *First!* by Individual Inc., a system developed by Thinking Machines [14] as well as the WAIS interface by Thinking Machines [15], and the method of choice in various academic systems such as WALT [16].

Although vector space models do not use natural language understanding for text classification, the use of words has been found to be quite effective. In vector space models, similarity of one piece of information to another is estimated on the basis of the separation between two vectors representing the two pieces of information. (Strictly speaking, words are not the basic building blocks of this approach, but rather *terms* which may be short expressions involving more than one word. In practice however, words, or just word *stems* are used to build the vectors.) The vectors may be weighted with term frequency weights, and the vectors may be normalized. A popular measure of similarity is the cosine of the angle between two vectors.

The vector space model can be used directly in retrieval operations by converting a query into a vector q, computing the cosine of the angle between q and vectors D_i for all documents in the collection, and ranking the documents in the collection in order of decreasing cosines. On a small document collection, this approach works fairly well. On large collections, our experience is that vectors do not really mean anything. In particular with collections of news stories, sports news stories tend to have very colorful language, and gets distributed all over the document vector space.

The vector space model can be used as described above to build an information filter by creating some categories for each individual and subsequently by populating the categories with pieces of information which are close to those already in the category. Here closeness is defined as having a high cosine value between vectors representing the information and a vector which represents the category. A user of this filtering system can send feedback to the system about the relevance of the selected articles. This feedback can be used to change representative categorizing vectors using algorithms described by Salton [22]. In our experience, if there are articles in the collection which are distributed somewhat uniformly over the information space, then large amounts of feedback tend to randomize the categorization vectors.

2.3 Latent semantic indexing

In vector space methods, an implicit assumption has to be made that words are pairwise independent in terms of the vector space. This is needed to ensure that the basis vectors of the space are orthogonal. However some words tend to occur frequently with some others. A way to overcome this difficulty is found in latent semantic indexing (LSI). LSI is a technique based on singular value decompositions of word frequency matrices. This reduces the dimension of the resulting information space. By finding the co-occurrence probabilities of terms, it is possible to use LSI to retrieve information which may not contain exactly the terms in the query.

Just as with vector space models, LSI can be used to filter documents by establishing categorizing vectors. Feedback from users is used to determine which categorizing vectors best represents a user's interest. An experiment was done at Bellcore to compare the accuracy of LSI-based filtering of technical reports with a key-word-based filtering approach. This experiment and results showed that the LSI method was the most successful of the methods which were tested [17].

2.4 Probability network methods

There are other approaches that extend the key-word and boolean logic model with artificial intelligence. An example of such a system is RUBRIC, written using EMYCIN [7]. This system was the forerunner of the popular commercial information retrieval and filtering system called TOPIC from Verity, Inc. TOPIC is used as the basis for a number of text retrieval and filtering applications, including such widely disseminated products as Microsoft's compact disc products.

There are several variations on the theme of networks. The basic idea however seems to be one of constructing a belief network, which may in some cases be a hierarchy. Topics may be nodes in this hierarchy and beliefs guide the placement of information at a certain topic. Rules of inference, which can be tempered with degrees of beliefs, link concepts so that a combination of concepts may result in a piece of information being associated with a particular topic.

2.5 Other methods

In addition to the methods described above, there are several other approaches to information filtering. For example, there are approaches involving inference networks [8], using hash functions [9], and using massively parallel machines [10]. We also have not considered the standard key-word and boolean logic approach, which work well when the queries are constructed by experts. Most commercial systems still use the key-word and logic approach to information retrieval. By carefully calibrating filters to retrieve relevant information from online sources, it is possible to construct very effective filters using only simple key-word and logic systems.

Although carefully crafted systems can work, it has two major disadvantages. One is the amount of work which goes into carefully specifying the topics of interest to a particular user. The other is the lack of flexibility in such a system. Since the types of information as well as a person's interests undergo changes with time, it is necessary for a filtering system to keep up with these changes to

Filter	References
Word disambiguation	Cotrell
Fuzzy retrieval	Petry *et.al.*
Probabilistic model	Kwok
Adaptive Retrieval	Belew
Adaptive Resonance	Carpenter *et.al.*

Table 2 Some Neural Filtering Methods

be truly effective. Carefully crafted systems cannot be changed easily because an apparently minor change in one place may affect the performance in some unanticipated fashion. Neural network-based models by contrast have some hope of modifying themselves with respect to changes in individual interests.

3 NEURAL NETWORKS FOR FILTERING

Information retrieval and filtering are terms usually applied in connection with text information, while most of the work in neural networks has been directed towards numerical data. The classification of numerical data can be relevant to some information retrieval tasks. For example, if auditory signals are classified using a neural network, and later this network is used to classify an unknown sample, this is not what we normally call filtering. However, if the same neural network is used to listen to a cacophony of voices, from which the net recognizes a familiar sound, then this network can be considered to be doing information filtering. Similar statements can be made about neural networks which recognize facial profiles for security systems, neural systems which recognize underwater sonar signals or even neural networks which look for anomalies in cardiac rhythms. Some previous work on connectionist models also are peripherally related to filtering. Work on language understanding is related to information filtering [18]. Some work in fuzzy information retrieval also has connections to neural networks, although generally information is retrieved from conventional databases [19]. There are several examples of neural networks applied directly to information retrieval, which are considered in detail below.

Most neural retrieval algorithms are inspired by the vector space model [20], [21]. For example, the descriptors used in [21], which is billed as a *probabilistic model*, are closely related to those used in the feedback version of Salton's model [22]. This approach assumes that the vector space model is the correct model of information retrieval, neural networks are used to build the model in an easier way. Even though the fundamental paradigm is that of a vector space of documents, the actual implementation details are different in the various models [23], [24]. Neural networks are used in this case to model probabilities or term weights which determine the relative placement of documents in the document vector space. The attempt in this case is not to improve on the conventional retrieval models but rather to emulate them, to show that neural networks can learn what the conventional methods can compute. Table 2 lists various neurally motivated retrieval methods.

3.1 Criteria for assessing neural net filters

Information filtering may be a reflection of the way people learn concepts. Information filters can be evaluated by seeing how well they work for people. This can be expressed in terms of two properties.

- Real-time performance

    It should be able to process information in real-time, i.e. in $O(N)$ time. Both learning and filtering should be in real time. In addition, the processing speed should not be adversely affected by the number of users. Since information is usually centralized, a system which requires a separate computation for each user is more computationally complex than one where some of the computation can be shared.

- Adaptive learning

    This criterion is harder to measure than real-time performance. Human adaptiveness has the twin qualities of stability and plasticity. Most neural networks are plastic, but they are sometimes plastic at the expense of stability. We will develop a test criterion to assess the adaptiveness of various neural network (and other) filtering methods based on the stability/plasticity criterion of adaptiveness.

Both supervised and unsupervised networks can adapt to inputs. Indeed so do most of the non-neural filtering systems. In supervised systems the adaptation occurs only when it is taught something. In an unsupervised system adaptation is part of the functioning of the system, and it learns constantly. Since supervised systems learn only when taught, careful teaching may make it more accurate and responsive. But some of the problems in this may be similar to problems with key-word and boolean retrieval, i.e. they require some care to ensure that stable classifications are formed. Unsupervised systems make no such assumptions about care. Hence they are generally more robust and can withstand arbitrary sequences of inputs. There are also systems which combine some unsupervised learning with supervised learning. ARTMAP [26] is an example of a system which combines supervised and unsupervised learning.

We will consider two neural filtering systems. The first considers *reinforcement learning*. In this model, the information about failure or success is used to guide the learning of the system. In contrast, standard backpropagation networks use the distance between the classification produced by the system and the correct classification to correct for errors. We will describe AIR, a well known system that uses reinforcement learning. Note that reinforcement learning is also a supervised method according to the definition we use in that learning takes place only in response to user inputs. Later we will consider a filtering system based on the Adaptive Resonance model of unsupervised learning.

4 REINFORCEMENT LEARNING - AIR

The best known connectionist system for information retrieval is probably Belew's Adaptive Information Retrieval (AIR) system. This is a supervised system in the sense that learning does not take place unless the user provides some feedback. In experiments reported in [25], a connectionist network was used to relate words, documents and authors. Modifications of the network took place as a result of feedback given by users based on their judgement of the relevance of documents retrieved in response to a particular query. AIR is related to an earlier system by Mozer [27]. Mozer's system used a standard parallel distributed architecture, while Belew's connectionist networks were designed specifically to aid in information retrieval tasks.

AIR uses a three layer network consisting of layers for words, documents and authors. Experiments described used small sets of words belonging to documents, usually words in titles [28]. AIR does not start with a "network from scratch" but rather, from a network constructed from the initial documents contained in the system. Thus links are not created at random, but exist only where a word is contained in a document or where a document is authored by a certain author. Except for standard noise word removal and word stem creation, all words in the document (or title as described in experiments) become one of the nodes in the top layer of the network. The links in the network as weighted using *inverse term frequency* as in [22]. The total weights of all links going out of a node is forced to be a constant, similar to the normalization done in [22]. As for the size of the networks considered, an example cited in [28] consisted of an initial collection of 1600 documents which formed a network of about 5000 nodes.

Learning in AIR is based on relevance feedback. Although the learning rules differ, the feedback is similar in spirit to the one done by Salton in [22]. The system responds to a query by retrieving documents, in fact it retrieves parts of the network. The user marks some of the nodes in the retrieved part (however many he feels like marking) using markers on a four-part scale, indicating whether a node is relevant or not relevant to the particular query. Weights of the links are changed based on responses to the queries. Nodes representing words, documents and authors can be represented in the response. After the user marks some of these nodes, weights of links connected to the marked nodes are changed according using a learning rule. Several rules for changing the weights of links are considered in [28]. The learning rule which was found to be best was the *correlation learning rule*. According to this rule, the change in weight of a link connecting say, a document i to a word node j is based on the conditional probability that node j is relevant, given the fact that node i is relevant. This conditional probability is judged from the user's response.

Although AIR contains the feedback mechanism just as in Salton's relevance feedback, Belew does not directly extend this to the construction of a filter. Indeed in [29], he and John Rentzepis considers the problem of creating information filters for email, but does not offer a solution. They state that "... information filters suffer from an even more fundamental limitation, viz., the inability to characterize *a priori* what constitutes interesting, relevant information." They do not see AIR as a way to do this type of a priori classification. This goes along with the assessment that systems like AIR are supervised since they do only post hoc learning, after the user has provided some input. Nevertheless, information filters can be constructed using AIR along the lines that Salton's relevance feedback can be used to construct information filters.

While the connectionist approach used by Belew in AIR is probably the best known information retrieval application of this kind, one that may appear to be similar is the approach used by Stanfill and Kahle [10]. As Belew points out in [25], the similarity here is superficial. The Stanfill and Kahle implementation uses relevance feedback to refine a user's query. By comparison, in AIR, the system learns from relevance feedback to modify the user's connectionist network representation for ever.

Although AIR does not use the standard backpropagation network, its use of connectionism is not unique in terms of concept formation networks. Other neural network paradigms can also be used for information retrieval and information filtering. Various researchers have applied networks such as Kohonen's feature maps and adaptive resonance networks to information retrieval and classification. We will consider one such approach, using adaptive resonance networks, as a typical example of an unsupervised filter construction.

5 AN UNSUPERVISED CLASSIFIER - ART

The basic requirements of filters described here are to operate in real-time and to be adaptive. Both of these requirements are qualities of human information classification. Adaptive Resonance Theory (ART) models many qualities of human perception. Of particular interest here is the fact that the categories formed by an ART 1 system can be made finer by adjusting the vigilance criterion. The ART 1 neural architecture has the property of stable self-organization of binary patterns. An ART 1 network forms its classification categories based on bottom-up and top-down matching as mediated by an attentional mechanism. The attentional mechanism will determine whether bottom-up input patterns and top-down expectations match closely enough as determined by a parameter.

For background on the theory and applications of Adaptive Resonance Networks see [30]. For a brief survey article see [31]. Application of ART networks to word and letter classification are considered in several articles, see for example [32], [33]. The networks consider here have two layers with the neurons in different layers contained in the sets F_1 and F_2.

In the nets we consider input to the net is presented at layer F_1. These inputs are propagated to the next layer via bottom-up links. As result of this propagation, the neurons within F_2 will be activated. On the other hand, F_2 will send a top-down expectation to F_1. In addition to the top-down and the bottom-up transmissions, there is an attentional mechanism. The attentional mechanism determines whether there is sufficient agreement between the top-down expectation and the bottom-up presentation. In cases where there is insufficient agreement, the mismatching top-down expectation is inhibited by the attentional mechanism so that some other expectation may match the bottom-up pattern.

ART 1 classification has a continuity property which is not shared by many other classifiers. For a particular collection of binary input patterns presented to the ART system, the classification becomes finer as the vigilance is increased. When a classification becomes finer, larger classes are broken into smaller classes. Note that such a property is not true, for example, for a nearest neighbor classifier. This continuity property can be utilized in an ART-based filter to tune filters based on user responses, varying the granularity of classification at different levels depending on the level of user interest.

A filtering network M 1 is a simple extension of the basic ART 1. The main difference between the mapping version M 1 and ART 1 is that M 1 has an "output" stage which associates a function value with each ART 1 category [34]. This output stage can also influence the formation of categories, by not allowing resonance to occur at a particular node in the second ART 1 layer F_2 when the value associated with this node is not the one which is suggested by the current example from the graph of the function. The M 1 network is able to learn a continuous mapping in real time (assuming proper hardware) without the aid of an external teacher. Therefore an M 1 network can be hooked to a data stream without regard to the complexities of sequences generated by this stream.

6 EXPERIMENTAL DETERMINATION OF FILTER VALIDITY

While there are theoretical ways to study the asymptotic performance or accuracy of filters, a fairly rudimentary experiment can be set up to simulate the behavior of filters over a period of time. A realistic behavior monitor must deal with a large enough volume so that we can realistically predict the performance of these filters when dealing with massive amounts of information.

To get an idea of the massiveness of information, the *Net News* system available on the internet typically has the volume of five megabytes of information per day. By comparison, the amount of information dealt with in the AIR experiment described in [25] is probably of the order of a few thousand bytes. Therefore, realistic experiments must be conducted with much larger data sets.

We need to measure the twin requirements of real-time performance and adaptiveness. Real time performance is relatively easy to determine, we can do complexity analysis both for worst and average cases, and we can measure the

Information Filtering 317

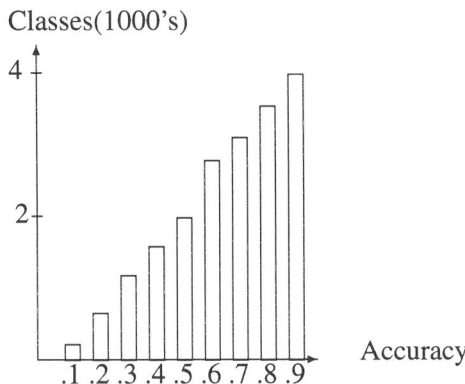

Figure 1   Accuracy(0.0-1.0) vs. Classes(0-4000) for an ART model

amount of time it actually takes to do filtering. There are actually two aspects to this computational complexity. If a filter is not trained properly then it may have fewer neural codes to deal with and thereby may have faster filtering. But a poorly trained filter will not be able to classify all of the information it is seeing, or it may misclassify this information. Thus there is often a tradeoff between speed of filtering and accuracy of filtering.

Figure 1 shows how there is a tradeoff between speed and accuracy. In this example, an ART-based model is used to classify sets of words in documents. As vigilance increases, the number of classes created also increases. The number of classes affect filtering performance by increasing the time required for search. In this case there is an approximately linear increase in the number of classes with the level of accuracy, therefore we can predict an approximately linear increase in computational load with accuracy. If a massively parallel computer is used however, so long as the entire neural network can be implemented, there may be no perceivable difference in time. For example, on a MasPar MP-1 with 4094 processors as tested there was no difference in filtering using vigilance level of 0.1 vs 0.9 since the MasPar was able to match approximately 4000 neural codes obtained at vigilance 0.9 just as easily as it was able to match approximately 180 codes with vigilance 0.1. Note that this computational load does not increase with the number of users.

Similar computational load requirements can be measured for different filters. Filters however differ in terms of their computational load depending on a number of factors. For example, with unsupervised neural networks there is a lot of up-front cost involved with training, but filtering can be real-time. With supervised systems there is little initial training but filtering and retraining can be expensive (this is mainly due to the need to perform separate computations for each user.) Some filters, such as ART, can be implemented using primarily integer arithmetic, while some others require extensive floating point computations. In practice we have found Salton's filters to be the most time-consuming, mainly due to the fact that extensive floating point computations are performed for each user.

Adaptiveness of a filter is much harder to measure. This is due to the subjective nature of human judgement, which means that an extremely tolerant person would find even the simplest filter adaptive enough, while a demanding person would find almost no filter adequate. To lend some objectivity to this exercise however, we can rely on preclassifications which are provided in many news feeds. For example the ClariNet news feed contains categories called *slugwords* which can be used to determine how a committee of people would tend to classify a particular story.

In general, the set up for testing an information filter involves a information source and a user interface where a user may interact with the system [10], [25]. In the case of human testing, a tester indicates interests to the filtering system through the user interface. For text filters the user indicates whether categorization performed by the filter matches the user's expectations. For adaptive text filtering system, the user may go through a phase of indicating category choices to the filter so that the filter may adapt to the user. Subsequently at some point, the user would stop indicating category choices and would instead indicate whether the category choices made by the filter are correct according to the user's interpretation of expectations built up during the process of training.

In automatic testing, there is no user to indicate category choices. Hence category information has to be derived from somewhere else. In this connection, category information provided by publishers becomes useful. Many information providers add some category information to their feeds. For example, UPI/Clarinet adds category information at several levels through the directory structure of NetNews newsgroups. When this type of information is not available, the origin of a news story, such as whether it is a business, sports or political story, can be used as a categorizer. We will call a categorizer of this form as a *preclassifier*. Note that preclassifiers are objective, even if they are not particularly correct.

Information Filtering

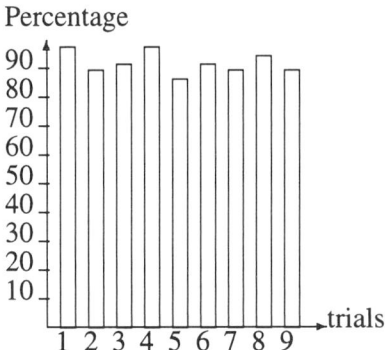

Figure 2 Progressive Accuracy for an ART filter (see text)

The procedure for testing filters now is as follows: Let F indicate a filter. For an article a, $F(a)$ will be a category, which we may think of as a number. Initially, for all articles a, $F(a) = 0$ indicating that there is no classification of this article which is meaningful for a user. For a fixed, presumably large, collection of articles, the articles belong to several categories identified by the preclassifier. Group many of these categories, at random, to form a small (of the order of ten) "user" categories. Having made the assignment of articles to "user" categories, divide the collection of articles into two collections; the first part will be the training part and the second part will be the testing part. The relative sizes of these two parts can be varied during testing to obtain results on the amount of training necessary to achieve certain results. Start by running the filter on all the articles in the training collection. Let a be an article whose preclassifier category is c, and hence whose user category is u. Instruct the adaptive filter that its output $F(a)$ should be u. Perform this action with all the articles in the training collection. Different filters may have differing requirements for stabilization, or in other words, criteria may be different for knowing when the filter is fully trained. But after the training phase is over, the testing phase can begin. During the testing phase, articles b in the testing collection are run through the filter to compute $F(b)$. The filter is considered to be accurate on article b if $F(b) = u$ which is the class that will be assigned by the preclassifier.

Figure 2 shows the result of using this test on an ART filter. First the ART network was trained at vigilance 0.9 on a collection of approximately twenty thousand articles from various news sources. After training, vigilance was set to 0.2 and the filter was tested on batches of 300 articles. Initially all articles were unclassified by the filter since no user categories were present. However, for each article a, its unsupervised neural class $N(a) = n_a$ was recorded. An automatic classifier examines each article and determined what user category it should belong to, i.e. it determines u such that $F(a) = u_a$. The ART filter records this information by constructing an association between the unsupervised class and the user category, i.e. a function A such that $A(n_a) = u_a$. On the next filtering run, if a new article b is such that $N(b) = n_a$ then the assignment $F(b) = u_a$ is assumed. The filter has made an error if

$$\begin{aligned} N(a) &= n_a \\ F(a) &= u_a \\ N(b) &= n_a \\ F(b) &= u_b \\ u_a &\neq u_b \end{aligned}$$

i.e. if the assumption about the value of $F(b)$, made on the basis of a previously learned value of $A(n_a)$, is incorrect.

Figure 2 shows that the classification is approximately ninety percent accurate on subsequent trials (some of the inaccuracy here can also be ascribed to the inaccuracy of the automatic preclassifier which used the news agency as a way to classify.) One observable behavior with ART classification is that accuracy does not degrade with time. This is due to a stability property of ART. We may conjecture, based on results on ART vs vector space clustering reported in [35] that vector space methods will not exhibit the same level of stability.

The tests here can be expanded in several directions. For example, there is a tradeoff in some filters between the time it takes to train a system and the accuracy of the resulting classification. Another direction in which filters may be tested involves setting up an ongoing filtering process to see if changes made by the filter on all incorrect answers causes the filter to become unstable. This stability is one good measure of the adaptiveness of a filter.

7 FUTURE DIRECTIONS

We have considered a number of information filtering approaches. These included those motivated by natural language processing, statistical similarity measures and database management. We have also considered a few methods for applying neural networks to information filtering. In neural filtering we have described both supervised and unsupervised methods.

While there have been various efforts directed at information filtering, only a few have been motivated directly by models of how people think. Even in neural information processing models, often properties of human information processing such as stable learning and adaptability are not found. A neural network filter which has some of these qualities can be expected to integrate well with our own information processing methods. We have introduced a method here for testing information filtering models to see if they conform to human filtering criteria. Future work will examine how various filters compare under these criteria.

Refinement of filtering methods will happen over a period of time. Meanwhile there are a number of applications of various filtering methods which exist now. These applications may penetrate further into all information processing activity. Future applications will result from this penetration. For example, there are no facilities at present for choosing desirable features from an entertainment network with hundreds of channels. Information filtering which learns user preferences is one way to approach this problem. Most of the information concerning financial transactions and market movements are now filtered by professionals. This may penetrate further into information monitored by ordinary individuals if efficient information filtering tools are put into place. Information filtering may also help with some of our social problems. For example, it is estimated that over fourteen million tons of newspaper is produced in the U.S. every year. An electronic newspaper could significantly alleviate this problem and in addition provide people with the news that fits their interests.

REFERENCES

[1] Blair, D.C and M.E.Maron, "An evaluation of retrieval effectiveness for a full-text document-retrieval system," Communications of ACM, 28 (1985) pp.289-299

[2] Pacific Telesis, Annual Report, 1987.

[3] Elmer-Dewitt, Philip, "Taking a trip to the future on the electronic superhighway", TIME, April 22 (1993) pp.50-58.

[4] Belkin, N.J. and W.B.Croft, "Information filtering and information retrieval: Two sides of the same coin ?" Communications of ACM, 35 (1992) pp.29-38.

[5] Salton, G. Automatic Text Processing, Addison-Wesley, 1989.

[6] Dumais, S.T., G.W.Furnas, T.K.Landauer and R.A. Harshman, "Using latent semantic analysis to improve information retrieval," CHI '88 Conference Proceedings: Human factors in computing systems, New York, 1988, pp.281-285.

[7] McCune, B.P., R.M. Tong, J.S.Dean and D.G. Shapiro, "RUBRIC: A system for rule-based information retrieval," IEEE Trans. Software Engineering, SE-11 (1985) pp.939-945.

[8] Turtle, H. and W.B.Croft, "Evaluation of an inference network-based retrieval model," ACM Trans. Inf. Systems, 9 (1991) pp.187-222.

[9] Fox, E.A., Q.F. Chen, A.M. Daoud and L.S. Heath, "Order-preserving minimal perfect hash functions and information retrieval," ACM Trans. Inf. Systems, 9 (1991) 281-308.

[10] Stanfill, C. and B. Kahle, "Parallel free-text search on the connection machine system," Communications of ACM, 29 (1986) pp.1229-1239.

[11] DeJong, G. "Prediction and substantiation: A new approach to natural language processing," Cognitive Science, 3 (1979) pp.251-273.

[12] Rau, L.F. and P.S. Jacobs "$NL \cap IR$: Natural language for information retrieval," International Journal of Intelligent Systems, 4 (1989) pp.319-344.

[13] Mauldin, M.L., Information Retrieval by Text Skimming, Ph.D. Thesis, Carnegie Mellon University, 1989.

[14] Stanfill, C., R. Thau and D. Waltz, "A parallel indexed algorithm for information retrieval," Proc. 12th SIGIR Conference, (1989) 88-97.

[15] Kahle, B., "Wide area information concepts," Thinking Machines Technical Memo, No. DR-89-1.

[16] Frisse, M.E., "Searching for information in a hypertext medical handbook," Communications of ACM, 31 (1988) 880-886.

[17] Folz, P.W. and S.T. Dumais, "Personalized Information delivery: An analysis of information filtering methods," Communications of ACM, 35 (1992) pp.51-60.

[18] Cotrell, G.W., "Parallelism in inheritance hierarchies with exceptions," Proc. 9th IJCAI, (1985) pp.194-202.

[19] Petry, F.E., B.P. Buckles, A. Yazici and R. George, "Fuzzy information systems," Proc. 1st IEEE Intl. Conf. Fuzzy Systems, (1992) pp.1187-1202.

[20] Bordogna, P. Carrara and G. Psai, "Extending boolean information retrieval: A fuzzy model based on linguistic variables," Proc. 1st IEEE Intl. Conf. Fuzzy Systems, (1992) pp.769-778.

[21] Kwok, K.L., "Application of neural network to information retrieval," Proc. IJCNN, Volume 2 (1990) pp.623-626.

[22] Salton, G. and C. Buckley, "Parallel text search methods," Communications of ACM, 31 (1988) pp.202-215.

[23] Kowk, K.L., "A neural network for probabilistic information retrieval," Proc. 12th SIGIR Conference, (1989) pp.21-31.

[24] Yu, C.T. and G. Salton, "Precision weighting - an effective automatic indexing method," J. of ACM, 23 (1976) 76-86.

[25] Belew, R.K., "Adaptive information retrieval," Proc. 12th SIGIR Conference, (1989) pp.11-20.

[26] Carpenter, G.A., S. Grossberg and J.H. Reynolds, "ARTMAP: Supervised real-time learning and classification of nonstationary data by a self-organizing neural network," IEEE Expert,, 6 (1991).

[27] Mozer, M.C., "Inductive information retrieval using parallel distributed computation," Technical Report, Institute of Cognitive Science, UCSD, La Jolla, CA. 1984.

[28] Belew, R.K., "Designing appropriate learning rules for connectionist systems," Proc. 1st ICNN, Volume II (1987) pp.479-486.

[29] Belew, R.K. and J. Rentzepis, "Hypermail: Treating electronic mail as literature," SIGOIS Bulletin, 11 (1990) pp.48-54.

[30] Grossberg, Stephen, Neural Networks and Natural Intelligence, MIT, 1988.

[31] Carpenter, Gail A., and Stephen Grossberg, "The ART of Adaptive Pattern Recognition by a Self-Organizing Neural Network", Computer, March 1988.

[32] Grossberg, S. and G.O. Stone, "Neural dynamics of word recognition and recall: Attentional priming, learning and resonance," Psychological Review, 93 (1986) pp.46-74.

[33] Grossberg, S., "Competitive learning: From interactive activation to adaptive resonance," Cognitive Science, 11 (1987) pp.23-63.

[34] John, Thomas, "Relationship of ART 1 models to continuous functions", BP Research, Topical Report Number 8094 (1990).

[35] Moore, B., "ART1 and pattern clustering," Proc Connectionist Models Summer School, 1988, pp.174-185.

17

EMPIRICAL COMPARISONS OF NEURAL NETWORKS AND STATISTICAL METHODS FOR CLASSIFICATION AND REGRESSION

Diane Duffy, Ben Yuhas,
Arvind Jain and Andreas Buja

Bellcore

1 INTRODUCTION

In this chapter we empirically compare neural networks and statistical methods on two telecommunications problems – one regression problem and one classification problem. Our goal is to explore the relative performance of neural networks and several statistical methodologies on moderate-sized regression and classification problems. A regression problem is one in which the goal is to estimate a numerical quantity based on partial input information; a classification problem is one in which the goal is to estimate the class to which an item belongs, again based on partial input information. We did not use a systematic method to choose problems for comparison, but rather chose one significant complex problem of each type. By significant, we mean that the underlying scientific or engineering question was important. By complex, we primarily mean that the dimensionality of the problem was large (roughly, greater than 20 measurement dimensions per observation or case).

When regression and classification problems are very low dimensional, say less than 4 dimensions, it may be possible to *see* the relationships between the covariates and the response. Graphical tools, therefore, can play a critical role in guiding model building for low dimensional problems. However, as the dimension increases, we are faced with the need for automatic model building strategies. Neural networks are naturally poised to tackle high dimensional problems. In statistics, several model building methodologies, notably CART (Classification and Regression Trees; Breiman et al. 1984) and MARS (Multivariate Adaptive Regression Splines; Friedman 1991) have been developed to automate model building in high dimensions. The relative performance advantages and disadvantages of these methods are the subject of this chapter.

The remainder of this chapter is organized as follows. In Section 2 we briefly describe the problems we have studied, and in Section 3 we summarize CART, MARS and neural networks. Section 4 presents the results of our comparisons, and Section 5 has concluding remarks.

2 TWO PROBLEMS

2.1 Modeling Switch Processor Memory

The regression problem we chose for this study involves modeling the size (in bytes) of switch processor memory for a digital switching system. This problem is of interest to telecommunications engineers who provision new switching systems. The covariate or input information that is available to engineers and that can be used to predict memory size is of the following form: total number of lines served by the switch processor, number of lines with speed-dialing, number of lines with ISDN capability, number of trunks, and so on. The data consist of records for 239 switch processors; each record contains the memory size in bytes and 26 covariate measurements such as the above. This problem is considered important because of the potential to reduce costs by improving the efficiency with which switch memory is ordered, and because of the extensibility of memory analysis of this type to other switching systems and to other digital computer systems.

In the initial analysis of the switch memory data, we found it important to set up sample re-use methods for comparing different models. Throughout this chapter we consider a cross validation set-up in which the original 239 switch processors are (randomly) divided into 11 groups of size 20 and one group of size 19. This random partition is generated once and thereafter remains fixed. For each modeling technique we do the following: omit one of the 12 groups from the set of data, build a model on the remaining 11 groups, estimate the responses for the omitted group, and repeat. Combining the resulting fits gives an *honest* picture of the model's performance in that it is always estimating data that were not used in the model building. These estimates are called cross validation estimates.

In order to assess the quality of the (cross validation) estimates we will rely both on numerical summaries (the mean squared error between the true sizes and the estimates and the mean absolute deviation between the true sizes and the estimates) and on a graphical display which the switch engineers found useful. Figure 1 illustrates the graphical display: the true sizes of the switch processors are plotted on the horizontal axis and the cross validated estimates on the vertical axis. (These cross validated estimates come from a MARS model. The details of this model are discussed in Section 4. Here we refer to the figure in order to illustrate the engineering display.) The line $y = x$ (which denotes perfect estimates) is drawn, as are two curved lines called *goal bands*. Points lying inside the goal bands are points for which the discrepancy between the true and cross validated sizes are less than the maximum of 100 kilobytes or 10% of the true size. This criterion was put forth by the engineers as desirable. The percentage of points outside the goal bands is sometimes used to summarize the extent to which the engineering criterion is not met.

For regression data such as these, CART, MARS, and neural networks are all directly applicable.

2.2 Characterizing DS0-rate Traffic

The classification problem in this chapter involves the automatic identification of the type of traffic being carried on a DS0-rate line. The motivation for this problem, the laboratory set-up used to collect the traffic data, and the preprocessing of the data are all described in Yuhas and Humphries (1992).

Four classes of traffic were used in these experiments: speech and modem data at baud rates of 2400, 4800 and 9600. For each traffic type, six seconds of digital data were collected. The data were then transformed by computing its 128-point Fast Fourier Transforms (FFTs). This produces a representation giving the relative amplitude of the frequency components of the signal. Since the original data were real, the resulting power-spectra are symmetric and therefore only 64 spectral coefficients are needed to uniquely identify the short-term power spectrum (STPS). Each STPS is then normalized to have a maximum amplitude one. The STPS spans approximately 14 milliseconds (ms) of the data. This translates into 430 STPSs of each traffic type, for a total of 1720 STPSs when all data types are combined. Since we have a sizable number of data points (STPSs), we use a half sample technique – models are built on half the data and their performance is tested on the other half. The data were divided into two equal sized subsets; each subset contained the same

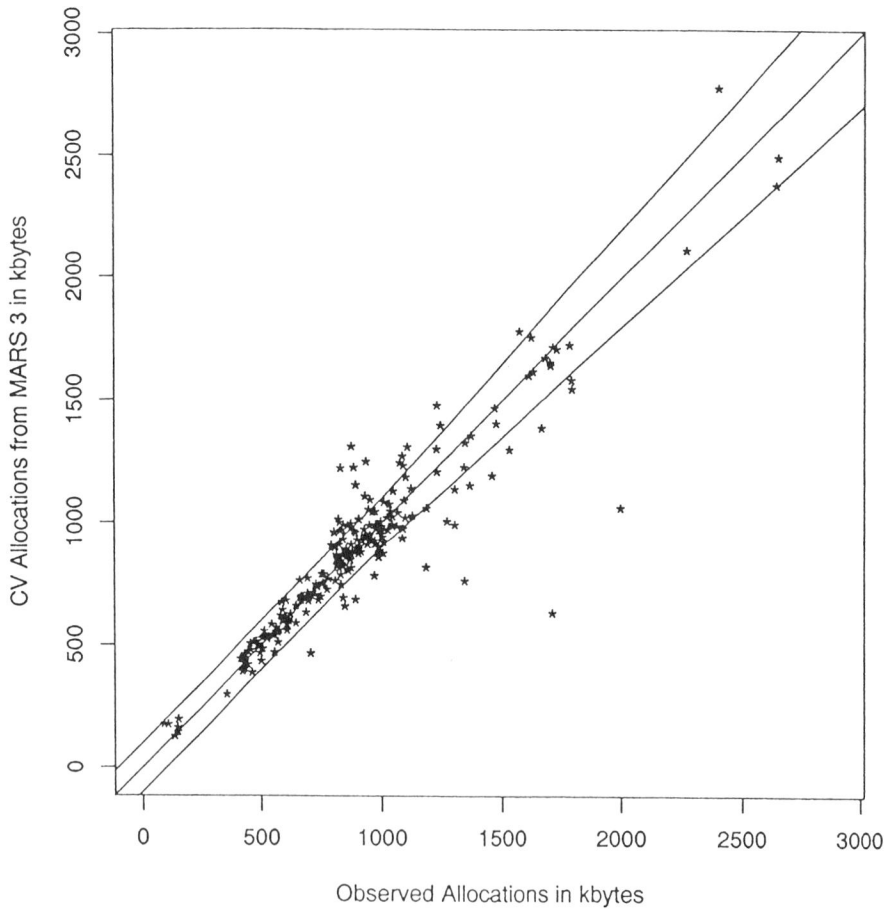

Figure 1 Engineering goals for MARS 3.

equal distribution of data points across the four traffic classes. These subsets were chosen once and remained fixed thereafter. The subset used to build the models is referred to as the *training data* and the subset used to test the model is referred to as the *test data*.

We derive two separate classification tasks from these data. In the two-class problem, the task is to differentiate speech from data (at any baud rate). In the four-class problem, the task is to identify each of the four types of traffic separately. In order to summarize how well a model does, we simply calculate the percentage of misclassifications made by the model on the test data. (Note that this is an honest picture of a model's ability to classify future traffic, since the test data are not used to build the model.) We also comment on how these misclassifications are distributed; that is, in the two-class problem, we examine how often voice is misclassified as data and how often data is misclassified as voice, and in the four-class problem, we describe the prevalence of the different types of misclassifications.

For the two-class problem, CART, MARS and neural networks are all directly applicable. For the four-class problem, however, the MARS technique requires extension, so we compared CART and neural networks. (It is worth noting that the required extensions to MARS for handling classification problems with more than two classes are under development in the statistics community. We expect that modified public domain MARS software incorporating these extensions will be available from the StatLib server at Carnegie Mellon University next year.)

3 REGRESSION AND CLASSIFICATION METHODOLOGIES

As outlined earlier, regression and classification are estimation problems, the former of an unknown quantitative variable, the latter of an unknown class membership. In both problems, estimation is based on the known values of covariate or input variables. The result of a regression analysis is a regression model, i.e., a function that maps a set of covariate values to a value of the unknown quantitative variable (the estimate). The result of a classification analysis is a classification rule, typically implemented as a vector function of the covariates with as many components as there are classes; the components of the vector give the estimated (relative) probabilities of falling in the different classes. In either case – regression or classification analysis – the problem is that of finding functions of the covariate variables suitable for the given task. The various regression and classification methodologies differ 1) in the class of functions used and 2) in the criteria and algorithms by which these functions are fitted to training data. In the following subsections, we give short descriptions of both the permissible function classes and the fitting procedures in the case of CART, MARS and neural networks. The terms *function* and *model* will be used interchangeably.

3.1 CART

The functions used by CART are given by repeatedly thresholding the covariate variables. As an example, part of a CART model for the switch memory problem could be a rule of the form: IF pu_port (a covariate) is less than 299.5 AND if $ad2s\_idp$ (another covariate) is less than 5 THEN the estimated memory size is 688700 bytes. A CART model is best summarized by a tree-diagram as shown in Figure 2, where each non-terminal node of the tree stands for an inequality condition characterized by a covariate and a threshold. Each terminal node stands for a predicted value. For a given set of covariate values, one descends down the tree according to the satisfied inequalities, till one finds a predicted value in a terminal node (here the estimated memory size in bytes). Note that the IF-THEN rule given above corresponds to the left-most terminal node in Figure 2. Since the predicted values are constant on areas characterized by a set of inequalities, we may characterize CART mathematically by saying that it uses step functions which are constant on generalized rectangles in covariate space.

CART's method for fitting such step functions proceeds in two phases: first a forward stepwise phase which adds a new threshold condition to the model at each step (essentially splitting a generalized rectangle into two), and then a backward phase which simplifies the model. The goal of the forward phase is to build a model that is rich enough to capture the key features of the data. The goal of the backward phase is to simplify the rich model so that it gains generalizability and parsimony. The backward phase is essentially a smoothing operation and can be guided by cross validation techniques.

At each step in the forward phase, the choice of which covariate to split on and which threshold to use is made by greedy minimization. The quantity being minimized is the lack of fit between the true responses and the estimates produced by the current model. The minimization is performed over all possible choices of covariates and thresholds.

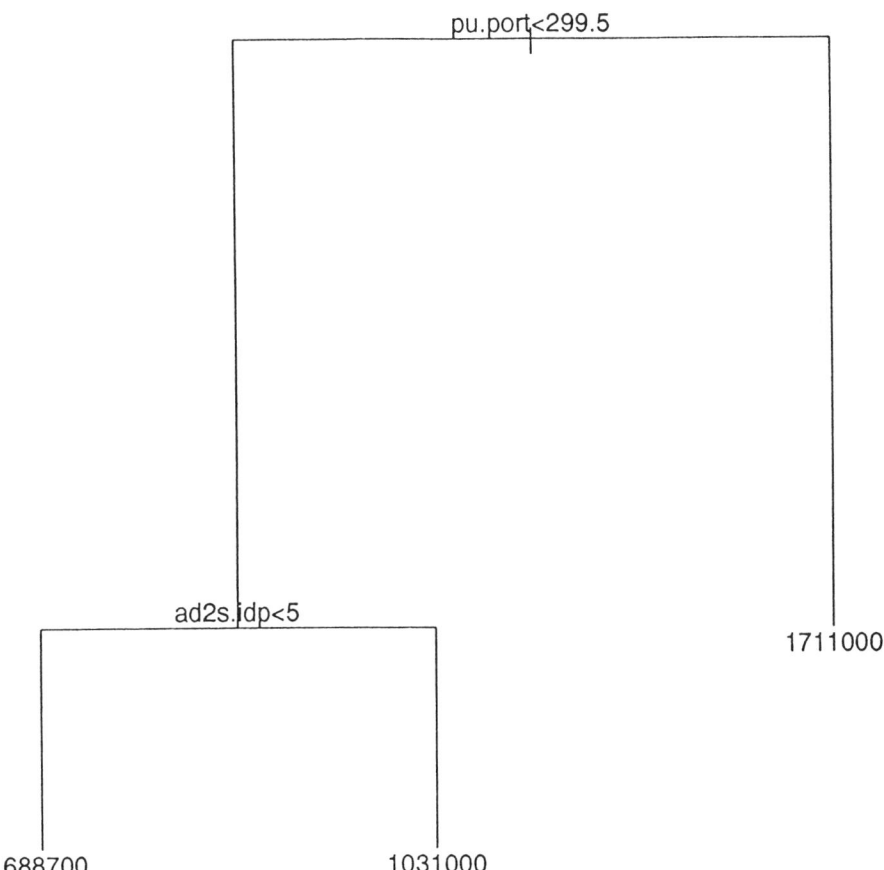

Figure 2 Small example of a CART tree.

In the regression context, each node of a CART tree is associated with the average response value over all training observations that lie in that node. The model estimate for a test observation is then the value of the terminal node in which the observation lies. Lack of fit is measured by the sum of squared errors between these model estimates and the true responses.

In the classification context, each node of a CART tree is associated with a vector whose components are the observed class proportions of training observations that lie in that node. These proportions are interpreted as the estimates of the probabilities for falling in each class. Lack of fit is measured with an entropy-like criterion involving these estimated class probabilities and the observed class assignments of the training data. A test observation is classified by finding the observation's terminal node in the CART tree and selecting that class which has the highest estimated probability in that terminal node.

There are two key parameters that govern the forward phase of CART – these are the minimum node size (mn) and the minimum deviance (md). The minimum node size is the smallest size that a node can be; the larger this value, the sooner the forward phase will stop. The minimum deviance governs how homogeneous a node can be. Node deviance or homogeneity is calculated differently in regression and classification (i.e., with sum squared error in regression, and with an entropy criterion in classification). The minimum deviance is expressed as a fraction of the deviance of the whole training set; the larger this fraction, the sooner the forward phase will stop.

The tree in Figure 2 was kept small for illustrative purposes. In actual practice, a tree model is purposely grown too large in the forward phase and then simplified in the backward phase. There are several possible implementations for the backward phase of CART. We use the shrunken tree approach described in Chapter 9 of Chambers and Hastie (1992) which preserves the structure of the tree but alters the values at the leaves; note that these values are simple numbers in the regression context and are probability vectors in the classification context. The alteration is done by recursively shrinking the value of each node toward that of the parent node. The degree of shrinkage is governed by a crucial parameter (k) whose choice is guided by a graphical display. The shrinking process imposes constraints on the values of each node and reduces the effective number of parameters. Since the values associated with the terminal nodes are used for estimation, the shrinkage process serves to smooth the estimates.

The paragraphs above give only the briefest description of CART. For more details and extensions of this approach see Breiman et al. (1984) and Chapter

9 of Chambers and Hastie (1992). We used the software provided in New S (Becker et al. 1988) for our CART computations. The forward phase takes up most of the computation. For a given problem, the computation time is largely a function of the minimum size and minimum deviance parameters. CART modeling is very fast, unlike both MARS and neural networks. Computation time is not really an issue with CART – even for our largest problem, the line monitoring data with 860 cases and 64 dimensions, CART models can easily be run interactively.

3.2 MARS

The functions used by MARS are based on building blocks of the form

$$max\{0, x - t\} \quad and \quad max\{0, t - x\},$$

where t a threshold value called a *knot*. As functions of x, they are linear with slope $+1$ or -1, respectively, on one side of the knot and constant zero on the other side. MARS uses the class of functions which are sums of products of these building blocks applied to the covariates. As an example, a simple fictitious MARS model for the switch memory problem would estimate the amount of required memory in bytes by a function such as

$$\begin{aligned}100450 \ &+ \ 2540 * max\{0, pu\_port - 229.5\} \\ &+ \ 754 * max\{0, 511 - pu\_port\} \\ &+ \ 34.5 * max\{0, ad2s\_idp - 4.5\} \\ &+ \ 10.5 * max\{0, pu\_port - 229.5\} * max\{0, 4.5 - ad2s\_idp\},\end{aligned}$$

where pu_port and $ad2s\_idp$ are again covariates. Product terms can have more than two factors, but the factors must be functions of different covariates. This class of functions is called first order interaction splines, where *first order* refers to the fact that these splines are piecewise linear as functions of one covariate, holding (all) the other covariates fixed.

There are some obvious similarities and differences between the functions used by CART and MARS: both rely on thresholding the covariates, but the splines of MARS are continuous functions of the covariates, while the step functions of CART are discontinuous at the thresholds. CART models may be easier to interpret, but MARS models may be more appropriate when the unknown quantity is expected to depend continuously on the covariates. In particular, MARS is better able to model low order structure such as linear dependencies which are hard to approximate by step functions.

As in CART, fitting of MARS models to training data consists of a forward stepwise model building phase, followed by a backward simplification phase. Again, the choices of covariates x and thresholds t and whether to use the $x - t$ or the $t - x$ form, are all made by greedy minimization.

The criterion being minimized is a sum of squares; the MARS algorithm critically relies on this form of the criterion since it makes extensive use of fast updating formulae for least squares problems. In the regression context, the sum of squares is naturally between the observed values and the model estimates. For classification, recall first that MARS is directly applicable only to two-class problems. The two classes are given values of 0 and 1, and MARS builds the model by treating the data as if it were regression data. After the model has been built, the MARS algorithm contains an additional step for classification problems which will be described later.

There are two important parameters that govern the forward phase of MARS – these are the number of terms (nk) and the maximum interaction level (mi). The number of terms is exactly the number of terms at the end of the forward phase; the smaller this number, the faster the forward phase can be completed. The maximum interaction level is the maximum number of factors in a product term; the smaller this number, the faster the forward phase can be completed.

The backward phase in MARS consists of deleting terms sequentially from the model. Terms are chosen to minimize the degradation of the fit of the MARS model to the training data (i.e., to minimize the increase in the sum of squared errors). The crucial decision is when to stop deleting, and the critical parameter governing this decision is a degrees of freedom parameter (df). The default value for df is 3, but in our experience a considerably larger value is often needed. We can use cross validation within the training data to choose df. The way this works is the following: we give the MARS algorithm a random seed and an integer value for the number of partitions; the algorithm randomly divides the training data into the desired number of partitions and then chooses df so that the sum of squared errors between cross validation estimates and true responses is minimized; a final model is then fit on all the training data with the chosen value of df. Note that if the number of partitions is p, then this approach requires the fitting of $p + 1$ MARS models.

Finally, for (two-class) classification problems, after a MARS model has been built the parameters are re-estimated using a linear logistic formulation. This formulation considers the logit of the probability that an observation is of class one. More formally, if π is the probability that an observation is of class one, then $logit(\pi) = log(\pi/(1 - \pi))$; note that while π lies between 0 and 1, $logit(\pi)$

is real valued. The functional form of the MARS model is considered fixed, but the parameters are chosen to maximize the linear logistic likelihood.

The preceding paragraphs introduced the basic ideas in MARS modeling. For a more thorough treatment, see Friedman (1991). Public domain FORTRAN code for fitting MARS models was written by Friedman and is available from the StatLib server at Carnegie Mellon University. We wrote an S interface to this code. Almost all of our MARS computations were done with version 3.5 of MARS; a couple of simpler runs for the switch memory data were done with version 2.5. Computationally, MARS is orders of magnitude more demanding than CART. Like CART, the forward phase requires the bulk of the work. The computation time is greatly affected by the parameters p, nk and mi. Roughly, for our largest problem, the line monitoring data with 860 observations and 64 dimensions, several hours of elapsed time were needed to fit one MARS model on a SPARC 2; several days were needed when p (the number of cross validation partitions) was 20.

3.3 Neural Networks

Neural networks include a wide class of models and various algorithms to fit the models, both of which are loosely motivated by the style of *computing* or processing performed in biological systems. These biological systems are characterized by a collection of simple processing units whose collective behavior is capable of performing complex processing. In a crude sense, neural network models and algorithms try to capture this characteristic. Here we present a very limited description of neural networks covering a methodology relevant for the classification and regression problems of interest in this chapter. Our description is confined to feed-forward networks without recurrence, a narrow class of models, and to back-propagation, one specific fitting algorithm. For a more complete treatment, see Hertz et al. (1991).

The neural network models considered here use a class of functions which are built up from building blocks or *units* of the form

$$f(a_0 + a_1 * x_1 + a_2 * x_2 + ...), \tag{1}$$

where f denotes the so-called sigmoid function, $f(u) = (1+e^{-u})^{-1} = 0.5\,(1+ tanh(u/2))$ (a monotonically increasing function on the reals, approaching one and zero at $\pm\infty$, respectively). Other functions could be used in principle; in particular, we will use linear f's in one instance. The way the building blocks or units of the form (1) are combined to form neural network functions can again be illustrated in terms of a fictitious model for the switch memory problem: estimate the amount of required memory in bytes by

$$3400 + 1{,}145000 * f(1.1 \ + \ 0.8 * f(0.1 + 0.012 * pu\_port + 0.2 * ad2s\_idp)$$
$$+ \ 3.5 * f(-0.3 + 0.340 * pu\_port + 1.1 * ad2s\_idp)$$
$$+ \ 1.9 * f(1.1 + 0.005 * pu\_port - 0.3 * ad2s\_idp) \ ).$$

The idea is to apply several units (three in the example) to the covariates and pass their results on as arguments to other units (just one in the example). Further nesting can be obtained by adding more *layers* of units. At the top layer of the hierarchy of units are one or more *output units* which furnish regression estimates (one output unit) or classification estimates (one output unit per class). To complete the graphical intuition behind a neural net architecture, the covariates are usually called *input units*, although they do not perform a computational function other than holding a covariate value. Those units layered between the input and output units are called *hidden units*. This collection of interconnected units is called a feedforward network without recurrence. It can be represented as a graph as shown in Figure 3. While the neural network paradigm allows for any number of units and any connectivity between these units, in this chapter we only consider layered feed-forward networks with at most one layer of hidden units.

The weights a_i in equation (1) are the free parameters that can be adjusted to fit training data. The process of adjusting these weights, called model fitting by statisticians, is referred to as *learning* or *training* in the areas of neural nets and pattern recognition. The algorithms that control these adjustments are called *learning algorithms*. There are a number of learning paradigms available in the literature (e.g., Hertz et al. 1991). One such popular algorithm is *back-propagation*, a version of steepest-descent adapted to neural net architectures. Prior to running the learning algorithm, the weights are initialized to small random values. [1]

A neural network model is parametrized not only by its weights but its architecture as well. While a learning algorithm will find the weights for a given architecture, it requires an *a priori* architecture selection. It is not uncommon to explore multiple architectures to determine an appropriate one. Fahlman and Lebiere (1990) and others have explored learning algorithms that construct

---

[1] If a feed-forward neural network with full connectivity between layers were to have all of its weights initialized to the same value, then the back-propagation learning algorithm would produce sets of weights that were identical going into every hidden unit. This would greatly reduce the computational power of the resulting models – in fact, any fully connected layered network with such an initialization is equivalent to a network with one hidden layer containing only one unit.

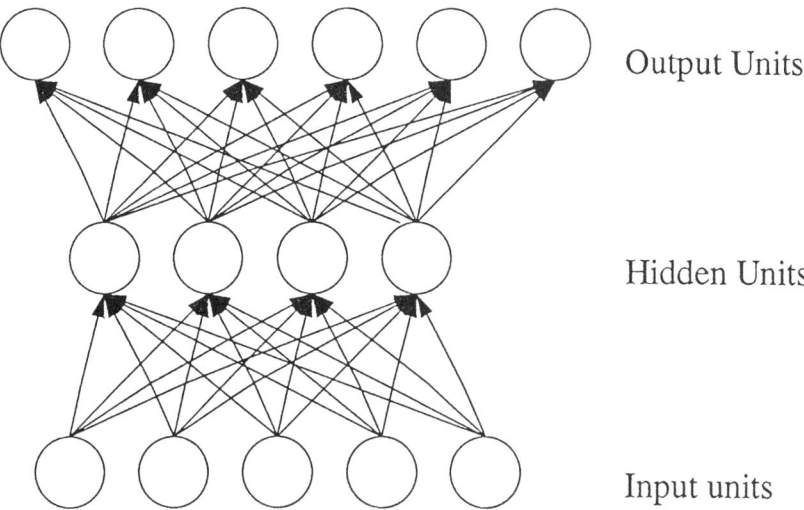

Figure 3 A graphical representation of a feed-forward neural network with three layers. Each unit is connected to all of the units in the layer immediately above. The inputs are applied to the units at the bottom and the information flows to the units at the top, whose values are then read out.

the architecture as they learn. The need to find flexible neural net architectures that fit the training data well, is balanced by the need to guard against too much flexibility that can result in overfitting the training data. This problem is particularly acute when the number of training samples is small.

The learning method used here is essentially defined in terms of minimization of the sum of squared errors across all output units:

$$E = \sum_n \sum_j (y_{nj} - o_{nj})^2,$$

where y_{nj} denotes the observed/desired training value of sample n and output unit j, and o_{nj} is the corresponding predicted or output value. We will see in later sections that this error measure is not optimal for all problems. For *regression*, where there is just one output unit, the criterion E measures the overall deviation between desired and predicted values, and the network is trained to approximate the observed values y_n of the training sample as a function of the observed covariate values. For *classification*, the observed class membership of sample n is coded as a vector $(y_{nj})_j$ of 0's and 1's, with a single 1 indicating the unit j whose associated class is observed. Thus, the criterion E forces the network to produce values near 0 and 1 in the output units indicating class membership as a function of the observed covariate values.[2] For given test covariates, the estimated class is taken to be that class associated with the output unit having the largest output value.

An issue common among neural net researchers, but less so among statisticians,[3] concerns the problem of sequential or *on-line* learning versus *batch* or off-line learning. By sequential learning one means the presentation of training samples one at a time and adjustment to the weights as the samples are presented; by batch learning one means that all the data are available at once and weight adjustment can be based on the entire set of data. Here we train the neural networks with a version of the back-propagation algorithm that uses a conjugate-gradient technique. This method uses a gradient that is calculated from the entire training set, and so as earlier with CART and MARS, this is a batch learning algorithm.

4 RESULTS ON REGRESSION FOR SWITCH MEMORY

Method. In this section we compare the performance of the various methods on the switch-memory regression problem.[4] As detailed in Section 2, a cross-validation scheme was used in which models were built on 11/12ths of the data and used to estimate the omitted 1/12th of the data. Note that this means 12 (different) models were built with each methodology. Three performance criteria are used: mean-squared error (MSE), mean absolute deviation (MAD) and Percent outside of Engineering Goals. All performance assessments are based on the combined cross-validated estimates.

CART. CART models were built on the 26 available covariates. The two parameters that govern the forward phase of CART, the minimum node size (mn) and the minimum deviance (md), have default values of 10 and .01, respectively. We explored values in the vicinity of the default values. The

---

[2] For numeric reasons, we actually use the values .1 and .9. Since the non-linear sigmoid function approaches 0 and 1 asymptotically, if the values 0 and 1 were used as target values, then the network would have to drive its weights to $\pm\infty$ in an attempt to obtain those target values.

[3] There are areas within statistics which deal explicitly with sequential problems, e.g., sequential design, sequential testing, and stochastic approximation.

[4] The classical statistical approach to a regression problem like this is multiple linear regression. This approach was tried but is not discussed here; it did not achieve the desired results primarily due to the high dimensionality of the covariate space.

twelve models we built all had $mn = 8$; ten had $md = .005$ and two had $md = .10$.

For the backward phase, the choice of the shrinkage parameter k can be guided by a graphical cross-validation scheme. Using these plots, values of k in the range .3 to .45 were chosen. The final models used between 9 and 12 of the 26 available covariates.

The CART model performs poorly with over 40% of the cross-validated estimates lying outside the engineering goals. It also has the highest value of MAD, but the MSE value is competitive with the other models.

MARS. Three different (sets of) MARS models are presented in the table. The first model (MARS 1) uses parameter values $mi = 3$, $nk = 40$ and (the default value of) $df = 3$. Recall that mi controls the maximum interaction level, or the maximum number of factors in a product term, nk is the number of terms in the forward phase, and df is the degrees of freedom associated with an individual term. This first model is proposed in TM-TSV-015632. It was built using MARS 2.5 and considerable experimentation was involved in guiding the choices for mi and nk.

The second MARS model (MARS 2) uses parameter values $mi = 3$, $nk = 60$ and $df = 8$. This model was developed later using MARS 3.5 and reflects the view that the default value of 3 for df may be too low.

The third MARS model (MARS 3) uses parameter values $mi = 3$, $nk = 60$ and $p = 10$. This was the last model built; MARS 3.5 was used. Here (internal) cross-validation was used to choose df. The parameter p, which gives the number of cross-validation partitions to be used in choosing df, was set to 10. The twelve df values that were chosen by the cross-validation were: 11.74, 6.38, 6.30, 10.39, 10.94, 10.37, 26.86, 18.91, 6.05, 7.98, 6.14 and 11.13. It is disturbing to note the wide variability in these selected df's. Figure 1 shows the performance on the engineering goals for MARS 3.

The three MARS models have roughly similar performance in terms of the engineering goals. It is interesting to note that while MARS 1 has higher values than MARS 2 and MARS 3 for both MSE and MAD, MARS 1 does very slightly better in terms of achieving the engineering goals.

Neural Networks. The covariates were presented to the neural networks via 26 input units. One output unit was used to represent the scaled memory size associated with those inputs. The number of hidden units was systematically varied between zero and twenty.

The networks were trained with 1000 iterations of a modified back-propagation algorithm. The algorithm attempted to minimize the mean-squared error (MSE) criterion described earlier. Prior to training, each architecture was initialized with small random weights. All results given are based on the final state of the network after training was completed.

The impact of the nonlinear squashing on the final estimates was also examined by retraining each architecture from the same initial weights, but with the output unit's function replaced with a linear f.[5]

In Table 1, we provide results for six neural net models. Included are two networks with no hidden units: one with a linear output unit (NN 0L) and one with a nonlinear output unit (NN 0N). Architectures with between two and twenty hidden units were also trained. Those networks we report on here were selected based on the minimum number of hidden units above which there did not appear to be appreciable improvement with respect to MSE on the training set. With a nonlinear output unit, an architecture with ten hidden units was selected (NN 10N), while a linear output unit required sixteen hidden units (NN 16L). For comparison, NN 10L and NN 16N give results for the corresponding architectures with the alternative output transfer functions.

The architectures that we selected to report on were not not necessarily those that gave the best performance on the test set. Moreover, since the performance on the test set did not vary monotonically with the number of hidden units, it is not obvious whether a better model would contain more or less hidden units. For the linear output unit, the worst test performance was seen with four hidden units, where 28% of the estimates fell outside of the engineering goals. The best test performance had only 21% outside of the engineering goals and was obtained by networks with two, twelve and fourteen hidden units. For the nonlinear output unit, the worst test performance was seen with two hidden units, where 29% of the estimates fell outside of the engineering goals. The best test performance had only 21% outside of the engineering goals and was obtained by a network with eighteen hidden units. A more sophisticated cross-validation technique might make it possible to identify those architectures that provide the best test performance.

---

[5] We did not replace other sigmoidal units with linear units, because if all of the hidden units are linear then the architecture is mathematically equivalent to a neural network with no hidden units.

The neural network models displayed systematically different performance depending upon whether or not hidden units were used. The neural net models without hidden units had lower values for MSE, but higher values for both MAD and Percent outside of Engineering Goals. This is most likely caused by overfitting the training data, since the MSE error on the training set was an order of magnitude less when using hidden units. To judge whether the network was overfitting the training data, we observed the performance on the test set during the training procedure. We found that while MSE error would begin to rise for the test set, the final network usually performed within 1% of the best performance seen on the Engineering Goals. Again, a more sophisticated use of cross-validation to determine when to stop training the neural network might reduce the overfitting, yielding models with comparable performance on the Engineering Goals and better performance on MSE.

The discrepancy between the MSE and the performance on the Engineering Goal can be attributed to the fundamental difference between the MSE criteria being minimized and the engineering goals. The networks with no hidden units provide many estimates just outside of the engineering goals, but with very few far outside. In contrast, the networks with hidden units have more estimates inside of the goals, but make much larger errors on their bad estimates. The use of a linear versus a nonlinear output unit has little effect when there are no hidden units (compare NN 0L with NN 0N).

Method	MSE	MAD	Outside Eng Goals
CART: mn=10	31020	123	43%
MARS 1: nk=40, mi=3, df=3	34233	90	20%
MARS 2: nk=60, mi=3, df=8	22873	83	22%
MARS 3: nk=60, mi=3, p=10	21531	81	22%
NN 0L: 0HU, linear output	22257	102	33%
NN 0N: 0HU, nonlinear output	22838	106	34%
NN 10L: 10HU, linear output	37260	103	25%
NN 10N: 10HU, nonlinear output	25160	90	22%
NN 16L: 16HU, linear output	26832	89	23%
NN 16N: 16HU, nonlinear output	32460	101	25%

Table 1 Performance assessments (mean-squared error (MSE), mean absolute deviation (MAD) and percent outside of engineering goals) of cross-validated estimates for switch processor memory data. MSE and MAD are given in kilobytes.

Comparisons. The best results from the point of view of the engineering goals are obtained with MARS 1, MARS 2, MARS 3, NN 10N, and NN 16L; all achieve values for Percent outside Engineering Goals in the $20-23\%$ range. These five models capture complex interactions among the 26 covariates – via interaction terms in the case of the three MARS models, and via hidden units in the case of the two neural net models. This is in contrast to the three models (CART, NN 0L and NN 0N) which are restricted to simpler combinations of the inputs.

The best results on MSE are obtained by MARS 2, MARS 3, NN 0L, and NN 0N. The best results on MAD are obtained by MARS 2 and MARS 3.

The engineering goals are the most important performance criterion from the substantive point of view, so that one of the five methods which do well here should be used. Since MARS 2 and MARS 3 also achieve best results for MSE and MAD, they would be preferred. We would recommend MARS 2 overall: it has top performance across all criteria, but it does not have the disturbing variability in the underlying twelve models that MARS 3 has, as indicated by the wide variation in the twelve choices of df.

5 RESULTS ON CLASSIFICATION OF DS0-RATE TRAFFIC

Method. In this section we compare the performance of the various methods on two traffic classification tasks. In the first task, the short-term power spectra (STPS) are classified into four individual classes, including speech and three modem types. In the second, the data is classified into two classes: data and speech. Because there were many more data samples available for this problem, a half-sample cross-validation technique was used. The performance criterion was percent correct.

Simple baseline techniques. Prior to building more sophisticated models, we used two simple techniques commonly used on classification problems. **Template matching** is a method commonly used by engineers. In this approach, an unknown STPS is classified by comparing it to a set of template STPSs representing the different classes. The unknown STPS is then labeled with the class associated with the closest matching template. Here, the closest match is defined in terms of mean-squared error. In our implementation, the templates were formed by calculating the average STPS of each class from the data in the training set. These templates can be seen in Figure 4. Note that the average

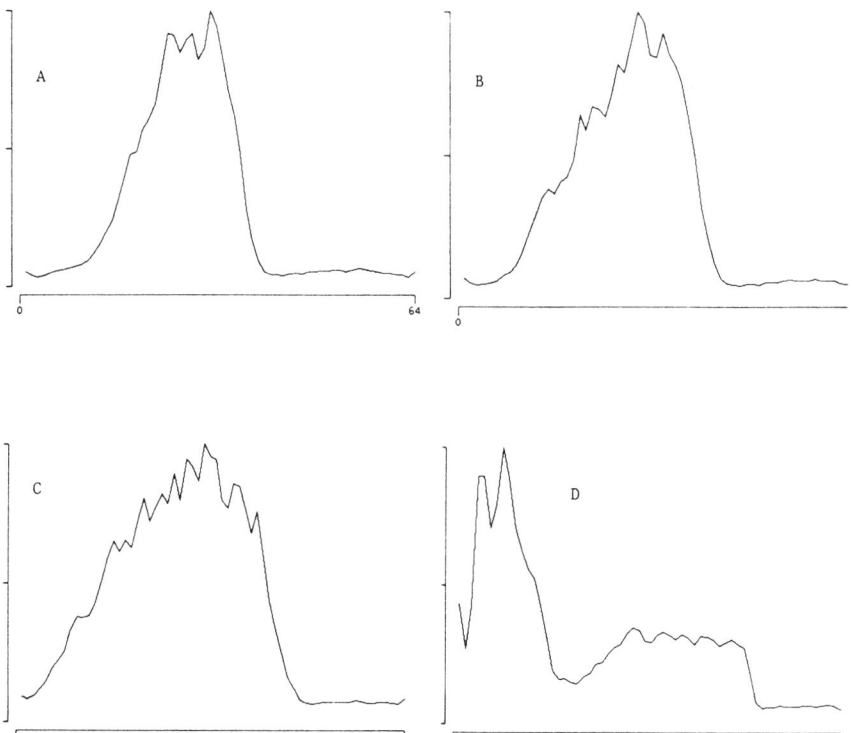

Figure 4 The average short-term power spectra produced for (a) 2400-baud, (b) 4800-baud, (c) 9600-baud and (d) voice. Each curve plots the power as a function of the individual frequency components of the short-term spectra. These spectra have been normalized to have the same peak value.

Method	Percent Correct	
	Train	Test
Template Matching	83.84%	83.48%
Linear Discriminant Analysis	90.47%	90.23%
CART: mn=7, md=.005, k=.192	97.44%	91.51%
NN 0L: 0HU, linear output	88.26%	88.72%
NN 0N: 0HU, nonlinear output	99.19%	95.12%
NN 10L: 10HU, linear output	97.33%	96.74%
NN 10N: 10HU, nonlinear output	100.00%	98.84%

Table 2 Percentage of the short-term power spectrum (STPS) correctly categorized into each of the **four classes** of data, including speech and three modem types.

Method	Percent Correct	
	Train	Test
Template Matching	97.91%	98.14%
Linear Discriminant Analysis	99.07%	98.95%
CART: mn=5, md=.005, k=.15	99.88%	97.79%
MARS (Logistic): nk=60, mi=3, df=3	100.00%	98.84%
NN 0L: 0HU, linear output	99.07%	98.95%
NN 0N: 0HU, nonlinear output	100.00%	99.77%
NN 10L: 10HU, linear output	100.00%	100.00%
NN 10N: 10HU, nonlinear output	100.00%	99.88%

Table 3 Percentage of the short-term power spectrum (STPS) correctly categorized into **two** classes, speech or data.

STPS for voice in panel (d) has a considerably higher proportion of its power at lower frequencies than the average STPSs of data in panels (a), (b) and (c). The average STPSs for data in panels (a), (b) and (c) are similar in overall shape, however there is an increase in power at the lower frequencies as the baud rate increases from 2400 in (a), to 4800 in (b) and to 9600 in (c). It is also worth noting that the individual spectra that are used to compute the average voice STPS vary considerably more than the individual spectra for data.

The second technique we used was **Linear Discriminate Analysis (LDA)**. This technique is commonly used by statisticians and is guaranteed to find the *best* linear separation of the classes under certain (rather restrictive) technical assumptions. (These assumptions do not hold here.)

Despite their simplicity (and in the case of LDA, the restrictive underlying assumptions) template matching and LDA did very well on the two class problem (see Table 3). This is because the two-class problem is not very difficult. For this problem, the only misclassifications made were on the speech spectra; no data spectra were misclassified. This asymmetry in the errors can be attributed to the inherent non-stationarity of the speech signal, which results in greater variability in its spectra.

In the four class problem, template matching did not perform well and LDA, although better, was also not competitive (see Table 2). Again, the errors that were made were not symmetric. The great majority of the confusions were among the different data types; only a few speech spectra were misclassified as data and no data was misclassified as speech.

CART. CART models were built using the 64 STPS coefficients. The two parameters that govern the forward phase of CART, the minimum node size (mn) and the minimum deviance (md), were set at $mn = 5$ and $md = .005$ for the four class labeling and $mn = 7$ and $md = .005$ for the two class labeling.

For the backward phase, the shrinkage parameter k was found by graphically plotting performance across a range of k's. The k's that provided the best performance on the training set were identified: $k = .192$ was used for the four class labeling, and $k = .15$ was used for the two class labeling.

CART performed slightly better than simple LDA on the four class problem and slightly poorer than LDA on the two class problem. Again, the misclassifications made were not symmetric. In the two class problem, speech was misclassified as data more often than data was misclassified as speech. In the four class problem, over half the misclassifications were among the different modem types. Of the remaining errors, speech was misclassified as data more often than data was misclassified as speech.

MARS. MARS was not used on the four class problem, as the currently available software does not support the building of MARS models for classification problems with greater than two classes. For the two-class problem, the default values for the forward building parameters of $nk = 60$ and $mi = 3$ were used, along with the backward parameter of $df = 3$. MARS performed well on the two class problem, with most errors being due to the misclassifications of speech as data.

Neural Networks. The covariates were presented to the neural networks via 64 input units. The number of output units depended upon the number of classes to be identified. For the four-class problem, four output units were used. The neural network was trained to identify the class associated with the input STPS coefficients by producing a value of .9 at the one output unit corresponding to that specific class and .1 at the other three output units. For the two-class problem, only one output unit was used and the neural network was asked to produce .9 for one class and .1 for the other.

The networks were trained as before using a modified back-propagation algorithm to minimize the mean-squared error (MSE). Training was stopped after 1000 iterations. As before, we looked for overfitting of the training data by observing the performance on the test set during the training procedure. In these runs, no evidence of overfitting was seen.

As before, a variety of architectures was tried, with the number of hidden units varying between zero and twenty. Each architecture was initialized with random weights before training and each architecture was trained multiple times. The impact of the nonlinear squashing on the final estimate was again tested by substituting the nonlinear output unit transfer function with a linear f.

Tables 2 and 3 provide results for both the linear (L) and nonlinear (N) output functions f and for networks with zero hidden units and with ten hidden units.

As with the other approaches, the neural networks had no problem solving the two class discrimination problem. Note that the neural network with a single linear output unit and no hidden units (NN 0L) provides the exact same mathematical solution as LDA, both performing at 98.95% on the test data. As before, the errors were due to the misclassification of speech as data.

On the four class problem, the neural networks performed better than the other techniques. When no hidden units were used, the network benefited from the nonlinear squashing function. A small improvement was also gained by adding the hidden units. With ten hidden units, there is only a slight difference between the network with the linear vs. the nonlinear output function. As seen with the other methods, the majority of the errors was caused by confusions among modem types. Of the remaining errors, more often speech was misclassified as data rather than the reverse.

Comparison across methods. All of the techniques performed well on the two-class task, making any substantial comparison difficult. However, there were differences in performance on the four class discrimination task, with the neural network models NN 0N, NN 10L and NN 10N performing best.

6 DISCUSSION

In general, both MARS and neural networks performed well in this study. These two approaches are powerful techniques for complex high-dimensional model building. MARS and neural networks both require various parameter settings which can greatly affect their performance (in particular, mi and df for MARS, and architecture and number of training iterations for neural networks). To use these methods effectively in applications, computing and analyst time are needed to carefully explore reasonable parameter options. The computing demands become particularly high when cross validation is used to guide the parameter choice, as in the MARS option for cross-validatory choice of df. One possible area for future work would be to automate an analogous cross-validatory choice for the number of training iterations in neural network models.

MARS and neural networks were trained to minimize MSE in both problems, but the comparisons focused on performance criteria relevant to the problems (the engineering goals in the switch memory problem and misclassifications in the DS0 traffic problem). While one expects MSE to be related to these other criteria, the relation need not be that strong. In particular, with the neural network models we have observed, while initial iterations tend to improve performance on both MSE and the other criteria, later iterations often improve performance only on MSE. MARS is intimately tied to MSE, but neural networks are not (in fact, any reasonably behaved differentiable function could be used). Another possible area for future work is to explore alternate error criteria for use during neural network training.

One weakness in our comparisons is that they are confined to a single cross validation partition in the switch memory problem and a single half sample partition in the DS0 traffic characterization problem. Hence it is difficult to calibrate the performance differences we see, particularly in the DS0 traffic problem where the absolute scale of the differences is so narrow. Repeated empirical studies that allowed one to calculate meaningful standard errors for comparing methods would be welcome. (We recognize, however, that the computing demands of such studies may be prohibitive.)

Finally, we are interested in the robustness of MARS and neural networks to contaminated observations, noisy input measurements, and highly collinear input. The MSE criterion is known to be non robust. In the case of neural networks, this may be somewhat alleviated by the squashing of the sigmoid function. In the case of MARS, this may be somewhat alleviated by the local nature of the fitting. For neural networks, alternate training criteria which have better robustness properties can be considered.

REFERENCES

[1] Becker, R.A., Chambers, J.M., and Wilks, A.R. (1988). *The New S Language: A Programming Environment for Data Analysis and Graphics.* Wadsworth: Pacific Grove, CA.

[2] Breiman, L., Friedman, J.H., Olshen, R.A., and Stone, C.J. (1984). *Classification and Regression Trees.* Wadsworth: Belmont, CA.

[3] Buja, A., Duffy, D.E., Hastie, T., and Tibshirani, R. (1991). "Discussion of 'Multivariate Adaptive Regression Splines", *The Annals of Statistics*, Vol. 19., No. 1., pp. 93-98.

[4] Chambers, J.M., and Hastie, T.J., eds. (1992). *Statistical Models in S.* Wadsworth: Pacific Grove, CA.

[5] Fahlman, S.E. and Lebiere, C. (1990). "The Cascade-Correlation Learning Architecture," in D.S. Touretzky (ed.), *Advances in Neural Information Processing Systems 2.* Morgan Kaufmann: San Mateo, CA.

[6] Friedman, J.H. (1991). "Multivariate Adaptive Regression Splines (with discussion)", *The Annals of Statistics*, Vol. 19. No. 1, pp. 1-141.

[7] Hertz, J., Krogh, A., and Palmer, R. (1991). *Neural Computation.* Addison-Wesley: Redwood City, California.

[8] Yuhas, B.P. and Humphries, C.M. (1992). "Characterizing traffic on DS0-rate lines using neural networks", *Proceedings of Globecom '92*, pp. 1319-1323, December 1992, Orlando, FL.

18

A NEUROCOMPUTING APPROACH TO OPTIMIZING THE PERFORMANCE OF A SATELLITE COMMUNICATION NETWORK

Nirwan Ansari

Center for Communications and Signal Processing Research,
New Jersey Institute of Technology

1 SATELLITE COMMUNICATIONS

Satellite communications plays an important role in both the military and the civilian sector. The properties that are intrinsic to satellite communications, such as broadcasting and geographical flexibility, explain the use of this system mainly in Wide Area Networks. With the commercial sector expanding, especially in terms of number of locations and distances between them becoming more widespread, there is potential for growth in satellite communication networks to take advantage of the relatively constant cost [1].

Here we introduce an approach for efficient traffic management for a satellite network of a geostationary orbital type that incorporates the idea of dynamically adapting the network as well as dynamically routing each arrival. We restrict our analysis to a virtual circuit model. The approach allows the network to change according to the input pattern, thus improving the grade of service (e.g., blocking rate) while maximizing network utilization. Two schemes are considered – the self-organization method and the cost minimization method.

The network is modeled as a mesh-connected topology where each node represents either a satellite or an earth station. The connections between nodes denote the links between stations. All connections and, therefore, the set of possible alternate routes for each origin-destination (O-D) pair are predefined.

2 THE SELF-ORGANIZATION METHOD

The scheme [2],[3] consists of two levels of management as shown in Figure 1. The first level adaptively configures maps for the satellite communication network, and the second level routes traffic under the fixed configuration.

2.1 FIRST LEVEL MANAGEMENT

The "map configuration" of the satellite communication network is posed as a pattern recognition and generation problem. During network operation, the state of the network (the number of circuits requested by users) fluctuates constantly. N exemplar maps to which the network can configure are initially selected. Each exemplar map is a map which can accommodate a typical demand of the network for a certain period of time during network operation. The exemplar maps are assumed either provided by an experienced operator of the network or derived from network statistics. The state of the network is constantly compared to each of the exemplar maps, which defines the pattern recognition task, the first stage of the map configuration task. The second stage analyzes the discrepancy between the chosen exemplar map and the state of the network, and generates intermediate maps which gradually deviate from the exemplar map and converge closely to the input data pattern. A new map is finally generated in the third stage.

Assume that the original N exemplar maps are given. The state of the network is assumed to be updated constantly from the Common Signaling Channel (CSC) and converted to a numerical value denoted by $R_i(t)$, that is, $R_i(t)$ is equal to the number of circuits in link i required by users at time t. To perform the recognition task, the state of the network, $R_i(t)$, is updated as input data to the first stage of the self-organization model. Then, the normalized distance (metric) between the input data $R_i(t)$ and each of the exemplar maps is computed:

$$D_j(t) = \sum_{i=1}^{L} |D_{ij}(t)|/R_i(t), \qquad (j = 1, 2, \cdots, N), \qquad (1)$$

where $D_{ij}(t) = R_i(t) - C_{ij}(t)$, $(i = 1, 2, ..., L$ and $j = 1, 2, ..., N)$. L is the total number of links in a network, $C_{ij}(t)$ is the total number of circuits assigned to link i of the jth exemplar map at time t.

$D_{ij}(t)$ indicates the load of link i of the network if the jth exemplar map is used. $D_{ij}(t) > 0$ implies that link i of the network is overloaded when map

Optimizing a Satellite Communication Network

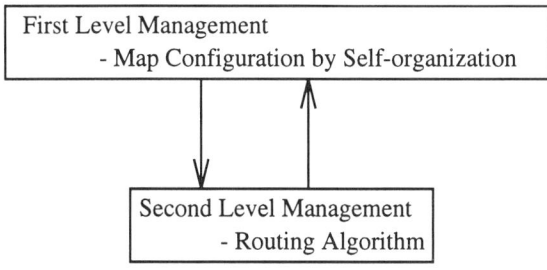

Figure 1 The self-organization method

j is used. $D_{ij}(t) < 0$ means that link i of map j provides more than enough circuits requested by the users in link i. Thus, $D_{ij}(t)$ indicates the resemblance between the input data $R_i(t)$ and exemplar map j. The exemplar map that best meets the requirement of the network is the one that yields the smallest normalized distance.

The exemplar map selected may not satisfy the requirement of the network. If the chosen exemplar map meets the network requirement, that is, $D_j(t)$ is small enough, we do not modify the map. If the chosen map does not satisfy the network requirement, we proceed to the second stage to modify the exemplar map to better meet the network demands. The process is governed by the following equation, similar to Kohonen's self-organization [4]:

$$C'_{ij}(t+1) = C'_{ij}(t) + \beta(n(j))D'_{ij}(t), \quad (i=1,2,\cdots,L;\ j=1,2,...,N), \qquad (2)$$

where $D'_{ij}(t) = R_i(t) - C'_{ij}(t)$. $C'_{ij}(t)$ is the number of circuits assigned to link i of map j which is being modified. $\beta(n(j))$ is a gain factor, $0 < \beta(n(j)) < 1$. The "Rprime"S is used to distinguish the map which is being modified from the original exemplar map chosen in the first stage.

$$\beta(n(j)) = k_1^{-(n(j)+k_2)}, \qquad (3)$$

$n(j)$ is the number of occurrences in which exemplar map j is selected. When a new exemplar map j is generated by the third stage, $n(j)$ is reset to 0. k_1 and

k_2 are non-negative constants. Note that $\beta(n(j))$ is a monotonically decreasing function of $n(j)$. In the third stage of the self-organization model, we compute a convergence factor, $\alpha(j)$, to decide whether a new map should be generated to replace the exemplar map.

$$\alpha(j) = \max_{i} |\beta(n(j))D'_{ij}(t)|, \qquad (i = 1, 2, \cdots, L). \qquad (4)$$

Obviously, $\alpha(j)$ is getting smaller when the jth map is selected more often, indicating that the modified map j is more closely approaching the network requirement. A new map is then generated based on the following conditions:

$$C_{ij}(t+1) = \begin{cases} C_{ij}(t) & \text{if } \alpha \geq r \\ \frac{C}{B}C'_{ij}(t) & \text{otherwise} \end{cases} \quad (i = 1, 2, \cdots, L), \qquad (5)$$

where $B = \sum_{i=1}^{L} C'_{ij}(t)$. B is the total number of circuits assigned to the modified map j at time (iteration) t. The threshold, r, is set to define whether a new assignment is made, and C is the capacity of the whole network.

From Equations (3) to (5), if the jth map has been selected often enough (i.e., $\beta(n(j))$ is small enough to make α less than r), the self-organization model generates a new map j to replace the originally chosen exemplar map.

2.2 SECOND LEVEL MANAGEMENT

Consider a set of nodes (switches) in a network interconnected by links, with each link consisting of a number of parallel circuits. Each circuit can carry one call. A call originating from node A and designated for node B requires for its completion a single-link or a multi-link path between nodes A and B, with at least one idle circuit on each link of the path at the time of call origination. The selected route consisting of one circuit on each link of the path is then occupied for the duration of the call, at the end of which the circuits become available for the connection of other calls [5]. The aim of a routing scheme is to influence, as much as possible, these random variations in the pattern of link occupancies so as to minimize network blocking.

A state-dependent scheme seeks to route each call so as to minimize the risk of blocking future calls, and, thus, it responds to the current state of the network on the basis of certain assumptions about future traffic demands.

The routing of a call arriving in a given configuration of link occupancies can thus be viewed as a choice of state transition from the given state of the network. Assuming that the cost of an additional call on a multi-link route is the sum of its cost on each link, a near-optimal routing strategy for a general communication network can be defined as follows [6]:

1. Route an $(A - B)$ call (originating from A and designated for B) over:
   - The direct route from A to B, if the (A, B) link (the link between node A and node B) has an idle circuit available.
   - The alternate route A-M-B path, if $C_{AM} + C_{MB} < C_{Ai} + C_{iB}$, and the (A, M) and (M, B) links have idle circuits available, where i is any other third node in alternate routes, and C_{AB} is the cost of using the (A, B) link.
2. Block the call in all other cases.

2.3 SIMULATION RESULTS

To evaluate the performance of the above scheme, a network consisting of 6 nodes, 10 links, and a capacity of 1000 circuits is used in the simulation. The cost for using a link is derived from the distance, time delay, financial charge for using that link or some other parameters regarding the link. In our simulation, the cost assigned for each link is assumed to be the combination of the load and the length of the link [2].

It is assumed that call arrivals follow a Poisson process, and the holding time (service time) follows an exponential distribution. Arrivals will either be accepted or be blocked immediately. That is, the link traffic follows the $M/M/N/N$ queueing model, where N is the link capacity.

The traffic of the network is characterized by the arrival rate and the service time of each link of the network. In our simulation, we assume that the network is characterized by seven traffic conditions corresponding to seven different time slots. Corresponding to each time slot, the traffic of each link is defined by a specific arrival rate λ_i (calls/time unit) and a specific mean service time $1/\mu_i$ (time units). The traffic condition of the network not only varies from time slot to time slot, but varies in different links. Table 1 shows the traffic parameters of the network corresponding to the seven different time slots. For example, if each time unit is equivalent to 1 minute, then the first time slot for the first link corresponds to a period of 120 minutes, the second time slot to 240 minutes, and likewise for the other time slots, as indicated in Table 1.

Link 1	Time units	120	240	180	120	120	180	480
	Arrival rate	1	0.5	1	0.3	0.5	0.2	0.5
	Service time	150	100	150	90	150	90	90
Link 2	Time units	120	240	180	120	120	180	480
	Arrival rate	1	0.5	0.5	0.3	0.5	0.2	0.25
	Service time	150	100	150	90	150	90	90
Link 3	Time units	180	120	240	120	120	120	540
	Arrival rate	1	0.25	1	0.5	1	0.25	0.3
	Service time	150	100	150	90	150	90	90
Link 4	Time units	180	120	240	120	120	120	540
	Arrival rate	1	0.5	1	0.5	0.5	0.25	0.3
	Service time	150	100	150	90	150	90	90
Link 5	Time units	120	240	180	120	120	60	600
	Arrival rate	1	0.5	0.5	0.5	0.3	0.3	0.5
	Service time	150	90	150	120	150	90	90
Link 6	Time units	120	240	180	120	120	60	600
	Arrival rate	1	0.3	0.5	0.5	1	0.25	0.5
	Service time	150	90	150	90	150	90	120
Link 7	Time units	120	240	180	120	120	180	480
	Arrival rate	1	0.5	1	0.3	0.5	0.2	0.3
	Service time	150	100	150	90	150	90	90
Link 8	Time units	120	240	180	120	120	180	480
	Arrival rate	1	0.5	0.5	0.3	0.5	0.2	0.25
	Service time	150	100	150	90	150	90	90
Link 9	Time units	180	120	240	120	120	120	540
	Arrival rate	1	0.5	1	0.5	1	0.25	0.3
	Service time	150	100	150	90	150	90	90
Link 10	Time units	180	120	240	120	120	120	540
	Arrival rate	1	0.5	1	0.5	0.5	0.25	0.3
	Service time	150	100	150	90	150	90	90

Table 1 Traffic parameters for a communication network, consisting of 10 links and 7 time slots. The columns correspond to the seven different time slots, and the rows to the ten links of the network.

Note that the seven time slots correspond to a one-day (24 hour) period. In the table, the columns correspond to the seven different time slots, the rows to the ten links of the network, and the traffic parameters are also indicated for each link. Note that the last time slot, which is the longest time slot, is characterized by lower arrival rates and shorter service times for each link of the network. Thus, the traffic in this time slot is low. In contrast, the third time slot is characterized by higher arrival rates and longer service times for each link of the network, thus exhibiting the highest traffic.

Although it appears that Poisson-distributed arrivals are directed towards nodes, for simplicity and equivalently, the arrivals are assumed designated for links rather than designated for nodes. Thus, an arrival to a link indicates that there is a call between the two nodes which make up that link. In the simulation, the traffic of the network is generated according to Table 1.

The performance is quantified by the average block rate. The average block rate at time t is defined by

$$\frac{1}{L} \sum_{i=1}^{L} \left| \frac{C_i - R_i(t)}{R_i(t)} \right|, \tag{6}$$

where L is the number of links in the network. Figure 2 shows the block rate as a function of time for the four cases considered where: (a) no management is done, i.e., calls to each link are blocked if all circuits in the link are occupied, (b) only map configuration by self-organization is employed, (c) only routing is employed, and (d) both map configuration by self-organization and routing are employed. As expected, the block rate is the lowest for case (d).

3 THE COST MINIMIZATION METHOD

The traffic management scheme that has been described in the preceding section serves as the basis for this model. This is a step forward in our attempt to realize a network that is not too specific for any service class. One of the major improvements in this model, as opposed to the one described in Section 2, is that here we will find a configuration that will result in an optimal network-wide performance. Further, given a configuration which is continuously updated by a better one, we perform the routing dynamically, again aiming at improving the performance.

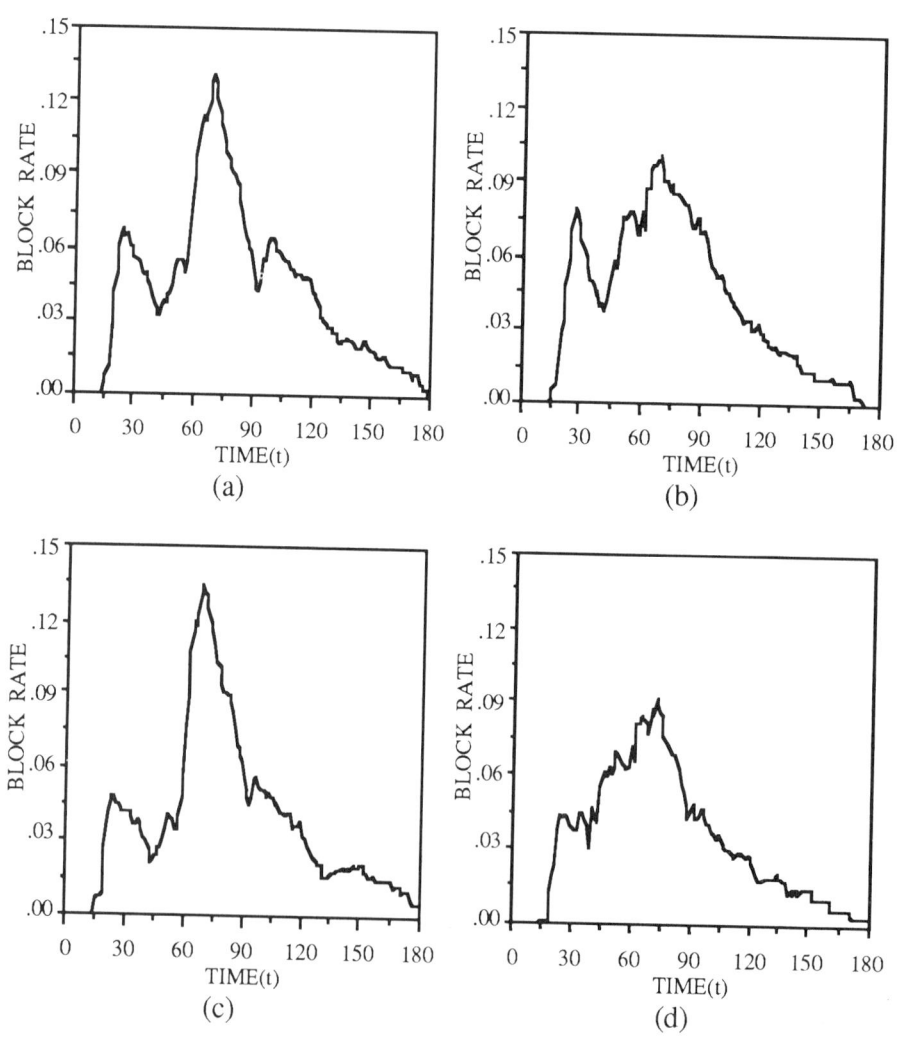

Figure 2 Block rate for the four cases described in the text.

In order to make the analysis simple, we choose the network-wide blocking rate as the only cost involved in this network. In the future we will extend the analysis to include other costs encountered in a communication network. Further details of the network model in Figure 3 are given below.

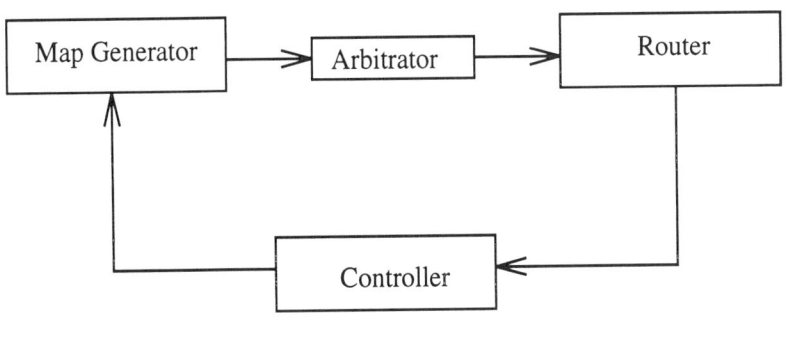

Figure 3 The traffic management scheme using the cost minimization method.

ROUTING

Routing is an integral part of a network. Here we consider the following variations of dynamic routing: 1) Alternate route by random choice; 2) Alternate route by minimum cost path; 3) Neural network approach to optimal routing. In the first case, alternately routed calls are allowed, and if no direct link is available, then one alternate route is chosen at random. In the second case, the choice of the alternate route is such that it minimizes a cost function. Neural network based routing finds the optimal path for each O–D pair, and then routes calls on this path. In what follows, we describe the routing of the first type specified above.

The routing is performed dynamically for every call, as follows: 1) An arriving call is first tried on the direct link; 2) If the direct link has no idle circuits then an alternate path is tried; 3) The choice of the alternate route is random; 4) If none of the alternate routes can accommodate this call, it is then blocked, and the call is lost. Here the routing module performs a simple routing function without much computation, which helps to reduce the processing delay of each call. The minimization of the blocking rate comes mainly from having a proper map in this module. There are other schemes, such as the least busy alternative (LBA) model [7], that try to dynamically compute the alternate route which minimizes the blocking rate. In our model, the proper choice of a map eliminates the need for computing the least busy alternate route. However, several of the dynamic routing schemes mentioned will be included in our future studies.

MAP GENERATOR

Arrival rates for each O–D pair and the total capacity of the network, along with the current status of the network, are given as input to this module. Based on this information, using a neural network based optimization technique, a map is produced as output. Here we use a simulated annealing technique [8] to search for the map that best suits the network conditions and constraints. In this module the map is searched by varying two parameters, namely, the link capacities **c** and the number of circuits that can be used by alternately routed calls **r** for each link. (**c**-**r**)circuits are reserved for direct calls.

CONTROLLER

The controller keeps track of network status and performance. The controller uses this information to decide when to request a new map. It is also the function of this module to give specifications to the Map Generator module. As mentioned before, link capacities and the reservation of circuits are the variables in a map. From previous studies we know that by properly assigning link capacities we can improve the performance. The optimal performance occurs when the capacity assignment is such that all links saturate simultaneously [9]. In this optimal case the capacity-to-flow ratio of all the links shall be equal to the average of the network. Also, we know that circuit reservation for first-routed calls is effective as a control mechanism against instabilities at high load conditions [10]. Based on these observations, we want to configure the map by properly choosing the values for link capacities and the number of reserved circuits.

ARBITRATOR

The function of the arbitrator is simple: to confirm that changing a network configuration will, in fact, be beneficial in reducing the cost or improving the performance. This is necessary since, when generating new maps, we do not take into account the cost associated with replacing a map by another one. This could be easily integrated into the map generation module.

3.1 ANALYTICAL MODEL

The proposed analytical model are generalized to have two distinct components: 1) Derivation of cost function; 2) Optimization of the network performance.

Many efficient nonlinear optimization techniques, such as simulated annealing [8] and mean field annealing [11], are considered for the traffic management scheme for circuit-switched and packet-switched networks, respectively. Results based on simulated annealing for circuit-switched networks are presented here.

Similar to the self-organization method, arrivals of any O–D pair follow a Poisson process. Call holding times are exponentially distributed. Each link is represented by an $M/M/N/N$ queueing model, where N is the number of circuits in that link. In order to simplify our analysis, we make the following assumptions.

1) All arrivals to any link form a Poisson process and are independent.

2) Link blocking probabilities are independent.

3) Processing and propagation delays are negligible.

The last of these three assumptions lets us study the performance of the network based solely on the proposed scheme. Furthermore, this is close to the real situation in circuit-switched networks. Obviously, calls arriving at an alternate route are not of Poisson distribution, but this is found to be a reasonable assumption especially when each link derives traffic from many end users. Similarly, a link blocking independence assumption is found to be reasonable, which greatly reduces the complexity of the analysis. Justifications for these assumptions and the extent to which they approximate real network traffic measurements are explained well in [12].

The notations below are used in discussion which follows.

(i, j) Link from node i to node j

$(i - j)$ Call originating from node i and destined for node j

λ_{i-j} External (new) arrival rate of $(i - j)$

M_{ij} Set of tandem nodes that forms alternate routes for $(i - j)$

B_{i-j} Probability that $(i - j)$ is blocked from the network

B_{ij} Probability that any call is blocked in (i, j)

B_{ij}^R Probability that an alternately routed call is blocked in (i, j)

C_{ij} Capacity (in number of circuits) of (i, j)

Suppose the average network blocking probability is denoted by $\bar{B}$. This is obtained by summing all of the call blocking probabilities, and normalizing the sum by total arrivals to the network.

$$\bar{B} = \frac{1}{\gamma} \sum_{(i-j)} \lambda_{i-j} B_{i-j}, \tag{7}$$

where γ, the total input traffic to the network, is given by

$$\gamma = \sum_{(i-j)} \lambda_{i-j}. \tag{8}$$

Arrival rates λ_{i-j} are known quantities. Under the previously stated assumption of link independence, B_{i-j} can be expressed in terms of B_{ij} and B_{ij}^R. A call $(i - j)$ is blocked from the network only when all available routes are occupied. Further, each alternate route is busy when either or both of the links constituting that route are busy. Here it is assumed that any alternate route is formed by, at most, two links. Hence the call blocking probability is,

$$B_{i-j} = B_{ij} \prod_{m \in M_{ij}} \left[1 - (1 - B_{im}^R)(1 - B_{mj}^R)\right]. \tag{9}$$

Therefore, if the link blocking probabilities B_{ij} and B_{ij}^R are known we can find the average network blocking rate by substituting Equations (9) and (8) into Equation (7). These probabilities can be derived from the birth-death process of the $M/M/N/N$ model [9].

When solving for blocking rates of each link, arrival rates of all links must be known. Since only the arrival rates to each O-D pair are given, this and the arrival rates for each link are related via the flow on each link. That is, the flow on each link is found by two methods, first by using the $M/M/N/N$ model, and second, by finding the overflow of arrivals from other links.

Once this network blocking rate is defined, the cost function is, in this example, simply the network blocking rate as a function of vectors **r** and **c**. Vector **c** is the set of available capacities for each link and vector **r** is the set of values which specify the amount of circuits available for alternate routes on each link. Once the cost function is specified, the next task is to optimize the network performance. Here, simulated annealing is used to minimize the cost function.

3.2 CONFIGURING MAPS BY SIMULATED ANNEALING

A map differs from another by two parameters, namely, r and c. The parameter r is also known as the reservation parameter. So the problem can be stated as:

$$E : S \rightarrow R, \qquad (10)$$

where E is the cost function and S is the solution space. E maps each state in the solution space to a real number. This mapping is subject to the constraint

$$\sum_{(i,j)} C_{ij} = C. \qquad (11)$$

Then the optimal solution S_{opt} is,

$$S_{opt} \in S : E(S_{opt}) \leq E(S_i) \qquad \forall S_i \in S \qquad (12)$$

where S_i is one of the states, a possible solution. In circuit-switched networks the grade of service is measured in terms of the rate at which calls are blocked from the network. Given the input statistics to each O-D pair, a closed form solution for this function can be obtained [13].

In reality there are many parameters, such as power, the available bandwidth of the individual links, and the block rate of the network, that need to be taken into account when optimizing a satellite network. In this work, only the blocking probability of the network is chosen. Since simulated annealing only requires a proper cost function and the careful choice of an annealing schedule, by choosing an appropriate cost function, any parameter that affects the network performance can be included in optimizing the network performance. The cost function $E(s)$ can be written as

$$E(s) = \bar{B}(r, c). \qquad (13)$$

When choosing values for the control parameter that corresponds to the "temperature," two things need to be given close attention. One is the distance between two successive temperature levels and the other is the value of this parameter in relation to the difference in the possible range of the costs of two states in the solution space, $\triangle(E) = E(S_2) - E(S_1)$. Since the size of the network affects $\triangle(E)$, and the constraint set forth in the beginning is the total capacity of the network, the control parameter T (temperature) is related to C, the total capacity of the network. More temperature levels are needed as the annealing process approaches a solution and also larger steps at the beginning. One such function used in the simulation is

$$T \propto \frac{C}{\log_e(j)}, \qquad (14)$$

where j is the iteration index incremented linearly.

3.3 FUNCTION OF THE CONTROLLER

The controller is fed with information about the current status of the network, including the arrival rates of the O-D pairs and the load balance, at regular time intervals. Let $\triangle$ denote the ratio of the network's total flow to the total capacity and let δ_{ij} denote the ratio of flow to capacity of link (i, j). At each update a measure of the network's load imbalance, denoted by d, is computed from these ratios using a sum-square-error method, as given below.

$$\triangle = \frac{F}{C}$$
$$\delta_{ij} = \frac{f_{ij}}{c_{ij}}$$
$$d = \frac{1}{C} \sum_{(i,j)} (\triangle - \delta_{ij})^2, \qquad (15)$$

where C and F are the total capacity and the total flow of the network, respectively. Therefore, the parameter d is a measure of load imbalance of the network. There is a threshold value, d_t, defined with respect to the network load balance $\triangle$. When the measure of load imbalance d is larger than the threshold value d_t, then the network is considered to be operating in an inefficient state for the current traffic condition. In the simulations done here, the threshold value is defined as follows.

$$d_t = 0.1 \times \triangle.$$

When d_t is small, even a slight deviation of network load balance from the ideal condition is not tolerated. In this situation, the Map Generator is called frequently. Thus, this parameter d_t can be set according to the specific application. In the simulations, the network status is updated at the end of every time unit and the parameter d is computed after 10 updates. 100 previous time units are used in measuring the traffic pattern.

3.4 SIMULATION RESULTS

Similarly, the network chosen for the simulation is mesh-connected. The rates of arrival into the nodes are non-symmetric. There are 11 nodes and 47 links in the network, and also 47 O–D pairs in this example. Each O-D pair has predefined alternate routes. The number of alternate routes is different from one O-D pair to the other, ranging from 0 to 4, and they are arbitrarily chosen.

As an example, the routing was performed with 20% of the link capacities reserved for direct calls, and only c is varied to optimize the network performance. The results are plotted in Figure 4. The annealing is done for different arrival rates between 0.5 and 1.3 calls/time unit (normalized). For these arrival rates, the network is simulated and the normalized throughput is again measured when the routing is done with the map that was selected by the Map Generator module. In all cases the throughput showed improvement, especially in terms of high arrival rates.

4 DISCUSSION

Two schemes have been developed, supported by simulation results, to optimize the performance of the circuit-switched satellite network. Improved network performance has been achieved. Future efforts include designing and analyzing a traffic management module similar to the one presented in this chapter for the packet-switched local area network (LAN), and integrating the traffic management schemes for both satellite and LAN networks into one model. Control parameters will be defined to make it possible to use the model for a satellite network as well as for a LAN.

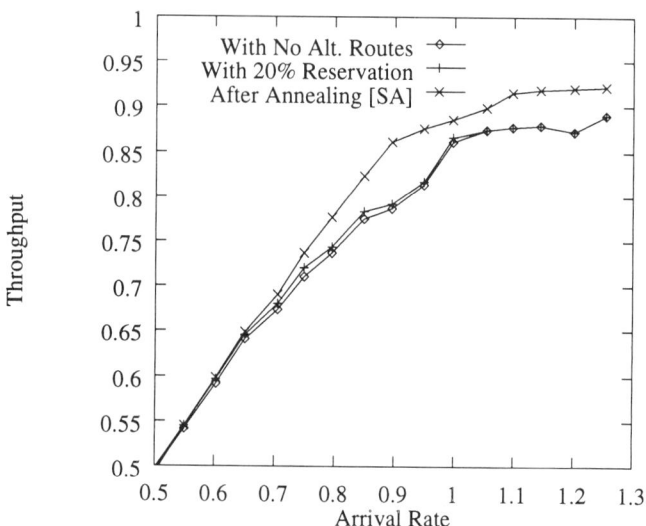

Figure 4 Network performance with simulated annealing.

Acknowledgments

This work was partially supported by the NASA Lewis Research Center, Cleveland, OH, contract number NAG3-1244. Special thanks are extended to S. Balasekar, L. Fitton and Z. Siveski for their helpful comments and suggestions that have improved the quality of this manuscript.

REFERENCES

[1] Rees, D. W. E., Satellite Communications: The First Quarter Century of Service. New York: John Wiley & Sons, 1990.

[2] Ansari, N. and D. Liu, "The Performance Evaluation of A New Neural Network Based Traffic Management Scheme For A Satellite Communication Network," Proc. 1991 IEEE Global Telecommunications Conference, December 2–5, 1991, Phoenix, AZ, pp. 110-114.

[3] Ansari, N. and Y. Chen, "Configuring Maps for a Satellite Communication Network by Self-organization," Journal of Neural Network Computing, vol. 2, no. 4, pp. 11-17, Spring 1991.

[4] Kohonen, T., Self-Organization and Associative Memory. Berlin: Springer-Verlag, 1984.

[5] Krishnan, K. R., "Markov Decision Algorithms for Dynamic Routing," IEEE Communication Magazine, vol. 28, no. 10, pp. 66-69, 1990.

[6] Regnier, J. and W. H. Cameron, "State-Dependent Dynamic Traffic Management for Telephone Networks," IEEE Communication Magazine, vol. 28, no. 10, pp. 42-53, 1990.

[7] Mitra, D., R. J. Gibbens and B. D. Huang, "State-dependent Routing on Symmetric Loss networks with Trunk Reservations I," AT&T Bell Labs, Murray Hill, NJ, Technical Report, November 1990.

[8] Van Laarhoven, P. J. M. and E. H. L. Aarts, Simulated Annealing: Theory and Applications. Dordrecht, Holland: D. Reidel Publishing Co., 1987.

[9] L. Kleinrock, Queuing Systems. Volume 1. New York: John Wiley, 1976.

[10] Akinpela, J. M., "Over-load Performance of Engineering Networks with Non-hierarchical and Hierarchical Routing," AT&T Bell Labs Tech J., vol. 63, no. 7, pp. 1261-1281, 1984.

[11] Peterson, C. and B. Söderberg, "A New Method for Mapping Optimization Problems onto Neural Networks," International Journal of Neural Systems, vol. 1, no. 1, pp. 3-22, 1989.

[12] Gerard, A. and M. Bell, "Blocking Evaluation for Networks with Residual Capacity Adaptive Routing," IEEE Trans. on Communications, vol. 37, no. 12, pp. 1372-1380, 1989.

[13] Balasekar, S. and N. Ansari, "Adaptive Map Configuration and Dynamic Routing to Optimize the Performance of A Satellite Communication Network," to be presented at GLOBECOM'93.

INDEX

adaptive information retrieval (AIR) 313-315
Adaptive Resonance Theory (ART) 315-320
additive white Gaussian noise (AWGN), see Gaussian noise
admission control 72-78, 106, 109, 127-141
analytic models 78, 361-363
ANNA 277,278
annealing schedule (also see cooling schedule) 363-364
ARTMAP 313
asynchronous transfer mode (ATM) 26, 63-74, 81-83, 87, 91, 92, 101, 105, 106, 109-125, 127-129,132-134,141
automatic gain control (AGC) 161
backpropagation 77, 111, 114, 116, 118,131
Bayesian detection 148-151
Bayesian estimation 151-153
belief network 307, 310
binomial sources 132,133
B-ISDN 63, 64
bit error rate (BER) 161, 165
Boltzmann machine 58, 205,206
call/connection admission control(CAC) 72-79, 109, 127-141
cascade-correlation 336
case-based reasoning
cell/connection blocking probability 128, 130, 134, 138, 139
cell loss probability 92,104
cell loss rate 76, 77, 81, 86, 87, 103, 110, 117, 121, 123, 124
cell scheduling 91
channel assignment 191-210
channel equalization 143-171
character recognition 271-285
circuit routing 15, 24
classification 3-5, 235, 244, 245, 248, 305, 308, 311, 313-316,325, 329

Classification and Regression Trees (CART) 330-333, 338, 345
clustering 8, 260, 295, 296, 320
CMAC 98
codebook replenishment 295-297
combinatorial optimization problems 70, 194 (see also optimization)
competitive learning 289
congestion control 4, 69, 92, 109, 110, 124
conjugate-gradient optimization 244, 248, 338
connectivity reduction 21, 23, 24, 31
constant bit-rate (CBR) 68
cooling schedule (also see annealing schedule) 198,200
cost function 47, 194-198, 361-363
decision boundary 148-150, 153, 155
decision feedback equalization 156, 158, 165, 166
designing neural network solutions 6, 15-24
deterministic switches 27
digital microwave radio (DMR) 143, 160
discrete cosine transform (CDT) 293, 298
distributed neural networks 71
document analysis 271, 305
dynamic programming 93, 95, 96, 97, 100, 105, 106
EMYCIN 310
estimation 3, 325, 329
equivalent bandwidth method 134
exciser 173-176
exhaustive search 37-38, 41-42, 50
expert system 235
filtering 1,7, 157, 173,177,305-324
fluid flow model 133
frequency sensitive competitive learning (FSCL) 289,291
function approximation 98,105,112
fuzzy logic 311, 192
GRAND 192
Gaussian noise 58, 152, 157, 164, 179, 184, 186

genetic algorithms 308
graph coloring 193-194, 198
greedy routing 31, 37-38,41-42,50
guaranteed constraint satisfaction 17-22, 24, 31
hardware 15, 272,277
Hidden Markov Model(HMM) 235, 251
Hopfield networks 5, 23, 45, 49, 194
hybrid system 257, 293
ISDN 64, 326
image coding 296
image compression 287
information filtering/retrieval 305-324
interconnection network 27, 37-61
intersymbol interference (ISI) 143, 144, 145, 146, 158, 164
Kohonen map 8, 156, 215, 290, 315, 351
language identification 233-254
latent semantic indexing (LSI) 309
Linde-Buzo-Gray (LBG) algorithm 261, 292-297
least-mean-square (LMS) 144, 154, 176-188
LeNet 273-275
link capacity control 79-87
load balance 364,365
local area network (LAN) 231, 365
M/M/1 91, 118
M/M/K 120
M/M/N/N 353, 361, 362, 363
Madaline 156
maximum flow 31
maximum-likelihood estimation (MLE) 144, 147
mean field annealing 361
mesh-connected 85, 349, 363
microwave radio, see digital microwave radio (DMR)
mobile switching center (MSC) 212-213
modularized 105
motion compensated luminance differences (MCLD) 293,298
multimedia traffic 70
multipath fading 143, 146, 161
multiple overlapping winner-take-all 18-21
multiplexing 67

Multivariate Adaptive Regression Splines (MARS) 93, 325, 333-335, 339, 345
optical character recognition (OCR) 271-285
optimization 5, 47, 70,83, 95,97, 137, 194-195, 205-207, 358-359
packet arbitration 26
packet routing 14, 26
peak reservation method 134
phase-shift keying (PSK) 160, 165
Poisson process 101,102, 133 353 ,355, 359
pseudo-random noise (PN) 178
pulse amplitude modulation (PAM) 145, 152, 158
quadrature amplitude modulation (QAM) 158-160, 163-164
quadrature phase shift keying (QPSK) 165
quality-of-service (QOS) 64, 67-70, 72, 73, 77, 81, 83, 92, 93, 97, 103, 104, 110, 111, 117, 128, 130, 134
queueing 26, 91, 92, 93, 94, 95, 101, 102, 103, 110, 118, 120, 124, 353, 359
queue models
   M/M/1 91, 118
   M/M/K 120
   M/M/N/N 353, 359-361
queueing delay 92, 93, 94, 101, 102, 103
radial basis functions 156
radio network 5, 8, 143, 160, 191, 196, 205, 208, 211,213,215,216
real-time learning 312
recurrent neural network 116
regression 3-5, 325, 329
reinforcement learning 7, 91, 93, 313-315
routing 14, 26, 37-61, 64, 65, 67, 117, 352-353, 357-358
RUBRIC 310
satellite 287, 349-365
self-organizing feature maps 215-219, 223,231, 315-316

signal-to-noise ratio (SNR) 148, 152, 161, 288, 292
simulated annealing 5, 58, 191, 194-208, 220, 361-362
soft-decision 151
speech processing 233-254, 255-270, 288, 327
spread spectrum 173-189
stochastic algorithm 202
supervised learning 7, 288, 313-315
switching 12, 24, 38, 65, 105, 109,111, 128, 197, 212, 326
tapped-delay line 144, 154, 155, 157, 167
TCP-IP 101,102,104
temporal differences 96
throughput 30, 37, 49, 363
unsupervised learning 8, 288, 289, 315-316
urgency scheduling 97, 103, 104, 105, 106
variable bit-rate (VBR) 68, 132
vector quantization 235, 260-261, 287-303
vector space 308
Viterbi algorithm 144, 244
wavelet transform 175-176, 188
Widrow-Hoff 177, 179, 180, 184, 186
winner-take-all circuit 17-22
word recognition 244